# FUEL PRODUCTION FROM NON-FOOD BIOMASS

## Corn Stover

# FUEL PRODUCTION FROM NON-FOOD BIOMASS
## Corn Stover

*Edited by*
**Barnabas Gikonyo, PhD**

Apple Academic Press Inc.
3333 Mistwell Crescent
Oakville, ON L6L 0A2
Canada

Apple Academic Press Inc.
9 Spinnaker Way
Waretown, NJ 08758
USA



First issued in paperback 2021

*Exclusive worldwide distribution by CRC Press, a member of Taylor & Francis Group*



ISBN 13: 978-1-77463-544-5 (pbk)
ISBN 13: 978-1-77188-123-4 (hbk)



**Library and Archives Canada Cataloguing in Publication**

Fuel production from non-food biomass: corn stover / edited by Barnabas Gikonyo, PhD.

Includes bibliographical references and index.
ISBN 978-1-77188-123-4 (bound)
1. Ethanol as fuel. 2. Corn stover as fuel. 3. Biomass energy. I. Gikonyo, Barnabas, editor

TP339.F83 2015 662'.88 C2014-907725-4

**Library of Congress Cataloging-in-Publication Data**

Gikonyo, Barnabas.
Fuel production from non-food biomass: corn stover / Barnabas Gikonyo, PhD.

pages cm
Includes bibliographical references and index.
ISBN 978-1-77188-123-4 (alk. paper)
1. Agricultural wastes as fuel. 2. Biomass energy. 3. Energy crops. 4. Corn.
5. Renewable energy sources. I. Title. II. Title: Corn stover.

TP339.G55 2015 333.95'39--dc23 2014045366

Apple Academic Press also publishes its books in a variety of electronic formats. Some content that appears in print may not be available in electronic format. For information about Apple Academic Press products, visit our website at **www.appleacademicpress.com** and the CRC Press website at **www.crcpress.com**

# ABOUT THE EDITOR

**BARNABAS GIKONYO, PhD**

Barnabas Gikonyo graduated from Southern Illinois University, Carbondale, Illinois (2007), with a PhD in organic and materials chemistry. Hc currently teaches organic and general chemistry classes at the State University of New York Geneseo, along with corresponding laboratories and the oversight of general chemistry labs. His research interests range from the application of various biocompatible, polymeric materials as “biomaterial bridging surfaces” for the repair of spinal cord injuries, to the use of osteoconductive cements for the repair of critical sized bone defects/fractures. Currently, hc is studying the development of alternative, non-food biofuels.

# CONTENTS

# ACKNOWLEDGMENT AND HOW TO CITE

The editor and publisher thank each of the authors who contributed to this book. The chapters in this book were previously published in various places in various formats. To cite the work contained in this book and to view the individual permissions, please refer to the citation at the beginning of each chapter. Each chapter was read individually and carefully selected by the editor; the result is a book that provides a nuanced look at the possibilities of corn stover as a biofuel. The chapters included are broken into three sections, which describe the following topics:

- *Part 1: Overview.* This article offers good evidence for the potential value of corn stover as a source of power.
- *Part 2: Technology.* The editor's own research is geared toward biofuel production from lignocellulosic biomass and non-human food sources. The papers contained in this section help put this into sharp focus.
- *Part 3: Considerations for Policy Formation.* These articles provide interesting economic, land-use, and environmental viewpoints that must be considered when considering the future of biofuel for corn stover.

# LIST OF CONTRIBUTORS

**Eva Albers**
Department of Chemical and Biological Engineering, Industrial Biotechnology, Chalmers University of Technology, Göteborg SE-412 96, Sweden and Taurus Energy AB, Ideon, Ole Römers väg 12, Lund, SE-223 70, Sweden

**Valdeir Arantes**
Forest Products Biotechnology/Bioenergy Group, University of British Columbia, 2424 Main Mall, Vancouver, British Columbia, V6T1Z4, Canada

**Venkatesh Balan**
DOE Great Lakes Bioenergy Research Center (GLBRC), Michigan State University, East Lansing, MI 48824, USA and Biomass Conversion Research Laboratory, Department of Chemical Engineering and Materials Science, Michigan State University, 3815 Technology Boulevard, MBI Building, Lansing, MI 48910, USA

**Jie Bao**
State Key Laboratory of Bioreactor Engineering, East China University of Science and Technology, 130 Meilong Road, Shanghai 200237, China

**Ye Chen**
Novozymes North America, Franklinton, NC, 27525, USA

**Bruce E. Dale**
DOE Great Lakes Bioenergy Research Center (GLBRC), Michigan State University, East Lansing, MI 48824, USA and Biomass Conversion Research Laboratory, Department of Chemical Engineering and Materials Science, Michigan State University, 3815 Technology Boulevard, MBI Building, Lansing, MI 48910, USA

**Jennifer B. Dunn**
Systems Assessment Group, Argonne National Laboratory, 9700 South Cass Avenue, Argonne, IL 60439, USA

**Sandra Eksioglu**
Department of Industrial and Systems Engineering, Mississippi State University, 260Q McCain Engineering Building, Mississippi State, MS 39762, USA

**Xiadi Gao**
BioEnergy Science Center (BESC), Oak Ridge National Laboratory, Oak Ridge, TN 37831, USA, Department of Chemical and Environmental Engineering, Bourns College of Engineering, University of California (UCR), Riverside, CA 92521, USA, and Center for Environmental Research and Technology (CE-CERT), Bourns College of Engineering, University of California, Riverside, CA 92507, USA

**Peter D. Goldsmith**
Department of Agricultural and Consumer Economics, University of Illinois at Urbana-Champaign, Urbana, IL 61801, USA

**Donald L. Grebner**
Department of Forestry, Mississippi State University, 105 Thompson Hall, Mississippi State, MS 39762, USA

**Yanqing He**
State Key Laboratory of Bioreactor Engineering, East China University of Science and Technology, 130 Meilong Road, Shanghai 200237, China

**Jason Holmes**
Novozymes North America, Franklinton, NC, 27525, USA

**Jinguang Hu**
Forest Products Biotechnology/Bioenergy Group, University of British Columbia, 2424 Main Mall, Vancouver, British Columbia, V6T1Z4, Canada

**Rakesh Koppram**
Department of Chemical and Biological Engineering, Industrial Biotechnology, Chalmers University of Technology, Göteborg SE-412 96, Sweden

**Rajeev Kumar**
BioEnergy Science Center (BESC), Oak Ridge National Laboratory, Oak Ridge, TN 37831, USA and Center for Environmental Research and Technology (CE-CERT), Bourns College of Engineering, University of California, Riverside, CA 92507, USA

**Ho-Young Kwon**
Department of Natural Resources and Environmental Sciences, University of Illinois at Urbana-Champaign, W-503 Turner Hall, MC-047, 1102 South Goodwin Avenue, Urbana, IL 61801, USA

**Annika Lambert**
SEKAB E-Technology AB, P.O. Box 286, Örnsköldsvik, SE-891 26, Sweden

**Xuezhi Li**
State Key Laboratory of Microbial Technology, Shandong University, Jinan, China

**Jie Lu**
State Key Laboratory of Microbial Technology, Shandong University, Jinan, China and Dalian Polytechnic University, Dalian, China

**Steffen Mueller**
Energy Resources Center, University of Illinois at Chicago, 1309 South Halsted Street, MC 156, Chicago, IL 60607, USA

**Rita H. Mumm**
Department of Crop Sciences, University of Illinois at Urbana-Champaign, Urbana, IL 61801, USA

**Fredrik Nielsen**
Department of Chemical Engineering, Lund University, P.O. Box 124, Lund, SE-221 00, Sweden

**Lisbeth Olsson**
Department of Chemical and Biological Engineering, Industrial Biotechnology, Chalmers University of Technology, Göteborg SE-412 96, Sweden

**Joel O. Paz**
Department of Agricultural and Biological Engineering, Mississippi State University, 130 Creelman St., Mississippi State, MS 39762, USA

**Ryan Petter**
Department of Agricultural Economics, Purdue University, 403 West State Street, West Lafayette, IN 47907-2056, USA

**Amadeus Y. Pribowo**
Forest Products Biotechnology/Bioenergy Group, University of British Columbia, 2424 Main Mall, Vancouver, British Columbia, V6T1Z4, Canada

**Yinbo Qu**
State Key Laboratory of Microbial Technology, Shandong University, Jinan, China

**Selvarani Radhakrishnan**
Department of Agricultural and Biological Engineering, Mississippi State University, 130 Creelman St., Mississippi State, MS 39762, USA

**Kent D. Rausch**
Department of Agricultural and Biological Engineering, University of Illinois at Urbana-Champaign, Urbana, IL 61801, USA

**Jack N. Saddler**
Forest Products Biotechnology/Bioenergy Group, University of British Columbia, 2424 Main Mall, Vancouver, British Columbia, V6T1Z4, Canada

**Blake A. Simmons**
Deconstruction Division, Joint BioEnergy Institute (JBEI), Emeryville, CA 94608, USA and Sandia National Laboratories, Livermore, CA 94551, USA

**Seema Singh**
Deconstruction Division, Joint BioEnergy Institute (JBEI), Emeryville, CA 94608, USA and Sandia National Laboratories, Livermore, CA 94551, USA

**Hans H. Stein**
Department of Animal Sciences, University of Illinois at Urbana-Champaign, Urbana, IL 61801, USA

**Mark A. Stevens**
Novozymes North America, Franklinton, NC, 27525, USA

**Wallace E. Tyner**
Department of Agricultural Economics, Purdue University, 403 West State Street, West Lafayette, IN 47907-2056, USA

**Michael Q. Wang**
Systems Assessment Group, Argonne National Laboratory, 9700 South Cass Avenue, Argonne, IL 60439, USA

**Sune Wännström**
SEKAB E-Technology AB, P.O. Box 286, Örnsköldsvik, SE-891 26, Sweden

**Lars Welin**
Taurus Energy AB, Ideon, Ole Römers väg 12, Lund, SE-223 70, Sweden

**Charles E. Wyman**
BioEnergy Science Center (BESC), Oak Ridge National Laboratory, Oak Ridge, TN 37831, USA, Department of Chemical and Environmental Engineering, Bourns College of Engineering, University of California (UCR), Riverside, CA 92521, USA, and Center for Environmental Research and Technology (CE-CERT), Bourns College of Engineering, University of California, Riverside, CA 92507, USA

**Hui Xu**
Novozymes North America, Franklinton, NC, 27525, USA

**Ruifeng Yang**
Dalian Polytechnic University, Dalian 116034, China

**Fei Yu**
Department of Agricultural and Biological Engineering, Mississippi State University, 130 Creelman St., Mississippi State, MS 39762, USA

**Guido Zacchi**
Department of Chemical Engineering, Lund University, P.O. Box 124, Lund, SE-221 00, Sweden

**Jian Zhang**
State Key Laboratory of Bioreactor Engineering, East China University of Science and Technology, 130 Meilong Road, Shanghai 200237, China

**Longping Zhang**
State Key Laboratory of Bioreactor Engineering, East China University of Science and Technology, 130 Meilong Road, Shanghai 200237, China

**Jian Zhao**
State Key Laboratory of Microbial Technology, Shandong University, Jinan, China

**Yongming Zhu**
Novozymes North America, Franklinton, NC, 27525, USA

# INTRODUCTION

Corn stover is made up of the stalks, leaves, and cobs that are left after the edible kernels of corn are harvested. This biomass is composed of about 70 percent cellulose and hemicellulose, which can be converted to ethanol, and 15 to 20 percent lignin, which can be burned as a boiler fuel for steam electricity generation.

Concerns about fossil fuel's connection to climate change have triggered interest in using corn stover for energy production. A 2012 study from the United States Energy Department's Argonne National Laboratory found that biofuels made with corn residue produced 95 percent less greenhouse gas emissions than gasoline.

Although controversy still exists about the use of corn stover, with some critics saying that it will cause food shortages, particularly for developing nations, the research in this book focuses on using corn's already existing, non-food biomass, and argues that food and biofuel could potentially be produced from the same fields.

*Barnabas Gikonyo, PhD*

Combined heat and power (CHP) production using renewable energy sources is gaining importance because of its flexibility and high-energy efficiency. Biomass materials, such as corn stover and forestry residues, are potential sources for renewable energy for CHP production. In Mississippi, approximately 4.0 MT dry tons of woody biomass is available annually for energy production. In Chapter 1, Radhakrishnan and colleagues collected and analyzed 10 years of corn stover data (2001–2010) and three years of forest logging residue data (1995, 1999, and 2002) in each county in Mississippi to determine the potential of these feed stocks for sustainable CHP energy production. The authors identified six counties, namely Amite, Copiah, Clarke, Wayne, Wilkinson and Rankin, that have forest logging residue feedstocks to sustain a CHP facility with a range of capacity between 8.0 and 9.8 MW. Using corn stover alone, Yazoo and Washington counties can produce 13.4 MW and 13.5 MW of energy, respectively. Considering both feedstocks and based on a conservative amount of 30% available forest logging residue and 33% corn stover, they found that 20 counties have adequate supply for a CHP facility with a capacity of 8.3 MW to 19.6 MW.

Corn stover is a promising feedstock for bioethanol production because of its abundant availability in China. To obtain higher ethanol concentration and higher ethanol yield, in Chapter 2, by Li and colleagues, liquid hot water (LHW) pretreatment and fed-batch semi-simultaneous saccharification and fermentation (S-SSF) were used to enhance the enzymatic digestibility of corn stover and improve bioconversion of cellulose to ethanol. The results show that solid residues from LHW pretreatment of corn stover can be effectively converted into ethanol at severity factors ranging from 3.95 to 4.54, and the highest amount of xylan removed was approximately 89%. The ethanol concentrations of 38.4 g/L and 39.4 g/L as well as ethanol yields of 78.6% and 79.7% at severity factors of 3.95 and 4.54, respectively, were obtained by fed-batch S-SSF in an optimum conditions (initial substrate consistency of 10%, and 6.1% solid residues added into system at the prehydrolysis time of 6 h). The changes in surface morphological structure, specific surface area, pore volume and diameter of corn stover subjected to LHW process were also analyzed for interpreting the possible improvement mechanism.

Dry dilute acid pretreatment at extremely high solids loading of lignocellulose materials demonstrated promising advantages of no waste water generation, less sugar loss, and low steam consumption while maintaining high hydrolysis yield. However, the routine pretreatment reactor without mixing apparatus was found not suitable for dry pretreatment operation because of poor mixing and mass transfer. In Chapter 3, by He and colleagues, helically agitated mixing was introduced into the dry dilute acid pretreatment of corn stover and its effect on pretreatment efficiency, inhibitor generation, sugar production, and bioconversion efficiency through simultaneous saccharification and ethanol fermentation (SSF) were evaluated. The overall cellulose conversion taking account of cellulose loss in pretreatment was used to evaluate the efficiency of pretreatment. The two-phase computational fluid dynamics (CFD) model on dry pretreatment was established and applied to analyze the mixing mechanism. The results showed that the pretreatment efficiency was significantly improved and the inhibitor generation was reduced by the helically agitated mixing, compared to the dry pretreatment without mixing: the ethanol titer and yield from cellulose in the SSF reached 56.20 g/L and 69.43% at the 30% solids loading and 15 FPU/DM cellulase dosage, respectively, corresponding to a 26.5% increase in ethanol titer and 17.2% increase in ethanol yield at the same fermentation conditions. The advantage of helically agitated mixing may provide a prototype of dry dilute acid pretreatment processing for future commercial-scale production of cellulosic ethano

Liquid hot water (LHW) pretreatment is an effective and environmentally friendly method to produce bioethanol with lignocellulosic materials. In Lu and colleagues' previous study, high ethanol concentration and ethanol yield were obtained from water-insoluble solids (WIS) of reed straw and corn stover pretreated with LHW by using fed-batch semi-simultaneous saccharification and fermentation (S-SSF). However, high cellulase loading and the large amount of wash water possibly limit the practical application of LHW pretreatment. To decrease cellulase loading and the amount of wash water, in Chapter 4, the authors performed Tween 40 pretreatment before WIS was subjected to bioethanol fermentation. Results showed that the optimum conditions of Tween 40 pretreatment were as follows: Tween 40 concentration of 1.5%, WIS-to-Tween 40 ratio of 1:10 (w/v), and pretreatment time of 1 hour at ambient temperature. After Tween 40 pretreatment, cellulase loading could be greatly reduced. After Tween 40 pretreatment, the residual liquid could be recycled for utilization but slightly affected ethanol concentration and yield. The unwashed WIS could obtain a high ethanol concentration of 56.28 g/L (reed straw) and 52.26 g/L (corn stover) by Tween 40 pretreatment using fed-batch S-SSF. Ethanol yield reached a maximum of 69.1% (reed straw) and 71.1% (corn stover). Tween 40 pretreatment was a very effective and less costly method with unwashed WIS. This pretreatment could greatly reduce cellulase loading and save wash water. Higher ethanol concentration was obtained almost without reducing ethanol yield.

It is widely recognised that fast, effective hydrolysis of pretreated lignocellulosic substrates requires the synergistic action of multiple types of hydrolytic and some non-hydrolytic proteins. However, due to the complexity of the enzyme mixture, the enzymes interaction with and interference from the substrate and a lack of specific methods to follow the distribution of individual enzymes during hydrolysis, most of enzyme-substrate interaction studies have used purified enzymes and pure cellulose model substrates. As the enzymes present in a typical "cellulase mixture" need to work cooperatively to achieve effective hydrolysis, the action of one enzyme is likely to influence the behaviour of others. The action of the enzymes will be further influenced by the nature of the lignocellulosic substrate. Therefore, it would be beneficial if a method could be developed that allowed us to follow some of the individual enzymes present in a cellulase mixture during hydrolysis of more commercially realistic biomass substrates. In Chapter 5, Pribowo and colleagues developed a high throughput immunoassay that could quantitatively and specifically follow individual cellulase enzymes during hydrolysis. Using monoclonal and polyclonal antibodies (MAb and PAb, respectively), a double-antibody sandwich enzyme-linked immunosorbent assay (ELISA) was developed to specifically quantify cellulase enzymes from Trichoderma reesei: cellobiohydrolase I (Cel7A), cellobiohydrolase II (Cel6A), and endoglucanase I (Cel7B). The interference from substrate materials present in lignocellulosic supernatants could be minimized by dilution. A double-antibody sandwich ELISA was able to detect and quantify individual enzymes when present in cellulase mixtures. The assay was sensitive over a range of relatively low enzyme concentration (0–1 μg/ml), provided the enzymes were first pH adjusted and heat treated to increase their antigenicity. The immunoassay was employed to quantitatively monitor the adsorption of cellulase monocomponents, Cel7A, Cel6A, and Cel7B, that were present in both Celluclast and Accellerase 1000, during the hydrolysis of steam-pretreated corn stover (SPCS). All three enzymes exhibited different individual adsorption profiles. The specific and quantitative adsorption profiles observed with the ELISA method were in agreement with earlier work where more labour intensive enzyme assay techniques were used.

Previous research on alkaline pretreatment has mainly focused on optimization of the process parameters to improve substrate digestibility. To achieve satisfactory sugar yield, extremely high chemical loading and enzyme dosages were typically used. Relatively little attention has been paid to reduction of chemical consumption and process waste management, which has proven to be an indispensable component of the bio-refineries. To indicate alkali strength, both alkali concentration in pretreatment solution (g alkali/g pretreatment liquor or g alkali/L pretreatment liquor) and alkali loading based on biomass solids (g alkali/g dry biomass) have been widely used. The dual approaches make it difficult to compare the chemical consumption in different process scenarios while evaluating the cost effectiveness of this pretreatment technology. Chapter 6, by Chen and colleagues, addresses these issues through pretreatment of corn stover at various combinations of pretreatment conditions. Enzymatic hydrolysis with different enzyme blends was subsequently performed to identify the effects of pretreatment parameters on substrate digestibility as well as process operational and capital costs. The results showed that sodium hydroxide loading is the most dominant variable for enzymatic digestibility. To reach 70% glucan conversion while avoiding extensive degradation of hemicellulose, approximately 0.08 g NaOH/g corn stover was required. It was also concluded that alkali loading based on total solids (g NaOH/g dry biomass) governs the pretreatment efficiency. Supplementing cellulase with accessory enzymes such as α-arabinofuranosidase and β-xylosidase significantly improved the conversion of the hemicellulose by 6–17%. The current work presents the impact of alkaline pretreatment parameters on the enzymatic hydrolysis of corn stover as well as the process operational and capital investment costs. The high chemical consumption for alkaline pretreatment technology indicates that the main challenge for commercialization is chemical recovery. However, repurposing or co-locating a biorefinery with a paper mill would be advantageous from an economic point of view.

While simultaneous saccharification and co-fermentation (SSCF) is considered to be a promising process for bioconversion of lignocellulosic materials to ethanol, there are still relatively little demo-plant data and operating experiences reported in the literature. In Chapter 7, Koppram and colleagues designed a SSCF process and scaled up from lab to demo scale reaching 4% (w/v) ethanol using xylose rich corncobs. Seven different recombinant xylose utilizing *Saccharomyces cerevisiae* strains were evaluated for their fermentation performance in hydrolysates of steam pretreated corncobs. Two strains, RHD-15 and KE6-12 with highest ethanol yield and lowest xylitol yield, respectively were further screened in SSCF using the whole slurry from pretreatment. Similar ethanol yields were reached with both strains, however, KE6-12 was chosen as the preferred strain since it produced 26% lower xylitol from consumed xylose compared to RHD-15. Model SSCF experiments with glucose or hydrolysate feed in combination with prefermentation resulted in 79% of xylose consumption and more than 75% of the theoretical ethanol yield on available glucose and xylose in lab and PDU scales. The results suggest that for an efficient xylose conversion to ethanol controlled release of glucose from enzymatic hydrolysis and low levels of glucose concentration must be maintained throughout the SSCF. Fed-batch SSCF in PDU with addition of enzymes at three different time points facilitated controlled release of glucose and hence co-consumption of glucose and xylose was observed yielding 76% of the theoretical ethanol yield on available glucose and xylose at 7.9% water insoluble solids (WIS). With a fed-batch SSCF in combination with prefermentation and a feed of substrate and enzymes 47 and 40 g $l^{-1}$ of ethanol corresponding to 68% and 58% of the theoretical ethanol yield on available glucose and xylose were produced at 10.5% WIS in PDU and demo scale, respectively. The strain KE6-12 was able to completely consume xylose within 76 h during the fermentation of hydrolysate in a 10 m3demo scale bioreactor. The potential of SSCF is improved in combination with prefermentation and a feed of substrate and enzymes. It was possible to successfully reproduce the fed-batch SSCF at demo scale producing 4% (w/v) ethanol which is the minimum economical requirement for efficient lignocellulosic bioethanol production process.

Pretreatment is essential to realize high product yields from biological conversion of naturally recalcitrant cellulosic biomass, with thermochemical pretreatments often favored for cost and performance. In Chapter 8, by Gao and colleagues, enzymatic digestion of solids from dilute sulfuric acid (DA), ammonia fiber expansion (AFEX™), and ionic liquid (IL) thermochemical pretreatments of corn stover were followed over time for the same range of total enzyme protein loadings to provide comparative data on glucose and xylose yields of monomers and oligomers from the pretreated solids. The composition of pretreated solids and enzyme adsorption on each substrate were also measured to determine. The extent glucose release could be related to these features. Corn stover solids from pretreatment by DA, AFEX, and IL were enzymatically digested over a range of low to moderate loadings of commercial cellulase, xylanase, and pectinase enzyme mixtures, the proportions of which had been previously optimized for each pretreatment. Avicel® cellulose, regenerated amorphous cellulose (RAC), and beechwood xylan were also subjected to enzymatic hydrolysis as controls. Yields of glucose and xylose and their oligomers were followed for times up to 120 hours, and enzyme adsorption was measured. IL pretreated corn stover displayed the highest initial glucose yields at all enzyme loadings and the highest final yield for a low enzyme loading of 3 mg protein/g glucan in the raw material. However, increasing the enzyme loading to 12 mg/g glucan or more resulted in DA pretreated corn stover attaining the highest longer-term glucose yields. Hydrolyzate from AFEX pretreated corn stover had the highest proportion of xylooligomers, while IL produced the most glucooligomers. However, the amounts of both oligomers dropped with increasing enzyme loadings and hydrolysis times. IL pretreated corn stover had the highest enzyme adsorption capacity. Initial hydrolysis yields were highest for substrates with greater lignin removal, a greater degree of change in cellulose crystallinity, and high enzyme accessibility. Final glucose yields could not be clearly related to concentrations of xylooligomers released from xylan during hydrolysis. Overall, none of these factors could completely account for differences in enzymatic digestion performance of solids produced by AFEX, DA, and IL pretreatments.

Conventional fossil fuels dominate the marketplace, and their prices are a direct competitor for drop-in biofuels. Chapter 9, by Petter and Tyner, examines the impact of fuel selling price uncertainty on investment risk in a fast pyrolysis derived biofuel production facility. Production cost specifications are gathered from previous research. Monte Carlo analysis is employed with uncertainty in fuel selling price, biomass cost, bio-oil yield, and hydrogen price parameters. Experiments reveal that fuel price has a large impact on investment risk. A reverse auction would shift risk from the private sector to the public sector and is shown to be more effective at encouraging private investment than capital subsidies for the same expected public cost.

Although the system for producing yellow corn grain is well established in the US, its role among other biofeedstock alternatives to petroleum-based energy sources has to be balanced with its predominant purpose for food and feed as well as economics, land use, and environmental stewardship. In Chapter 10, Mumm and colleagues model land usage attributed to corn ethanol production in the US to evaluate the effects of anticipated technological change in corn grain production, ethanol processing, and livestock feeding through a multi-disciplinary approach. Seven scenarios are evaluated: four considering the impact of technological advances on corn grain production, two focused on improved efficiencies in ethanol processing, and one reflecting greater use of ethanol co-products (that is, distillers dried grains with solubles) in diets for dairy cattle, pigs, and poultry. For each scenario, land area attributed to corn ethanol production is estimated for three time horizons: 2011 (current), the time period at which the 15 billion gallon cap for corn ethanol as per the Renewable Fuel Standard is achieved, and 2026 (15 years out). Although 40.5% of corn grain was channeled to ethanol processing in 2011, only 25% of US corn acreage was attributable to ethanol when accounting for feed co-product utilization. By 2026, land area attributed to corn ethanol production is reduced to 11% to 19% depending on the corn grain yield level associated with the four corn production scenarios, considering oil replacement associated with the soybean meal substituted in livestock diets with distillers dried grains with solubles. Efficiencies in ethanol processing, although producing more ethanol per bushel of processed corn, result in less co-products and therefore less offset of corn acreage. Shifting the use of distillers dried grains with solubles in feed to dairy cattle, pigs, and poultry substantially reduces land area attributed to corn ethanol production. However, because distillers dried grains with solubles substitutes at a higher rate for soybean meal, oil replacement requirements intensify and positively feedback to elevate estimates of land usage. Accounting for anticipated technological changes in the corn ethanol system is important for understanding the associated land base ascribed, and may aid in calibrating parameters for land use models in biofuel life-cycle analyses.

The greenhouse gas (GHG) emissions that may accompany land-use change (LUC) from increased biofuel feedstock production are a source of debate in the discussion of drawbacks and advantages of biofuels. Estimates of LUC GHG emissions focus mainly on corn ethanol and vary widely. Increasing the understanding of LUC GHG impacts associated with both corn and cellulosic ethanol will inform the on-going debate concerning their magnitudes and sources of variability. In Chapter 11, Dunn and colleagues estimate LUC GHG emissions for ethanol from four feedstocks: corn, corn stover, switchgrass, and miscanthus. We use new computable general equilibrium (CGE) results for worldwide LUC. U.S. domestic carbon emission factors are from state-level modelling with a surrogate CENTURY model and U.S. Forest Service data. This paper investigates the effect of several key domestic lands carbon content modelling parameters on LUC GHG emissions. International carbon emission factors are from the Woods Hole Research Center. LUC GHG emissions are calculated from these LUCs and carbon content data with Argonne National Laboratory's Carbon Calculator for Land Use Change from Biofuels Production (CCLUB) model. Our results indicate that miscanthus and corn ethanol have the lowest (−10 g $CO_2$e/MJ) and highest (7.6 g $CO_2$e/MJ) LUC GHG emissions under base case modelling assumptions. The results for corn ethanol are lower than corresponding results from previous studies. Switchgrass ethanol base case results (2.8 g $CO_2$e/MJ) were the most influenced by assumptions regarding converted forestlands and the fate of carbon in harvested wood products. They are greater than miscanthus LUC GHG emissions because switchgrass is a lower-yielding crop. Finally, LUC GHG emissions for corn stover are essentially negligible and insensitive to changes in model assumptions. This research provides new insight into the influence of key carbon content modelling variables on LUC GHG emissions associated with the four bioethanol pathways we examined. Our results indicate that LUC GHG emissions may have a smaller contribution to the overall biofuel life cycle than previously thought. Additionally, they highlight the need for future advances in LUC GHG emissions estimation including improvements to CGE models and aboveground and belowground carbon content data.

# PART I

# OVERVIEW

## CHAPTER 1

# ASSESSMENT OF POTENTIAL CAPACITY INCREASES AT COMBINED HEAT AND POWER FACILITIES BASED ON AVAILABLE CORN STOVER AND FOREST LOGGING RESIDUES

SELVARANI RADHAKRISHNAN, JOEL O. PAZ, FEI YU, SANDRA EKSIOGLU, AND DONALD L. GREBNER

## 1.1 INTRODUCTION

Combined heat and power is a concurrent process, and generates heat and electricity from the same location where they need to be utilized. CHP can produce heat from any fuel source such as natural gas, biomass, biogas and coal. CHP can be applicable for various types of existing technologies for generating electricity, power, and waste-heat recovery for heating, cooling and thermal applications [1]. The two most common types of CHP systems are gas turbines and steam turbines. In a gas turbine or reciprocating engine system, the CHP system produces electricity by burning fuel and a heat recovery unit is used to capture heat from the gas turbine. In a steam turbine system, the CHP system produces steam by burning a fuel and heating water to produce steam and this is used to generate electricity in a turbine. Steam turbine-based CHP systems typically use solid fuels like coal, biomass and waste products that are readily available to fuel the boiler unit [2]. CHP is an efficient and clean way of producing power and thermal energy. The total CHP system efficiencies for producing electricity and thermal energy range from 60% to 80%, which are higher than the average efficiency of conventional power plants (33%) in the United States. These CHP efficiency gains improve the economics, as well as have other environmental benefits [2].

Forestry is the biggest source for Mississippi's economy [3]. It covers 18 million acres or 62% of the total land area in Mississippi. In Mississippi, about four million dry tons of woody biomass is available for energy production, distributed into four major types namely, logging residues (70%), small diameter trees (20%), urban wastes (7%), and mill residues (3%) [4]. Logging residues represent a significant feedstock that can be utilized for energy production, and it is less expensive than small diameter trees. In addition to forest residues, Mississippi has other types of biomass feedstock such as corn that can be used for CHP production. According to the National Agricultural Statistics Service [5] corn production in Mississippi increased significantly from 0.9 MT in 2006 to 3.4 MT in 2007. Corn acreages and corn productivity grew dramatically from 1940 to 2010 [5,6]. According to a Mississippi Agricultural and Forestry Experiment Station report [7], Mississippi farmers have grown about 280,852 ha of corn and produced 2.3 MT over the last five years. In the past 20 years corn yield doubled and are increasing faster than any other crop grown in Mississippi.

The main objectives of this study were to: (a) assess the use of corn stover and forest logging residues as distributed feedstock sources for combined heat and power facilities in Mississippi; (b) determine the potential capacity of CHP plants that can run with available biomass; and (c) quantify the sustainable amount of feedstock that can support two biomass-based CHP facilities in Mississippi.

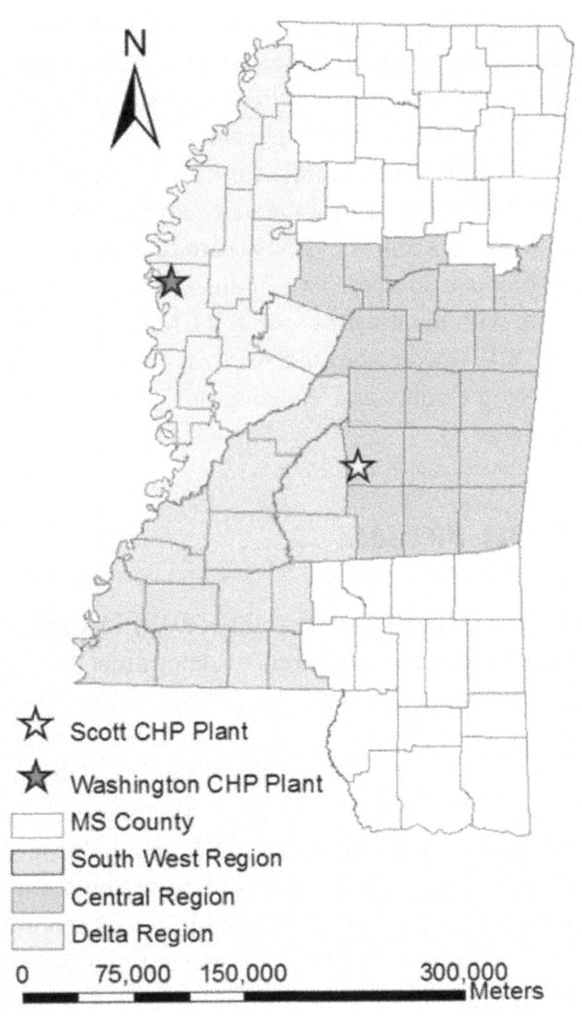


**FIGURE 1:** Map showing the location of the CHP plants considered in this study.

## 1.2 MATERIALS AND METHODS

### 1.2.1 STUDY AREA

The analysis of available biomass feedstock for CHP application focused on existing CHP facilities or plants in Mississippi. There are 22 CHP facilities in the state of Mississippi that generate a total of 570.4 MW [8–10]. The applications of these units range from dairy facilities with a capacity of 50 kW to oil refineries with a capacity up to 146.9 MW. Eight CHP facilities with lower capacities (<5 MW) are fueled by either wood, wood waste or other biomass sources. Two biomass-based CHP plants located in Mississippi, one with the lowest capacity (1 MW) and one with the highest capacity (5 MW), were selected for this study. The capacity of these CHP plants in Scott County and Washington County is 1 MW and 5 MW, respectively (Figure 1).

### 1.2.2 AVAILABLE BIOMASS

Ten years of corn production data (2001–2010) in each county in Mississippi were collected and analyzed to determine the potential of this feedstock for sustainable CHP energy production. Corn production values were obtained by averaging the county production of corn in terms of total bushels produced for the years 2001–2010 [5]. Corn production was summed across all Mississippi counties and converted from bushels to tons [11]. Dry weight of corn grain is 720.8 kg per cubic meter (56 pounds per bushel) [12]. For this study, one ton of corn stover was produced for every ton of harvested corn grain, based on values reported by [12,13]. The amount of corn stover that can be collected in a sustainable way depends on different factors such as type and sequence of collection operations, the efficiency of the collection equipment, environmental restrictions, such as the need to control erosion, maintain soil productivity and soil carbon levels. Wyman and Hinman stated that 58% of the corn stover could be collected on a sustainable basis [14]. Scechinger and Hettenhaus

[15] suggested lower rates ranging from 40% to 50%, while Brechbill [16] noted that a 53.5% utilization rate was sustainable. A conservative value of 33% collection rate of stover was used in this study, similar to the value used by Perlack [12]. The moisture content of corn stover was assumed to be 47% [17]. The available corn for energy production was calculated using Equation (1):

$$\text{Corn stover (dry tons)} = \text{Corn production (in wet tons)} \times 0.33 \times 0.53 \quad (1)$$

County level dry logging residue data from 1995, 1999, and 2002 were obtained from the Department of Forestry, Mississippi State University. The data were part of an inventory conducted by Periz-Verdin [4] who drew from two main sources, namely, the Mississippi Institute for Forestry Inventory (MIFI) satellite imagery-based data, and Forestry Inventory Analysis (FIA) Timber Products Output database (TPO) forest inventory data. The quantity of logging residues was based on the FIA–TPO forest inventory data as reported by the Southern Research Station. Periz-Verdin [4] converted the average volume of logging residues into dry tons using density values for pine and hardwoods of 0.507 and 0.61 dry t $m^{-3}$, respectively, for each county. The density values were obtained from a previous study by Gan and Smith [18]. According to Perlack [19], not all logging residues are available for bioenergy conversion. Based on this study, it was assumed 65% of logging residues are removed during the harvest of conventional products. The removal or collection of the logging residue impacts the soil nutrients and rate of soil erosion [20]. An optimal amount of biomass that varies between 0.8 and 2.2 t per ha per rotation period, depending on the region and local conditions, should be left on the soil to compensate for the extraction of essential nutrients (e.g., calcium, magnesium, potassium, and phosphorus) [21]. Periz-Verdin [4] found that the resulting biomass left on site was greater than the amount recommended by Borjesson [21]. After 65% of logging residues were recovered from timber harvesting, the percentage of biomass (35%) left in the field serves as soil nutrient.

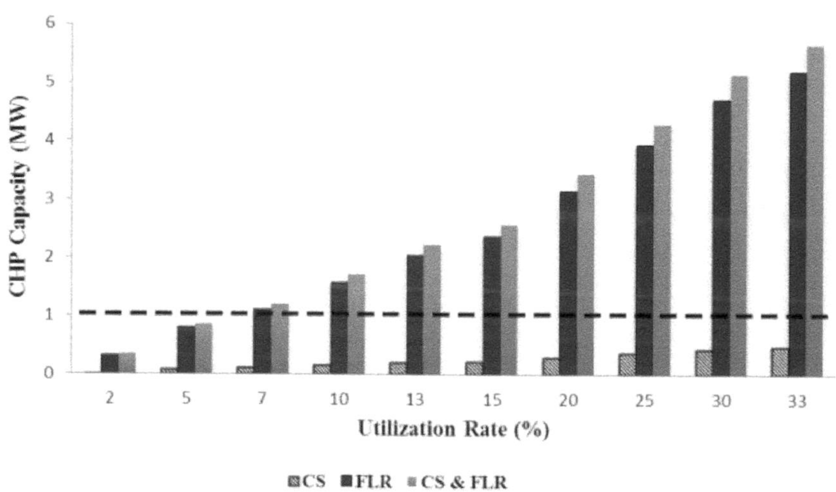


**FIGURE 2:** The capacities of the CHP plant in Scott county based on different utilization rates (%) of biomass feedstock. "---" Represents the existing capacity of the CHP plant in Scott county (1 MW).

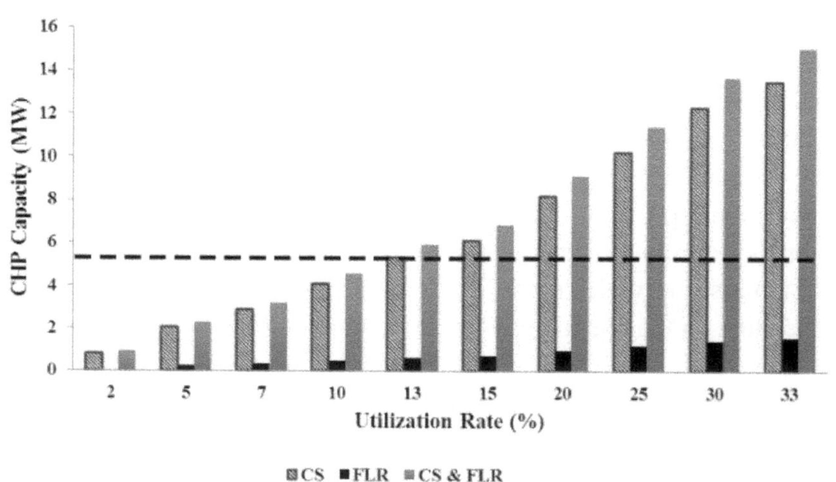


**FIGURE 3:** The capacities of the CHP plant Washington county calculated based on different utilization rates (%) of biomass feedstock. "---" Represents the existing capacity of the CHP plant in Washington county (5 MW).

### 1.2.3 BIOMASS CONVERSION

Different percentages of available corn stover and forest logging residue for CHP use were evaluated. Several corn stover (CS) and forest logging residue (FLR) utilization rates ranging from 2% to 33% were considered in this study (Figures 2 and 3). The maximum rates for available corn stover and forest logging residue were 33% and 30%, respectively. High Energy Heating Values (HHV) for corn stover and logging residue were set at 17.58 MJ/kg (7560 BTU/lb) and 19.93 MJ/kg (8570 BTU/lb), respectively [2,22]. The energy produced using corn stover and forest logging residue was calculated for each county. Viana [23] found that a typical CHP plant operated for a total of 8000 h a year or 333 days a year. For this study, the CHP plants in Scott and Washington counties were assumed to be in operation 24 h a day, 340 days a year and 8160 h per year. These values were slightly higher than the operational hours reported by Viana [23]. The total CHP energy efficiency was assumed at 70% [2].

Equation (2) was used to calculate the CHP power rating based on the energy content of feedstock (corn stover or forest logging residue):

$$\text{Power Rating} = \text{Feedstock Quantity} \times \text{Energy Content of feedstock} \times \text{CHP efficiency} \quad (2)$$

One dry ton per day of logging residue can produce 0.15 MW of CHP energy, while one ton per day of corn stover can produce 0.13 MW.

## 1.3 RESULTS AND DISCUSSION

The top five Mississippi counties in terms of available forest logging residue are listed in Table 1. Each of the five counties is covered by forest area that is 80% to 86% of the county area. Thus, these five counties are predominately covered by forest and the high percentage of forest land is reflected in the logging residue available in these counties (Figure 4). The top five counties can each produce at least 8.3 MW of CHP power from utilizing 30% of available forest logging residues (Table 1). The available logging residue tons ranged from 188 to 222 t per day.

**TABLE 1:** Top 5 Mississippi counties in terms of available forest logging residue and their potential CHP capacities.

| County | FLR (t/day) | FLR Utilization Rate at 30% (t/day) | FLR (MW) |
|---|---|---|---|
| Amite | 222 | 67 | 9.8 |
| Copiah | 219 | 66 | 9.6 |
| Clarke | 218 | 66 | 9.6 |
| Wayne | 212 | 64 | 9.3 |
| Wilkinson | 188 | 56 | 8.2 |

*Notes: Forest Logging Residue (FLR), utilization rate (UR).*

There are 20 counties in Mississippi that can each produce 8 MW or higher CHP power based on combined utilization of corn stover and forest logging residue (Table 2). The distribution of these counties is shown in Figure 5. Yazoo County had the highest amount of available feedstock for CHP production, and its potential CHP capacity was estimated at 19.6 MW. The potential county-level CHP capacities, based on combined utilization of corn stover and forest logging residue, range from 1.8 MW to 19.6 MW (Figure 6).

The CHP facility in Washington County has an existing capacity of 5 MW. In this study, we found that the CHP plant in Washington County can be operated with just a 13% corn stover utilization level (Table 3). The 1 MW capacity CHP plant in Scott County can be operated with a 7% forest logging residue utilization level within the county, but would require using corn stover at 7% utilization level from four adjacent counties (Table 3). In Washington County, it was possible to increase the capacity of the CHP plant from 5 MW to 13.5 MW by using a sustainable CS utilization level of 33% (Table 4). Similarly in Scott County, a 30% FLR utilization level could potentially increase the capacity of the CHP plant from 1 MW to

4.7 MW (Table 4). The capacities of CHP plants in Scott and Washington counties could increase to 5.2 MW and 14.9 MW, respectively, by combining both types of feedstock.

**TABLE 2:** Counties with potential CHP capacity of more than 8 MW based on combined use of corn stover and forest logging residue.

| County | CS (t/day) | CHP CS (MW) | FLR (t/day) | CHP FLR (MW) | CHP CS & FLR (MW) |
|---|---|---|---|---|---|
| Yazoo | 594 | 13.4 | 140 | 6.1 | 19.6 |
| + Washington | 597 | 13.5 | 32 | 1.4 | 14.9 |
| Noxubee | 279 | 6.3 | 124 | 5.5 | 11.8 |
| + Issaquena | 285 | 6.5 | 98 | 4.3 | 10.8 |
| Holmes | 198 | 4.5 | 139 | 6.1 | 10.6 |
| + Sharkey | 355 | 8.0 | 54 | 2.4 | 10.4 |
| Leflore | 393 | 8.9 | 34 | 1.5 | 10.4 |
| # Amite | 0 | 0.0 | 222 | 9.8 | 9.8 |
| Madison | 157 | 3.5 | 141 | 6.2 | 9.7 |
| Tallahatchie | 316 | 7.2 | 57 | 2.5 | 9.6 |
| # Copiah | 0 | 0.0 | 219 | 9.6 | 9.6 |
| # Clarke | 0 | 0.0 | 218 | 9.6 | 9.6 |
| Warren | 108 | 2.4 | 162 | 7.1 | 9.5 |
| # Wayne | 0 | 0.0 | 212 | 9.3 | 9.3 |
| Panola | 88 | 2.0 | 166 | 7.3 | 9.3 |
| + Bolivar | 315 | 7.1 | 49 | 2.1 | 9.3 |
| + Sunflower | 396 | 9.0 | 4 | 0.2 | 9.1 |
| Hinds | 96 | 2.2 | 156 | 6.9 | 9.0 |
| Rankin | 30 | 0.7 | 183 | 8.0 | 8.7 |
| + Wilkinson | 0 | 0.0 | 188 | 8.3 | 8.3 |

*Notes: Corn stover (CS) utilization rate was assumed to be 33% and forest logging residue (FLR) utilization level was assumed to be 30%; + Counties which are located in the delta region; # Counties which are produce CHP power more than 8 MW, utilizing only forest logging residue.*

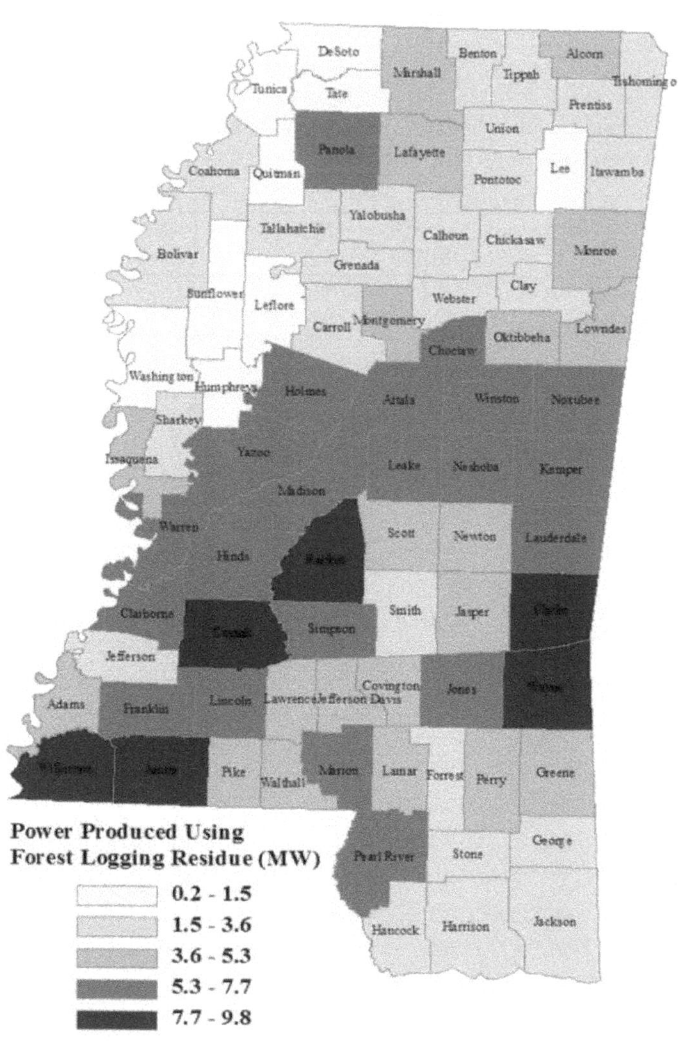


**FIGURE 4:** County-level potential CHP capacities (MW) for Mississippi based on 30% utilization of forest logging residue.

**FIGURE 5:** Top 20 counties in Mississippi with a potential capacity of more than 8 MW based on combined use of corn stover and forest logging residue.

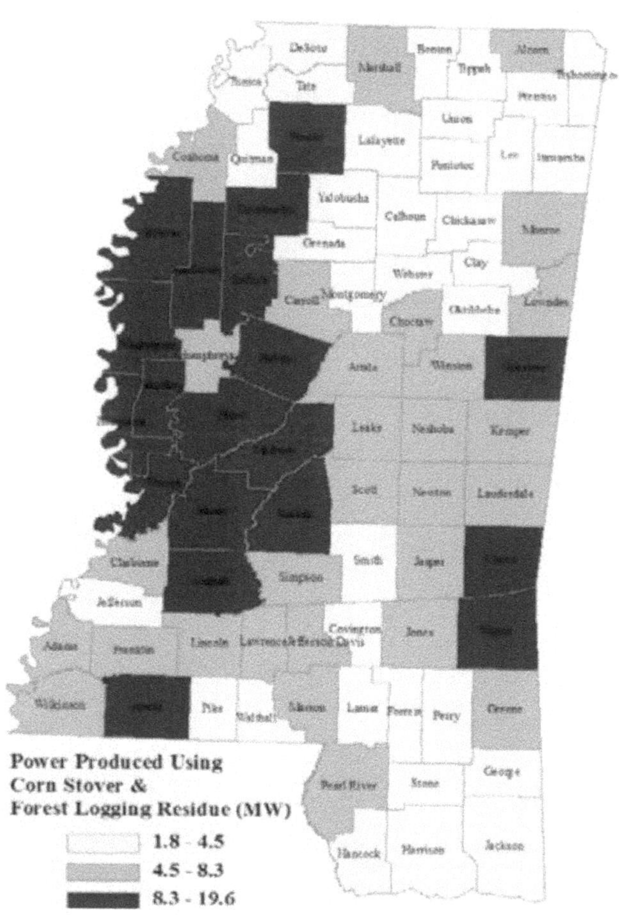


**FIGURE 6:** County-level potential CHP capacities (MW) for Mississippi based on combined use of corn stover (33%) and forest logging residue (30%).

**TABLE 3:** Counties with existing CHP facilities, available biomass and required utilization levels to support current CHP capacity.

| CHP plant (MW) | CS (t/day) | FLR (t/day) | RUR CS CHP capacity (%) | RUR FLR CHP capacity (%) |
|---|---|---|---|---|
| Washington (5 MW) | 597 | 32 | 13 | 15 * |
| Scott (1 MW) | 21 | 107 | 7 ** | 7 |

*Notes: Required utilization rate (RUR), Forest logging residue (FLR), Corn stover (CS); Forest Logging Residue supply was calculated as an average supply of 1995, 1999, and 2002; * There is not sufficient logging residue in Washington County to operate a 5 MW plant. However, combining the FLR in 4 adjacent counties at 15% utilization rate will be sufficient to support a 5 MW plant; Corn stover supply was calculated as an average supply of 2001 to 2010; ** There is not sufficient corn stover in Scott County to operate a 1 MW plant; However, combining the CS in four adjacent counties at 7% utilization rate will be sufficient to support a 1 MW plant.*

**TABLE 4:** Potential capacities of CHP plants in Washington and Scott based on higher utilization rates of corn stover and forest logging residue feedstock.

| CHP plant capacity (MW) | CS CHP capacity (MW) | FLR CHP capacity (MW) | CS & FLR CHP capacity (MW) |
|---|---|---|---|
| Washington (5) | 13.5 | 1.4 | 14.9 |
| Scott (1) | 0.5 | 4.7 | 5.2 |

*Notes: Based on 30% utilization of FLR and 33% utilization of CS.*

## 1.4 CONCLUSIONS

The potential increase in CHP capacity was assessed based on a sustainable utilization rate of available corn stover and forest logging residue in Mississippi. The results show that the available corn stover is 2.0 MT/year, and forest logging residues is 2.8 MT/year. The total amount of available biomass feedstock is 4.8 MT/year, which represents a significant amount of renewable resource that can be utilized for CHP production in Mississippi. In this study, a sustainable corn stover utilization rate of 33% can produce up to 126.9 MW and sustainable forest logging residue utilization rate of 30% can produce 332.7 MW of power using CHP. The Mississippi Delta region is the main source for corn stover, while the southwest region of Mississippi has more forested areas, which can be tapped for energy production.

## REFERENCES

1. Hinnells, M. Combined heat and power in industry and buildings. Energy Policy 2008, 36, 4522–4526.
2. Environmental Protection Agency. Catalogue of CHP Technologies; Environmental Protection Agency: Washington, DC, USA, 2012. Available online: http://www.epa.gov/chp/technologies.html (accessed on 10 May 2012).
3. Munn, I.A.; Tilley, B.K. The Impact of the Forest Products Industry on the Mississippi Economy: An Input-Output Analysis. In Forestry in Mississippi; Forest and Wildlife Research Center, Mississippi State University: Starkville, MS, USA, 2005.
4. Perez-Verdin, G.; Grebner, D.L.; Sun, C.; Munn, I.A.; Schultz, E.B.; Matney, T.G. Woody biomass availability for bioethanol conversion in Mississippi. Biomass Bioenergy 2009, 33, 492–503.
5. USDA-NASS. National Agricultural Statistics Service, 2012. Available online: http://quickstats.nass.usda.gov/results/012FF1F0-CB20–315E-9ED5–947848B06356. (accessed on 10 May 2012).
6. Woli, P.; Paz, J.O. Analyzing the Effect of Variations in Soil and Management Practices on the Sustainability of Corn Stover-Based Bioethanol Production in Mississippi; American Society of Agricultural and Biological Engineers: St. Joseph, MI, USA, 2011.
7. MSU Cares. Mississippi Corn Production, 2012. Available online: http://msucares.com/ crops/corn (accessed on 10 May 2012).
8. Energy and Environmental Analysis Inc. Combined Heat and Power Database, 2012a; Energy and Environmental Analysis Inc.: Washington, DC, USA, 2012. Available online: http://www.eea- inc.com/chpdata/index.html (accessed on 8 May 2012).
9. Energy and Environmental Analysis Inc. Combined Heat and Power Units Located in Mississippi, 2012b; Energy and Environmental Analysis Inc.: Washington, DC, USA, 2012. Available online: http://www.eea-inc.com/chpdata/States/MS.html (accessed on 8 May 2012).
10. Energy Information Administration, 2012. Available online: http://www.eia.doe.gov/ (accessed on 8 May 2012).
11. Larson, W.E.; Holt, R.F.; Carlson, C.W. Residues for Soil Conservation. In Crop Residue Management Systems; American Society of Agronomy Special Publication, Number 31; American Society of Agronomy-Crop Science, Society of America-Soil Science Society of America: Madison, WI, USA, 1978; pp. 1–15.

12. Perlack, R.D.; Turhollow, A.F. Assessment of Options for the Collection, Handling, and Transport of Corn Stover; Oak Ridge National Laboratory: Oak Ridge, TN, USA, 2002.
13. Lang, B. Estimating the Nutrient Value in Corn and Soybean Stover. In Fact Sheet BL-112; Iowa State University: Ames, IA, USA, 2002.
14. Wyman, C.E.; Hinman, N.D. Fundamentals of production from renewable feedstocks and use as transportation fuel. Appl. Biochem. Biotechnol. 1990, 24/25, 735–753.
15. Schechinger, T.M.; Hettenhaus, J. Corn Stover Harvest: Grower, Custom Operator, and Processor Issues and Answers Iron Horse Custom Farms; US Department of Energy: Oak Ridge, TN, USA, 1999.
16. Brechbill, S.C.; Tyner, W.E.; Ileleji, K.E. The economics of biomass collection and transportation and its supply to Indiana cellulosic and electric utility facilities. Bioenergy Res. 2011, 4, 141 152.
17. Frear, C.; Zhao, B.; Fu, G.; Richardson, M.; Chen, S.; Fuchs, M.R. An Evaluation of Organic Material Resources for Bioenergy Production in Washington State. In Biomass Inventory and Bioenergy Assessment; Washington State Department of Ecology: Spokane, WA, USA, 2005.
18. Gan, J.; Smith, C.T. Availability of logging residues and potential for electricity production and carbon displacement in the USA. Biomass Bioenergy 2006, 30, 1011–1020.
19. Perlack, R.D.; Wright, L.L.; Turhollow, A.F.; Graham, R.L.; Stokes, B.J.; Erbach, D.C. Biomass as Feedstock for a Bioenergy and Bio Products Industry: The Technical Feasibility of a Billion-Ton Annual Supply. In DTIC Document; U.S. Department of Energy: Oak Ridge, TN, USA, 2005.
20. Sanchez, F.G.; Carter, E.A.; Klepac, J.F. Enhancing the soil organic matter pool through biomass incorporation. Biomass Bioenergy 2003, 24, 337–349.
21. Börjesson, P. Economic evaluation of the environmental impact of logging residue recovery and nutrient compensation. Biomass Bioenergy 2000, 19, 137–152.
22. Gulf Coast CHP Applications Center. Combined Heat and Power Potential Using Texas Agricultural Wastes; Houston Advanced Research Center: The Woodlands, TX, USA, 2008.
23. Viana, H.; Cohen, W.B.; Lopes, D.; Aranha, J. Assessment of forest biomass for use as energy. GIS-based analysis of geographical availability and locations of wood-fired power plants in Portugal. Appl. Energy 2010, 87, 2551–2560.

# PART II

# PROCESS TREATMENTS AND TECHNOLOGIES

## CHAPTER 2

# CHARACTERISTICS OF CORN STOVER PRETREATED WITH LIQUID HOT WATER AND FED-BATCH SEMI-SIMULTANEOUS SACCHARIFICATION AND FERMENTATION FOR BIOETHANOL PRODUCTION

XUEZHI LI, JIE LU, JIAN ZHAO, AND YINBO QU

## 2.1 INTRODUCTION

The high demand for energy worldwide and fossil fuel reserves depletion have generated increasing interest in renewable biofuel sources [1]. The use of bioethanol produced from lignocellulosic material can reduce our dependence on fossil fuels [2]. Lignocellulosic material, for example, waste products from many agricultural activities, is a promising renewable resource for bioethanol production [3]. This generally cheap and abundant material does not compete with food production compared with agricultural crops [4]. The conversion of lignocellulosic material to bioethanol has been a research focus in China for the past decades [5]. In China, corn stover is an agricultural residue that is produced annually. Therefore, research on ethanol production from corn stover is of high importance in the new energy resource development [6]. The conversion process of lignocellulosic material to bioethanol generally includes four steps, namely, pretreatment, enzymatic hydrolysis, fermentation, and distillation [7]. Pretreatment technologies are necessarily applied to lignocellulosic material to decrease recalcitrance and to improve the yield of fermentable sugars [8], [9]. Many pretreatment methods have been proposed and investigated, such as alkaline [10], [11], steam explosion [12], [13], ammonia fiber expansion [14], [15], organic solvent [16], dilute acid [17], [18], and so on. Different pretreatment methods have different mechanisms, for example, they can decrease cellulose crystallinity and/or the polymerization degree, increase accessible surface areas, or selectively remove hemicellulose and lignin from the lignocellulosic material [19]. However, economic and environmental requirements limit the applicability of these methods. An effective pretreatment strategy should also minimize carbohydrate degradation and the production of enzyme inhibitors and toxic products for fermenting microorganisms [20]. One of the most promising pretreatment processes for lignocelluloses material is liquid hot water (LHW) pretreatment [21]–[23]. Some studies have been conducted on the mechanisms of LHW pretreatment [24]–[26]. However, different biomass types have different structures and show different reaction mechanisms.

In the process of ethanol production from lignocellulosic material, enzymatic hydrolysis and fermentation can be performed separately or simultaneously. In separate hydrolysis and fermentation (SHF), these two steps are separate, and SHF can coordinate the inconsistent contradiction between the temperatures for enzymatic hydrolysis and fermentation [27]. In simultaneous saccharification and fermentation (SSF), both steps occur in a single bioreactor where the glucose formed is rapidly converted to ethanol by the yeast. However, solid loading is limited by the higher effective mixing and high viscosity of the system in the SSF process [28]. Semi-SSF (S-SSF) of ethanol production is an operating mode between SSF and SHF. S-SSF consists of two phases, namely, pre-hydrolysis and SSF. To increase substrate concentration, fed-batch S-SSF process was carried out. Fed-batch S-SSF for ethanol production showed that higher substrate concentration and higher ethanol yield can be obtained compared with S-SSF and SSF when a suitable pre-hydrolytic period is selected [29]. In our previous study, LHW pretreatment was applied to corn stover to test the efficiency of enzymatic hydrolysis, and cellulose conversion rates of almost 100% were obtained [30]. In the present work, corn stover samples were subjected to a combination of LHW pretreatment and fed-batch S-SSF to obtain higher ethanol concentration and yield. The effects of different impact factors on the fermentation digestibility of LHW-pretreated corn stover in S-SSF and fed-batch S-SSF are discussed, and the chemical structures and morphological characteristics of corn stover during LHW pretreatment were presented.

## 2.2 MATERIALS AND METHODS

### *2.2.1 MATERIALS*

Corn stover was collected from a corner of field near Jinzhou New District (Dalian, China). It was stated that a permit was not required to collect the corn stover. It was also confirmed that the corn stover is not a protected or endangered species. Corn stover was manually cut into pieces, milled, and screened to collect 20 mesh to 80 mesh fractions. Samples were then homogenized and stored in a plastic bag for subsequent experiments. Corn stover was composed of the following: 10.9% benzene-alcohol (2:1) extractive, 38.8% glucan, 23.5% xylan, 15.6% acid-insoluble lignin, 2.4% acid-soluble lignin, and 3.7% ash in terms of oven-dried weight. The commercial cellulase used in the study was purchased from Imperial Jade Biotechnology Co., Ltd., Ningxia, China. Cellulase was derived from *Trichoderma longbrachiatum. Saccharomyces cerevisiae* was purchased from Angel Yeast Co., Ltd., China. The yeast was activated prior to fermentation. Approximately 1 gram of dry yeast was added to 20 mL of 5% sterilized glucose solution, activated at 38°C for 1 h, cooled to 28°C to 30°C, and used in the fermentation experiment. The fermentation medium contained 0.3% yeast extract, 0.5% peptone, 2.5% $KH_2PO_4$, 0.03% $MgCl_2$, and 0.025% $CaCl_2$.

### 2.2.2 LHW PRETREATMENT

LHW pretreatment was conducted in a 15 L digester with four small tanks (mechanical mill of Shanxi University of Science and Technology, China). Approximately 40 g of corn stover and 800 mL of deionized water were loaded into the small tanks. The start temperature for the pretreatment was 50°C, and the maximum temperature was controlled in the range 170°C to 210°C. The time to maximum temperature was maintained at 100±2 min, and the pretreatment reaction time was set to either 20 or 40 min. Severity factor, which was defined by Overend and Chornet, was used for measuring the pretreatment intensity in LHW. The severity factor provides a way to compare the combined effects of parameters on the changes in the composition to enable a better comparison of results and a better correlation with the compositional changes in the biomass after pretreatment [31]. The severity factors corresponding to different LHW pretreatment conditions are calculated using the following formula (1):

$$\log(R_0) = \log\left\{t\times \exp\left[\frac{T - T_{ref}}{14.75}\right]\right\} \quad (1)$$

where t is the reaction time (min), T is the pretreatment temperature (°C), and $T_{ref}$ = 100°C. After pretreatment, the solid residues and the prehydrolysates were separated by filtration with a Bŭchner funnel. The prehydrolysates were analyzed for pH and contents of glucose, xylan, acid-soluble lignin, furfural, and HMF. The solid residues were analyzed for yield and contents of chemical compositions. The solid residues were used for subsequent fermentation.

### 2.2.3 S-SSF

For S-SSF process, sample pre-hydrolysis was performed at 50°C for 6 h to 24 h prior to the main SSF phase. The weighted solid residue from LHW pretreatment was added into in 100 mL Erlenmeyer flask that contains pH 4.8 buffers. Cellulase loading was 25 to 50 filter paper unit per gram of oven-dried solid residues. After the pre-hydrolysis time, the medium temperature was adjusted to a constant fermentation temperature and maintained during the subsequent SSF. Then, approximately 1 mL of activated yeast was added into the medium. The fermentation experiments were performed in a constant-temperature incubator for 72 h. The flasks were sealed with rubber stoppers and equipped with syringe needles to remove the generated carbon dioxide. Samples were collected at 0, 12, 24, 36, 48, 60, and 72 h for glucose concentration and ethanol analyses. Glucose and ethanol were determined using the SBA-40D Biological Sensing Analyzer (Biology Institute of the Shandong Academy of Sciences, Jinan, China). Ethanol yield was calculated using the formula (2):

$$\text{Ethanol yield (\%)} = \frac{[EtOH]}{f \times biomass \times 1.111 \times 0.51} \times 100\% \tag{2}$$

where [EtOH] = ethanol concentration at the end of the fermentation minus any ethanol produced from the enzyme and medium (g/L); f = cellulose fraction of dry biomass (g/g); biomass = dry biomass concentration at the beginning of the fermentation (g/L); 0.51 = conversion factor for glucose to ethanol based on the stoichiometric biochemistry of yeast; and 1.111 = conversion factor of cellulose to equivalent glucose. Each experiment was performed using three parallel samples and the standard error was calculated using Microsoft Excel software in computer.

### 2.2.4 FED-BATCH S-SSF

Fed-batch was conducted in two ways. The first approach involved the feeding of solid residue (71% moisture content) at the pre-hydrolysis time of 6 h into the fermentation flasks to final substrate concentration of 16.1% (10% + 6.1%). The second approach involved the feeding of solid residue in batches at pre-hydrolysis times of 2, 4, 6, and 16 h, into the fermentation flasks. The final solid loadings in the second mode were 17.0% (initial loading of 10%, and 2.8%, 2.3%, and 1.9% at pre-hydrolysis times of 2, 4, and 6 h, respectively) for solid residue abstained at severity factor of 3.95 and 18.7% (10% + 2.8% + 2.3% + 1.9% + 1.7%, in which 1.7% was added at pre-hydrolysis time of 16 h) for that at severity factor of 4.54. The other conditions for fed-batch S-SSF were fermentation temperature of 36°C, pH 4.8, cellulase dosage of 40 FPU/g oven-dried solid residues, pre-hydrolysis time of 18 h, pre-hydrolysis temperature of 50°C, and initial solid loading of 10%. The required total cellulase was added before prehydrolysis. Approximately 1 mL of the activated yeast was added into the medium at the beginning of fermentation. The rest of the process steps were similar to those performed in the S-SSF pretreatment of solid residues.

### 2.2.5 ANALYSIS

The benzene–alcohol (2:1) extractive contents were determined using the Chinese National Standard method (GB/T2677.6-1994). The sample was extracted for 6 h with benzene–alcohol mixture (2:1), then the solvent mixture with extractives was distilled to recover solvent, and remaining residue was dried and weighed for calculating extractives content using formula (3):

$$\text{Extracives contents (\%)} = \frac{weight\ of\ remaining\ residue\ (g)}{sample\ weight\ (g)} \times 100\% \tag{3}$$

The content of acid-insoluble lignin (AIL) was determined according to the Chinese National Standard method (GB/T2677.8-1994). Extractives free sample was hydrolyzed with sulfuric acid of 72% ± 0.1% at 18~20°C for 2.5 h, then the system was diluted with distilled water to 3% of sulfuric acid concentration, and further hydrolyzed at 100°C for 4 h. After the hydrolysis, the hydrolysis residue was separated by filtration using the filtering crucible, and washed with fresh distilled water to about neutral. Dry the crucible and acid insoluble residue at 105 ± 3°C until a constant weight is achieved for calculating acid-insoluble lignin content, which is a percent of weight of the residue to weight of sample.

The content of acid-soluble lignin (ASL) was determined according to the Chinese National Standard method described in GB/T10337-1989. Using the hydrolysis liquor aliquot obtained in assay of acid-insoluble lignin, measure the absorbance of the sample at 205 nm on a UV-Visible spectrophotometer. 3% sulfuric acid was used to dilute the sample, and the same solvent was used as a blank. The amount of acid soluble lignin was calculated using formula (4)

$$\text{ASL (\%)} = \frac{A \times Dilution \times V}{1000 \times \varepsilon \times m} \times 100\% \tag{4}$$

where: A = absorption value at 205 nm; Dilution = dilution factor; V = filtrate volume (ml); ε = absorptivity of biomass [L/(g.cm)]; and m = weight of oven dry sample (g)

The glucan content and xylan content in solid biomass sample were determined according to National Renewable Energy methods. The glucan content and xylan content were calculated using formula (5) and formula (6) respectively:

$$\text{Glucan content (\%)} = \frac{glu\cos e \times 0.087 \times 0.9}{m} \times 100\% \tag{5}$$

$$\text{Xylan content (\%)} = \frac{\text{xylan} \times 0.087 \times 0.88}{m} \times 100\% \tag{6}$$

Where: glucose/xylan = glucose/xylan concentration (g/L); m = mass of oven-dried solid residues (g); 0.087 = volume of acid hydrolysis liquid (L); and 0.9/0.88 = conversion factor for glucose to glucan or xylose to xylan.

For the compositions of prehydrolysates, the contents of furfural, HMF, glucose and xylose were determined using HPLC and acid soluble lignin content was analyzed by ultraviolet-spectroscopy using method above.

### 2.2.6 SCANNING ELECTRON MICROSCOPY (SEM) ANALYSIS

The pretreated samples were washed with deionized water and then dried at 105°C for 4 h. The samples were then coated with gold in a Balzers SCD004 sputter coater and examined in a JEOL JSM-6460 LV SEM (Akishima, Japan).

### 2.2.7 SPECIFIC SURFACE AREA, PORE SIZE, AND DISTRIBUTION ANALYSES

The specific surface area and the pore size of samples were determined by a high-speed automatic surface area and pore size analyzer (NOVA2200e, Quantachrome Instruments Co., USA).

### 2.2.8 FT-IR ANALYSIS

FT-IR analysis was performed on both the untreated corn stover and the solid residues pretreated at 190°C and 210°C. All samples were dried and pressed into a KBr disc. IR spectra were obtained using a Spectrum One-B FT-IR spectrometer (PerkinElmer, USA) with a resolution of 0.5 $cm^{-1}$ in the range 4000 $cm^{-1}$ to 450 $cm^{-1}$.

## 2.3 RESULTS AND DISCUSSION

### 2.3.1 CHEMICAL COMPOSITION OF PREHYDROLYSATES AND SOLID RESIDUES FROM LHW PRETREATMENT

The chemical compositions of the prehydrolysates and solid residues from LHW pretreatment at the different temperatures and times assayed are presented in Table 1. Log ($R_0$) was the severity factor used to represent pretreatment severity. In the LHW pretreatment, a fraction of the lignocellulosic material was removed from the solid corn stover and transferred to the prehydrolysate. Approximately 24% to 45% of the original material was solubilized according to different pretreatment severity, which resulted in decreased solid residues yield. Acid-insoluble lignin increased with increasing severity factor for severity factors <3.66 because of the quick removal of xylan at lower severity factors. When the severity factor increased to 4.25, the acid-insoluble lignin content decreased and was even lower than the content of untreated raw materials. When the intensity factor was higher than 4.54, the acid-insoluble lignin content increased again. This result may be caused by the formation of "lignin-like" structures obtained as a result of condensation reactions between lignin and carbohydrate degradation products. The condensation substrate was adsorbed on the surface of solid residues, which increased the acid-insoluble lignin. The acetyl groups coupled with xylan were released as acetic acid in the prehydrolysates under the high severity factor, resulting in the decreased pH of the prehydrolysates. The decreased pH resulted in a decrease of the acid-soluble lignin amount of the solid residue, whereas the acid-soluble lignin amount in the prehydrolysates increased. Xylan (including xylose) and glucose were the main two sugars in the prehydrolysates. The xylan content in the prehydrolysates rose progressively with increasing severity factor (>3.95), and then a decrease was detected. This reduction was caused by xylan degradation at high severity factors. During LHW pretreatment, the sugar degradation products that are released into the prehydrolysate, for example, furfural and HMF, inhibit both yeast [32], [33] and enzymes [34]. In previous studies [30], the cellulose conversion rate in enzymatic hydrolysis was high at the severity factor of 3.95, whereas the cellulose conversion rate did not increase significantly with increased severity factor. Therefore, in the present work, the solid residues at the severity factor of 3.95 were selected as the substrates of subsequent fermentation because the degradation products in the prehydrolysates were lower at the pretreatment severity. For comparison, the solid residues at the severity factor of 4.54 were also used as the substrate.

**TABLE 1:** Chemical compositions of the prehydrolysates and solid residues obtained from LHW pretreatment of corn stover.

| Pretreatment conditions | | log (Rol | Prehydrolysate (g/L) | | | | | | Solid residues (%)** | | | | | |
|---|---|---|---|---|---|---|---|---|---|---|---|---|---|---|
| time (min) | tempera-ture (°C) | | pH | Xylan* | Glucose | ASL | Furfural | HMF | Yield | Xylan | Glucan | AIL | ASL | Extractives |
| 20 | 170 | 3.36 | 4.19 | 2.09± 0.13 | 1.21:+: 0.05 | 1.44± 0.05 | 0 | 0.10 ± 0.03 | 76.4 | 26.02 ± 0.16 | 42.33± 0.03 | 17.14± 0.07 | 1.7 | 8.38±0.10 |
| 20 | 180 | 3.66 | 3.95 | 3.09:+: 0.16 | 1.56± 0.03 | 1.58± 0.02 | 0 | 0.11:+: 0.03 | 67.7 | 15.29± 0.14 | 46.03:+: 0.04 | 17.57± 0.03 | 1.2 | 11.23±0.09 |
| 20 | 190 | 3.95 | 3.64 | 5.29± 0.15 | 2.46± 0.02 | 2.07:+: 0.04 | 0 | 0.73± 0.02 | 59.6 | 9.23± 0.15 | 53.67:+: 0.02 | 15.3± 0.04 | 1.1 | 15.83:+:0.06 |
| 20 | 200 | 4.25 | 3.37 | 2.10± 0.12 | 3.54± 0.04 | 2.76:+: 0.05 | 0.32± 0.04 | 1.96± 0.04 | 57.5 | 5.23± 0.13 | 53.31:+: 0.02 | 15.48± 0.06 | 0.5 | 18.24±0.10 |
| 20 | 210 | 4.54 | 3.31 | 0.55± 0.13 | 4.16± 0.01 | 2.72± 0.03 | 0.36± 0.06 | 2.46± 0.03 | 57.2 | 5.10± 0.12 | 54.33± 0.03 | 16.56± 0.03 | 0.5 | 19.78±0.07 |
| 40 | 170 | 3.66 | 3.98 | 3.65± 0.11 | 1.50± 0.04 | 1.53± 0.04 | 0 | 0.13± 0.02 | 70.3 | 17.13± 0.17 | 47.42± 0.04 | 18.11:+: 0.05 | 1.5 | 8.07±0.08 |
| 40 | 180 | 3.96 | 3.81 | 4.22± 0.12 | 1.98± 0.05 | 1.97± 0.02 | 0 | 1.16± 0.04 | 62.2 | 10.62± 0.19 | 50.16± 0.03 | 17.66± 0.07 | 1.1 | 12.17±0.08 |
| 40 | 190 | 4.25 | 3.45 | 2.55± 0.15 | 3.06± 0.03 | 1.31:+: 0.04 | 0.29± 0.09 | 1.77± 0.01 | 58.0 | 5.43± 0.12 | 54.60± 0.02 | 15.42± 0.05 | 0.8 | 17.85±0.09 |
| 40 | 200 | 4.55 | 3.34 | 0.65± 0.16 | 3.72± 0.03 | 1.63± 0.05 | 0.27± 0.02 | 2.75± 0.04 | 56.0 | 4.82 ± 0.16 | 54.73± 0.04 | 17.17± 0.04 | 0.3 | 19.25±0.07 |
| 40 | 210 | 4.84 | 3.34 | 0.47± 0.17 | 4.32± 0.05 | 1.68± 0.02 | 0.16± 0.05 | 2.51:+: 0.04 | 55.7 | 4.53± 0.18 | 54.78± 0.03 | 20.09± 0.06 | 0.3 | 22.40±0.10 |

**Based on all of xylose and xy lan in prehydrolysate. **Based on oven dry weight of solid residue, except yield that on basis of weight of untreated corn stover.*

### 2.3.2 S-SSF

In the work, *Saccharomyces cerevisiae* was used in S-SSF of solid residues obtained from LHW pretreatment of corn stover. In a previous work, *S. cerevisiae* exhibited stable viability and high fermentation efficiency in SHF and SSF [35], [36].

#### 2.3.2.1 MEDIUM.

During ethanol production, the fermentation medium composition affects the fermentation performance of the yeast [37], [38]. Pre-hydrolysis in S-SSF increases the fermentation sugar concentration in the fermentation system before yeast addition. The high sugar concentration in the fermentation system increases the osmotic pressure, which has a damaging effect on yeast cells [39]. A report suggested that the required nutrients, such as nitrogen and trace elements, are provided in adequate amounts to obtain high fermentation performance in the high sugar concentration medium using *S. cerevisiae* [40]. To obtain efficient ethanol fermentation with *S. cerevisiae*, numerous nutrients are required. Chemicals contribute significantly to the cost of large-scale production. On a laboratory scale, media are often supplemented with peptone and yeast extract. Magnesium, calcium, potassium, and phosphorus that influence the sugar conversion rate are required for the fermentation [41], [42]. For comparison, S-SSF without added nutrients was performed with the solid residues pretreated with LHW at the severity factors of 3.95 and 4.54. The results are shown in Figures 1a and 1b. The final ethanol concentration in media of the added nutrients did not increase significantly; but initial productivity of the fermentation process increased. The ethanol concentration when nutrients were added was higher than that when no nutrients were added before the fermentation time of 36 h (severity factor of 3.95) and 48 h (severity factor of 4.54). By contrast, the ethanol concentration when nutrients were added was lower than that when no nutrients were added after 36 h (severity factor of 3.95) and 48 h (severity factor of 4.54). This result suggests that additional nutrients are not necessary for ethanol fermentation of the pretreated corn stover with the commercial yeast, which decreases production cost.

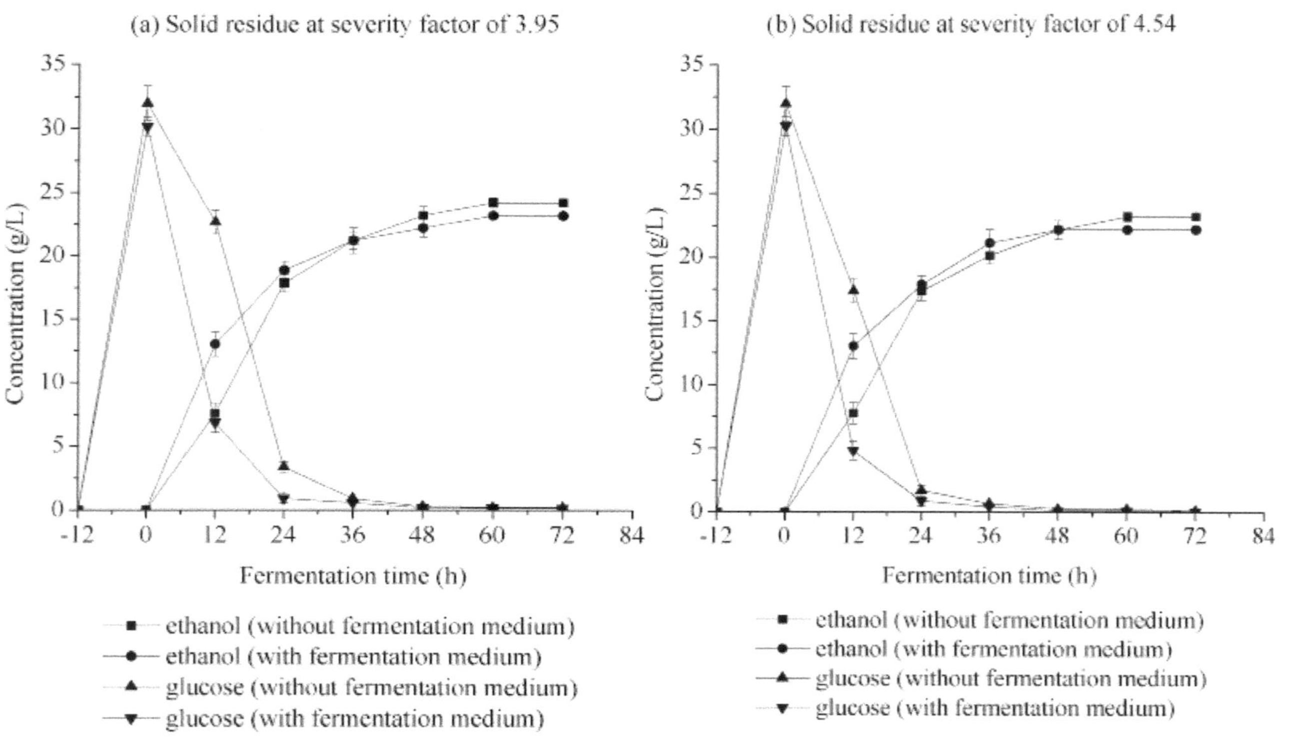


**FIGURE 1:** Concentration of ethanol and glucose obtained with and without fermentation medium. Other S-SSF conditions were cellulase loading of 50 FPU/g oven-dried solid residues, substrate concentration of 8.5%, pre-hydrolysis temperature of 50°C, pre-hydrolysis time of 12 h, pH 4.8, and fermentation temperature of 36°C.

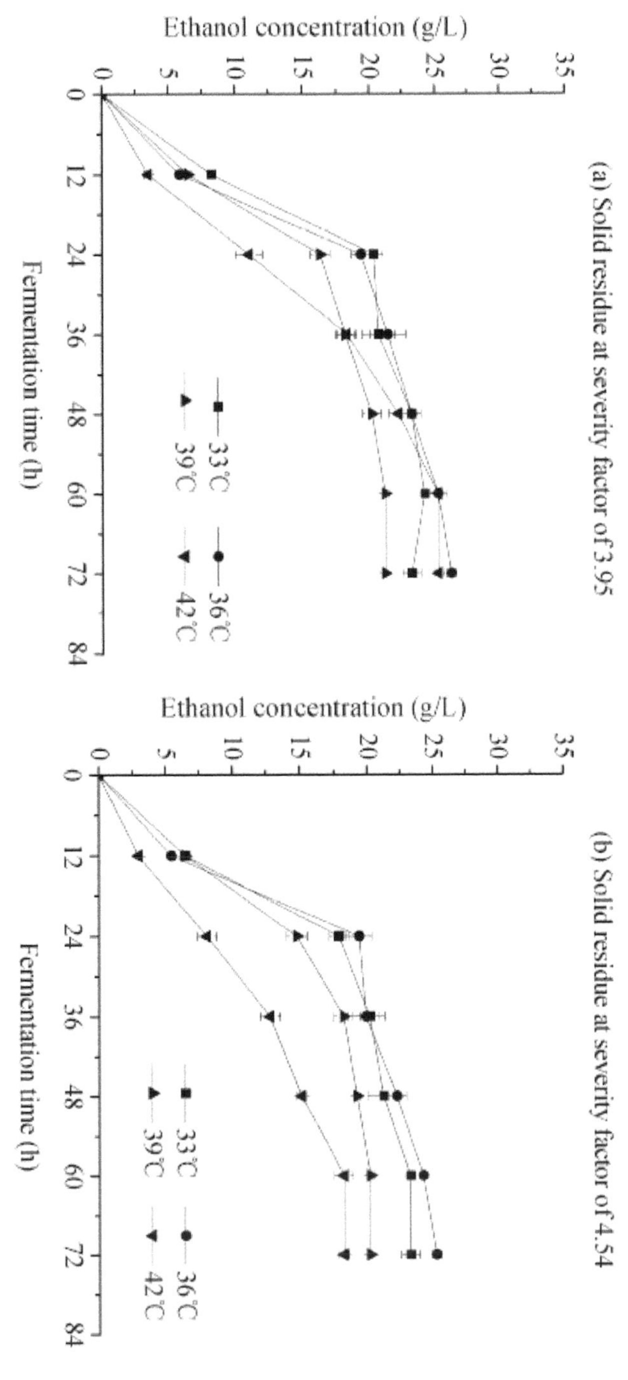


**FIGURE 2:** Effect of fermentation temperature on concentration of ethanol and glucose in S-SSF of pretreated corn stover with LHW. S-SSF conditions are same as that in Figure 1 except fermentation temperature.

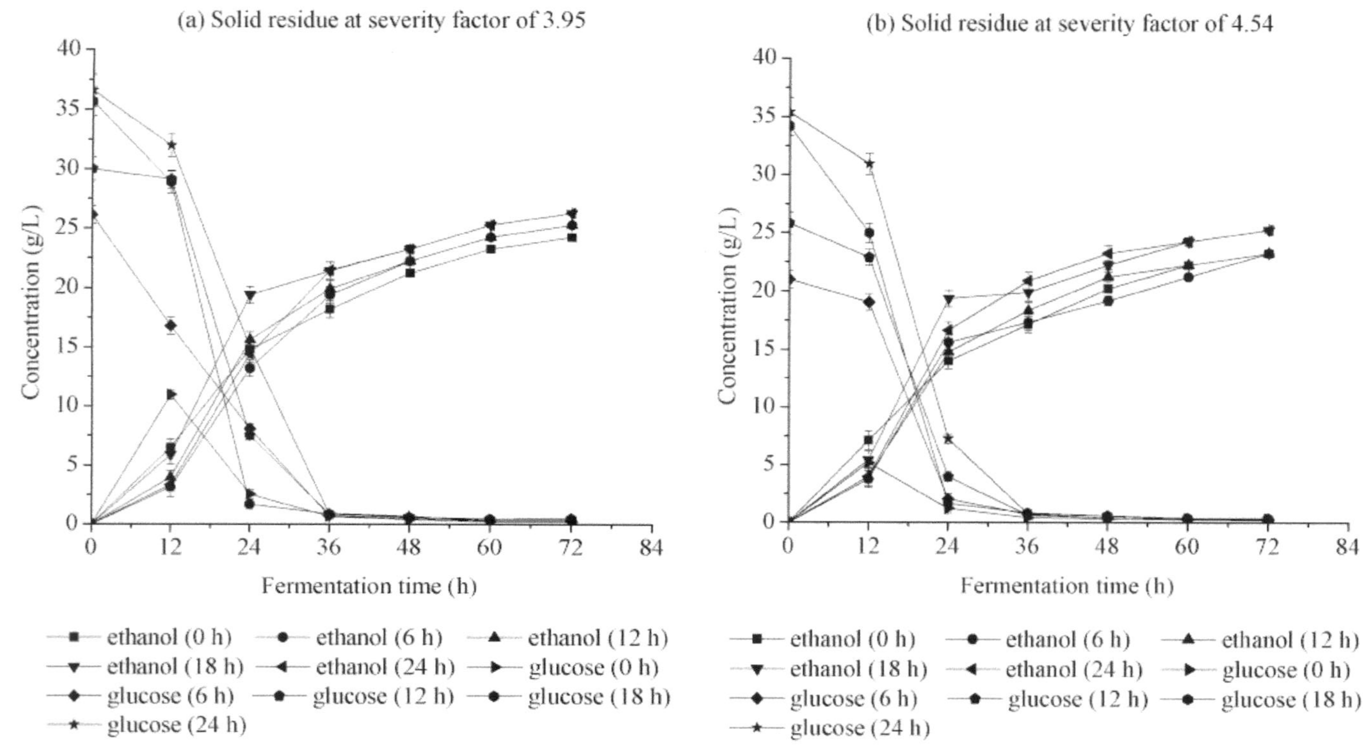


**FIGURE 3:** Effect of pre-hydrolysis time onconcentration of ethanol and glucose in S-SSF of pretreated corn stover with LHW. S-SSF conditions are same as that in Figure 1 except pre-hydrolysis time.

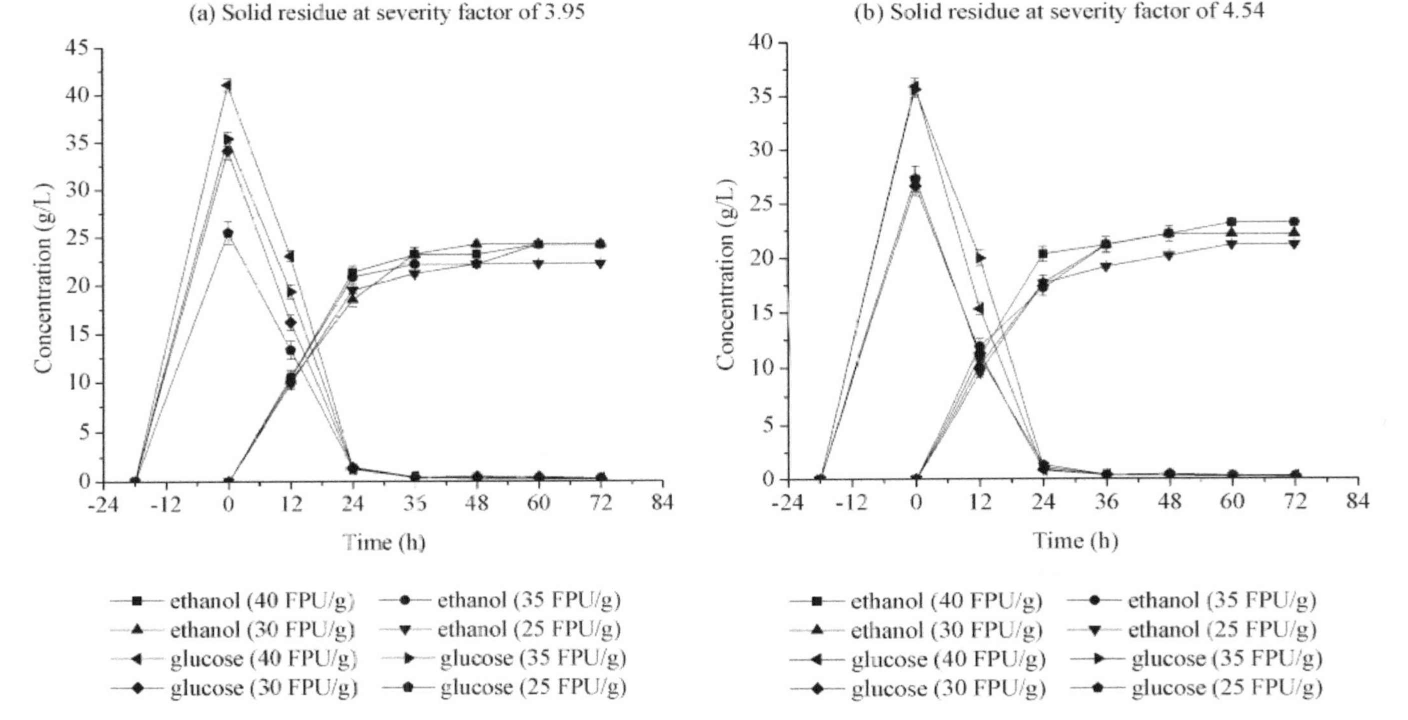


**FIGURE 4:** Concentrations of ethanol and glucose produced with different cellulase loadings. Other S-SSF conditions are same as that in Figure 1 except pre-hydrolysis time of 18 h.

**TABLE 2:** Ethanol concentration in S-SSF and Fed-batch S-SSF of solid residues from LHW pretreated corn stover with different solid loadings.

| Fermentation methods | S-SSF | | | | | | Fed-batch S-SSF | | | |
|---|---|---|---|---|---|---|---|---|---|---|
| Severity of preteatment of corn stover | Severity factor of 3.9S | | | Severity factor of 4.S4 | | | Severity factor of 3.9S | | Severity factor of 4.S4 | |
| Solid loading (%) | 8.5 | 10 | 17 | 8.5 | 10 | 17 | 16.1 (10+6.1) | 17 (10+2.8 +2.3+1.9) | 16.1 (10+6.1) | 18.7 (10 + 2.8+2.3+ 1.9 + 1.7) |
| Ethanol concentration (g/L) | 26.3 ± 0 | 33.3 ± 0 | 21.2 ± 1.0 | 25.2 ± 0.5 | 31.3 ± 0.7 | 31.3 ± 0.5 | 38.4 ± 0 | 28.3 ± 1.4 | 39.4 ± 1.4 | 39.4 ± 1.4 |

### 2.3.2.2 FERMENTATION TEMPERATURE.

Fermentation temperature is one of the main technological factors known to impact the activity of *S. cerevisiae* at industrial scale [43]. Optimal fermentation temperature can increase production ethanol yields using *S. cerevisiae*. The impact of fermentation temperature was investigated in S-SSF at 33, 36, 39, and 42°C after 12 h of pre-hydrolysis at 50°C, as shown in Figures 2a and 2b. For the two solid residues from LHW pretreatment of corn stover at the severity factors of 3.95 and 4.54, the highest ethanol yield occurred at temperature of 36°C after 72 h of fermentation time, Thus, S-SSF was performed in the present study at 36°C using *S. cerevisiae*.

### 2.3.2.3 PRE-HYDROLYSIS TIME.

The pre-hydrolysis time refers to the initial cellulose hydrolysis for a constant time prior to the main SSF phase. Pre-hydrolysis time, one of the important factors in S-SSF, influenced ethanol concentration. The concentrations of ethanol and glucose with respect to pre-hydrolysis time in S-SSF at 6, 12, 18, and 24 h and the SSF experiments are shown in Figures 3a and 3b. For S-SSF, the initial glucose concentration increased with increasing pre-hydrolysis time. The glucose concentrations gradually decreased from the initial higher values with the extension of fermentation time, which almost approached zero after 36 h. By contrast, the glucose concentration in SSF first increased because of the low ethanol production rate and the high enzyme concentration in the initial period, resulting in glucose accumulation during the initial 12 h. For S-SSF and SSF, the ethanol concentration rapidly increased within the first 24 h, and then increased slowly. This result was due to the exponential growth of yeast because of sufficient substrate supply, and glucose was quickly consumed during this period. As Figures 3a and 3b show, S-SSF at 18 h was different compared with other modes. The glucose concentration was reduced faster and the ethanol concentration increased faster than the other modes at the initial 24 h. The ethanol concentration in S-SSF at 18 h of pre-hydrolysis time is higher than that in S-SSF at 6 h, 12 h, and SSF at the same fermentation time. Ethanol concentration did not increase when pre-hydrolysis time was further extended to 24 h. In S-SSF at 18 h of pre-hydrolysis time, the final ethanol concentration reached 26.3 g/L (severity factor of 3.95) and 25.3 g/L (severity factor of 4.54) at 72 h of fermentation time. Thus, the optimal enzymatic pre-hydrolysis time to obtain the maximum ethanol concentration in S-SSF was 18 h when using pretreated corn stover with LHW as substrate.

#### *2.3.2.4 ENZYME LOADING.*

Enzyme cost has been recognized as a considerable contributor to bioethanol production cost. Therefore, the fermentation cost can be lowered by decreasing enzyme loading. Theoretically, ethanol yield increases with increasing enzyme loading. However, the complex structure of lignocellulosic material inhibits enzyme activity. From an economic perspective, higher enzyme loadings can result in the waste of a large number of enzymes. The optimization of enzyme loading dosage in S-SSF is a key requirement for large-scale bioethanol production. The results are shown in Figure 4. Ethanol concentrations showed a slight increase with increasing enzyme loading from 25 FPU/g to 30 FPU/g oven-dried solid residues after 72 h. The increase in enzyme loading above 30 FPU/g oven-dried solid residues did not increase ethanol concentration. Residual glucose concentration was less than 0.3 g/L in S-SSF with all enzyme loadings after 24 h. The appropriate cellulase loading was 30 FPU/g oven-dried solid residues.

#### *2.3.2.5 SOLID LOADINGS.*

The interest in high solid loading of enzymatic hydrolysis and fermentation is motivated by reduced liquor volume, resulting in lower operating cost [44]. As a drawback, high solid loading can result in difficulties in stirring the material. In addition, LHW pretreatment resulted in solid residues with higher moisture, complicating the further increase of high solid loading. Therefore, in the present work, S-SSF was carried out at the solid loadings of 8.5%, 10%, and 17% (w/v) of the solid residue from LHW-pretreated corn stover. Table 2 shows the ethanol concentrations and yields obtained from solid residues with LHW pretreated at 190 and 210°C subjected to S-SSF fermentation at different solid loadings. The highest ethanol concentration was obtained with 10% (w/v) solid loading. The ethanol concentration of 33.3 g/L corresponds to 98.4% of ethanol theoretical yield in the pretreated solid residues at the severity factor of 3.95. With the further increase to 17% (w/v) solid loading, the ethanol concentration and yield were even lower.

As the above analysis shows, S-SSF showed higher ethanol concentration and yield using corn stover pretreated with LHW as substrate. For possible commercial applications, ethanol concentration needs to be further increased to decrease the cost of follow-up distillation. The fed-batch S-SSF experiments results shown in Table 2 indicate that the ethanol concentration of the first feeding mode (feeding one time) were higher than those of the second feeding mode (feeding many times in batches). This difference indicates that the first feeding mode was appropriate for the corn stover pretreated with LHW. Compared with S-SSF, the ethanol concentration increased significantly in fed-batch S-SSF with the first feeding mode. At the same time, the ethanol yield reached almost 80%, which was a better result. In the optimum mode, the ethanol concentrations of 38.4 g/L (severity factor of 3.95) and 39.4 g/L (severity factor of 4.54) as well as ethanol yields of more than 78.6% (severity factor of 3.95) and 79.7% (severity factor of 4.54) were obtained. The optimum conditions for the fed-batch mode should be studied further to improve the ethanol production from corn stover. Table 3 shows a comparison of several ethanol productions using corn stover as the substrate found in the literature and in this work. Compared with the results of other studies, the ethanol concentration reached 39.4 g/L in the current study. The ethanol concentration in this work is higher than those in other studies. However, the high enzyme loadings needed to be reduced in the future studies.

**TABLE 3:** Comparison of various ethanol productions using pretreated corn stover as the substrate found in the literatures and in the current study.

| Pretreated method | Fermentation method | Hydrolysis Temp. (°C) | Temp. in SSF (°C) | Pre-hydrolysis time (h) | Fermentation time (h) | Enzyme loadings | Ethanol concentration (g/L) | Reference |
|---|---|---|---|---|---|---|---|---|
| LHW | Fed-batch S-SSF | 50 | 36 | 18 | 60 | Cellulase 30-35 FPU/g substrate | 39.4 | In this work |
| Alkaline | SSF | – | 33 | – | 72 | Cellulase 20 FPU/g substrate, β-glucosidase 10 CBU/g substrate | 27.8 | [58] |
| Dilute s lfuric acid | SHF | 45 | 35 | 72 | 72 | Cellulase 15 FPU/g substrate, β-glucosidase 9 CBU/g substrate | 30.6 | [59] |
| Dilute phosphoric acid | SHF | 45 | 35 | 72 | 48 | Cellulase 15 FPU/g, β-glucosidase 9 CBU/g substrate | 26.4 | [60] |
| Fungal | SSF | – | 37 | – | 72 | Cellulase 10 FPU/g solid | 25.0 | [61] |

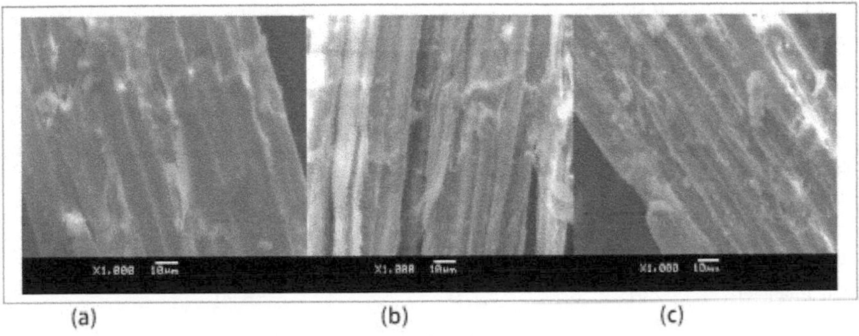


**FIGURE 5:** SEM micrographs of untreated and LHW-pretreated corn stover. a: untreated corn stover; b: pretreated corn stover at severity factor of 3.95; c: pretreated at severity factor of 4.54.

## *2.3.3 CHANGES IN STRUCTURE CHARACTERIZATION OF CORN STOVER AFTER LHW PRETREATMENT*

### *2.3.3.1 MORPHOLOGICAL CHARACTERIZATION.*

SEM was used to observe the changes of morphological characteristics of corn stover before or after LHW pretreatment. The SEM micrographs of untreated and pretreated corn stover are presented in Figure 5. LHW pretreatment significantly disrupted structure of corn stover with increased pretreatment severity and significantly decreased particle size. The surface status was also changed by LHW pretreatment. The untreated corn stover possessed a flat, smooth, rigid, regular, and compact surface structure (Figure 5a). After pretreatment with LHW, the smooth surfaces gradually became rough (Figure 5b and 5c), which was beneficial to reaction with enzyme.

**TABLE 4:** Changes in surface area, pore volume and pore diameter of corn stover before and after LHW pretreatment.

| Samples | Untreated corn stover | Pretreated at severity factor of 3.95 | Pretreated at severity factor of 4.54 |
|---|---|---|---|
| Specific surface area ($m^2/g$) | 8.55 | 17.08 | 10.55 |
| Pore volume ($cm^3/g$) | 0.001 | 0.028 | 0.020 |
| Pore diameter ($\times 10^{-9}$ m) | 2.91 | 6.77 | 8.90 |

### *2.3.3.2 SPECIFIC SURFACE AREA, PORE VOLUME AND PORE DIAMETER.*

As the aforementioned SEM analysis shows, LHW pretreatment could disrupt recalcitrant microstructure significantly to form a smaller average particle size, resulting in small pores in the pretreated sample surface. A greater amount of information related to cellulase action can be obtained from measurements of the pores or "interior" surface area of particles available for penetration by cellulase. In the present work, specific surface area, pore volume, and pore diameter were studied using the Brunauer–Emmett–Teller method by nitrogen adsorption to explain the mechanism involved in enhancing the enzymatic hydrolysis of corn stover with LHW pretreatment. The results are shown in Table 4. The specific surface area of corn stover pretreated with LHW was higher than that of the untreated sample, means that LHW pretreatment led to specific surface area increase. Generally, the specific surface area of particles is inversely proportional to their average particle size. However, in Table 4, the specific surface area decreased when the pretreatment severity factor increased to 4.54. This difference was due to the surface area of the particles being divided into exterior surface area and interior surface area. The specific surface area was affected mainly by the interior surface area. The interior surface area is influenced by pore volume. The pore volume of corn stover at the severity factor of 4.54 was lower than that of pretreated corn stover at the severity factor of 3.95 (Table 4).

Li C et al. reported that the difference of surface area and pore volume between untreated and AFEX treated corn stover was negligible, although SEM tomography have shown large increases in macroporosity after AFEX treatment. But there were a significant increase in the BET surface area (21.6 times greater) and the pore volume (26.6-fold greater) after IL pretreatment [45]. Yoon et al. found that ARP-treatment increased the BET surface area by 50% [46]. Our previous work showed that BET surface area of corn stover increased from 0.329 $m^2/g$ to 2.878 $m^2/g$, about 8.75 times greater, after dilute sulfuric acid pretreatment with acid consistency of 1 g/ml at 170°C for 60 min, and 1:15 of the ratio of corn stover weight (g) to liquor volume (mL) [47]. In this work, the specific surface area of corn stover increased from 8.55 $m^2/g$ to 17.08 $m^2/g$, which about 2 times greater, after LHW pretreatment at the severity factor of 3.95. The pore volume of pretreated corn stover with LHW increased 28 times compared with that of untreated corn stover. The increased surface area and pore volume provides easier enzyme access to cellulose.

Although the specific surface area of the substrate was provided by the decreased particle size, the pore volume has a significant function in facilitating hydrolysis by cellulase, the interconnecting function of other substrate factors such as pore diameters should also be considered. A report proposed that enzymatic hydrolysis is enhanced when the pore diameter of the substrate is large enough to accommodate both large and small enzyme components to maintain the synergistic action of the cellulase enzyme system [48]. Several extensive studies found that the rate-limiting pore diameter for lignocellulosic substrate hydrolysis was $5.1\times10^{-9}$ m [49]–[53]. The pore diameters of substrates, untreated corn stover, and pretreated corn stover at severity factors of 3.95 and 4.54 were $2.9\times10^{-9}$, $6.8\times10^{-9}$, and $8.9\times10^{-9}$ m, respectively (Table 4). The pore diameters of untreated corn stover were $<5.1\times10^{-9}$ m, whereas the pore diameter of pretreated corn stover was $>5.1\times10^{-9}$ m. The enlarged pore diameters after pretreatment of LHW enhanced action of enzyme on lignocellulosic substrate, and led to the enzymatic digestibility of corn stover pretreated with LHW increase.

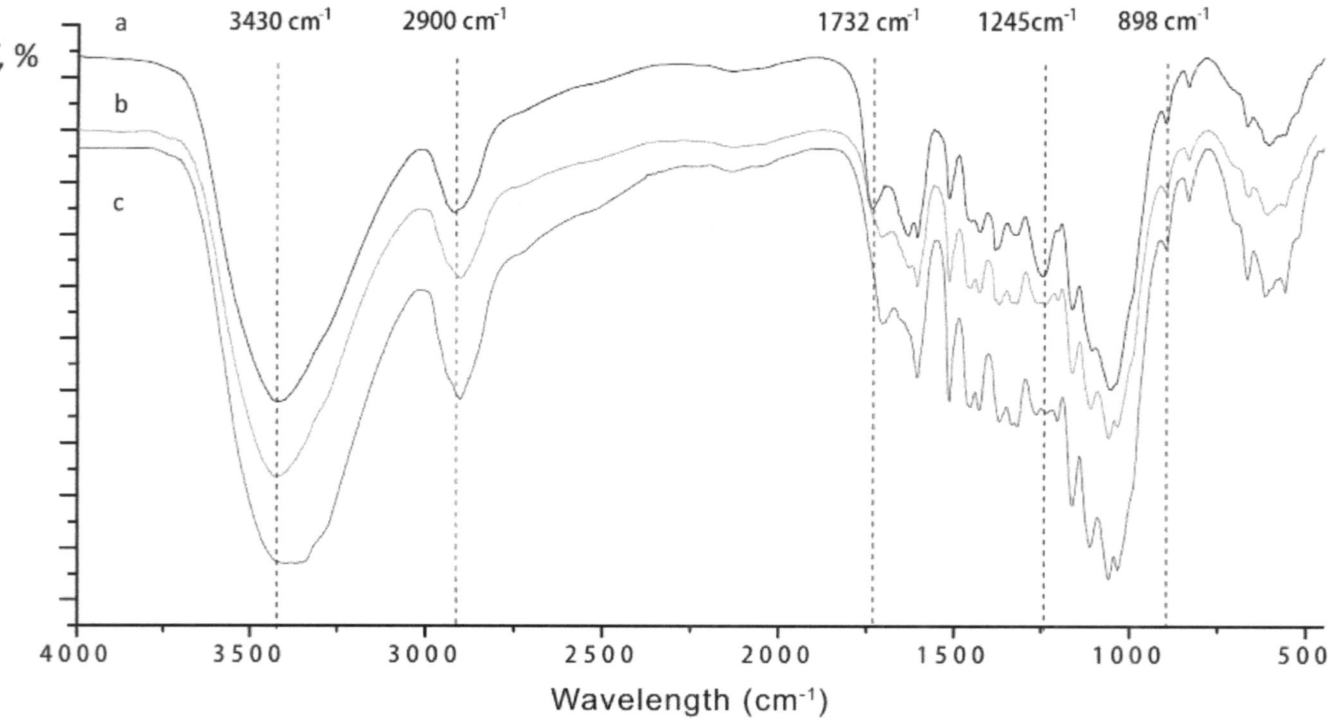


**FIGURE 6:** FTIR spectra of untreated and pretreated corn stover at different severity factor. a, b and c denote FTIR spectrum of untreated corn stover, pretreated at a severity factor of 3.95 and pretreated at a severity factor of 4.54, respectively.

### *2.3.3.3 FT-IR ANALYSIS.*

FT-IR spectra of the untreated and pretreated corn stover samples are shown in Figure 6. The band at 3430 $cm^{-1}$ is attributed to the O-H stretching of the hydrogen bonds of cellulose [54]. The peak exhibited reduction in intensity, indicating that hydrogen bonds in cellulose were disrupted during LHW pretreatment, and part of the crystalline cellulose in corn stover was disrupted during LHW pretreatment. The band position at 2900 $cm^{-1}$ is attributed to C–H stretching within the methylene of cellulose [55]. The relative absorbance decreased slightly, indicating that the methyl and methylene portions of cellulose were slightly ruptured. LHW pretreatments mostly reduced the intensity of the 1245 $cm^{-1}$ band attributed to the cleavage and/or alterations of acetyl groups, indicating that the acetyl groups were almost completely removed by LHW pretreatment [55]. The ester bond signal at 1732 $cm^{-1}$ was weaker in the spectra of LHW pretreated-samples than that of untreated samples, suggesting that some ester linkages between lignin and carbohydrates were cleaved during LHW pretreatment [56], led to some lignin fractions with low molecular weight partly dissolving out. The chemical compositions described above also shows that a small amount of soluble lignin were detected in the prehydrolysates from LHW pretreatment of corn stover. A small adsorption at 898 $cm^{-1}$ is characteristic of β-glycosidic linkages [57]. The relative adsorption decreased slightly, indicating that pretreatment disrupts the β-glycosidic linkages, led to part of carbohydrates was depolymerized.

## 2.4 CONCLUSIONS

The solid residues after LHW pretreatment of corn stover is suitable to be used as substrate for ethanol production, and the fed-batch S-SSF is one effective process for obtaining higher ethanol concentration and ethanol yield. The optimum feeding process in fed-batch S-SSF of the solid residues was that 6.1% of semi-weighted solid residues at pre-hydrolysis time of 6 h were added into the system. Ethanol concentrations of 38.4 g/L (severity factor of 3.95) and 39.4 g/L (severity factor of 4.54) and ethanol yields of 78.6% (severity factor of 3.95) and 79.7% (severity factor of 4.54) were obtained by the fed-batch S-SSF in the conditions of initial solid loading of 10% and pre-hydrolysis time of 18 h.

## REFERENCES


1. Lynd LR, Van Zyl WH, McBride JE, Laser M (2005) Consolidated bioprocessing of cellulosic biomass: an update. Curr Opin Biotechnol 16: 577–583.
2. Balat M (2011) Production of bioethanol from lignocellulosic materials via the biochemical pathway: A review. Energ Convers and Manage 52: 858–875.
3. Santos ALF, Kawase KYF, Coelho GLV (2011) Enzymatic saccharification of lignocellulosic materials after treatment with supercritical carbon dioxide. J Supercrit Fluids 56: 277–282.
4. Erdei B, Barta Z, Sipos B, Réczey K, Galbe M, et al. (2010) Ethanol production from mixtures of wheat straw and wheat meal. Biotechnol Biofuels 3: 16.
5. Fang X, Shen Y, Zhao J, Bao XM, Qu YB (2010) Status and prospect of lignocellulosic bioethanol production in China. Bioresour Technol 101: 4814–4819.
6. Zhao J, Xia LM (2009) Simultaneous saccharification and fermentation of alkaline-pretreated corn stover to ethanol using recombinant yeast strain. Fuel Process Technol 90: 1193–1197.
7. Limayem A, Ricke SC (2012) Lignocellulosic biomass for bioethanol production: current perspectives, potential issues and future prospects. Prog Energ Combust 38: 449–467.
8. Himmel ME, Ding SY, Johnson DK, Adney WS, Nimlos MR, et al. (2007) Biomass recalcitrance: engineering plants and enzymes for biofuels production. Science 315: 804–807.
9. Kumar P, Barrett DM, Delwiche MJ, Stroeve P (2009) Methods for pretreatment of lignocellulosic biomass for efficient hydrolysis and biofuel production. Ind Eng Chem Res 48: 3713–3729.
10. Kim I, Han JI (2012) Optimization of alkaline pretreatment conditions for enhancing glucose yield of rice straw by response surface methodology. Biomass Bioenerg 46: 210–217.
11. Chaudhary G, Singh LK, Ghosh S (2012) Alkaline pretreatment methods followed by acid hydrolysis of Saccharum spontaneum for bioethanol production. Bioresour Technol 124: 111–118.
12. Chen WH, Pen BL, Yu CT, Hwang WS (2011) Pretreatment efficiency and structural characterization of rice straw by an integrated process of dilute-acid and steam explosion for bioethanol production. Bioresour Technol 102: 2916–2924.
13. Yu ZD, Zhang BL, Yu FQ, Xu GZ, Song AD (2012) A real explosion: The requirement of steam explosion pretreatment. Bioresour Technol 121: 335–341.
14. Chundawat SPS, Vismeh R, Sharma LN, Humpula JF, Sousa LC, et al. (2010) Multifaceted characterization of cell wall decomposition products formed during ammo-

nia fiber expansion (AFEX) and dilute acid based pretreatments. Bioresour Technol 101: 8429–8438.

15. Li BZ, Balan V, Yuan YJ, Dale BE (2010) Process optimization to convert forage and sweet sorghum bagasse to ethanol based on ammonia fiber expansion (AFEX) pretreatment. Bioresour Technol 101: 1285–1292.
16. Sathitsuksanoh N, Zhu ZG, Zhang YHP (2012) Cellulose solvent- and organic solvent-based lignocellulose fractionation enabled efficient sugar release from a variety of lignocellulosic feedstocks. Bioresour Technol 117: 228–233.
17. Cara C, Ruiz E, Oliva JM, Sáez F, Castro E (2008) Conversion of olive tree biomass into fermentable sugars by dilute acid pretreatment and enzymatic saccharification. Bioresour Technol 99: 1869–1876.
18. Hsu TC, Guo GL, Chen WH, Hwang WS (2010) Effect of dilute acid pretreatment of rice straw on structural properties and enzymatic hydrolysis. Bioresour Technol 101: 4907–4913.
19. Rezende CA, Marisa Lima A, Maziero P, Azevedo ER, Garcia W, et al. (2011) Chemical and morphological characterization of sugarcane bagasse submitted to a delignification process for enhanced enzymatic digestibility. Biotechnol Biofuels 4: 54.
20. Taherzadeh MJ, Karimi K (2008) Pretreatment of lignocellulosic wastes to improve ethanol and biogas production: a review. Int J Mol Sci 9: 1621–1651.
21. Wan CX, Zhou YG, Li YB (2011) Liquid hot water and alkaline pretreatment of soybean straw for improving cellulose digestibility. Bioresour Technol 102: 6254–6259.
22. Mosier N, Hendrickson R, Ho N, Sedlak M, Ladisch MR (2005) Optimization of pH controlled liquid hot water pretreatment of corn stover. Bioresour Technol 96: 1986–1993.
23. Wang W, Yuan TQ, Wang K, Cui BK, Dai YC (2012) Combination of biological pretreatment with liquid hot water pretreatment to enhance enzymatic hydrolysis of Populus tomentosa. Bioresour Technol 107: 282–286.
24. Yu Q, Zhuang XS, Lv SL, He MC, Zhang Y, et al. (2013) Liquid hot water pretreatment of sugarcane bagasse and its comparison with chemical pretreatment methods for the sugar recovery and structural changes. Bioresour Technol 129: 592–598.
25. Yu G, Yano S, Inoue H, Inoue S, Endo T, et al. (2010) Pretreatment of rice straw by a hot-compressed water process for enzymatic hydrolysis. Appl Biochem Biotechnol 160: 539–551.
26. Mosier N, Hendrickson R, Ho N, Sedlak M, Ladisch MR (2005) Optimization of pH controlled liquid hot water pretreatment of corn stover. Bioresour Technol 96: 1986–1993.
27. Tomás-Pejó E, Oliva J, Ballesteros M, Olsson L (2008) Comparison of SHF and SSF processes from steam-exploded wheat straw for ethanol production by xylose- fermenting and robust glucose-fermenting Saccharomyces cerevisiae strains. Biotechnol Bioeng 100: 1122–1131.
28. Ask M, Olofsson K, Felice TD, Ruohonen L, Penttilä M, et al. (2012) Challenges in enzymatic hydrolysis and fermentation of pretreated Arundo donax revealed by a comparison between SHF and SSF. Process Biochem 47: 1452–1459.
29. Shen JC, Agblevor FA (2010) Modeling semi-simultaneous saccharification and fermentation of ethanol production from cellulose. Biomass Bioenerg 34: 1098–1107.

30. Lu J, Li XZ, Zhao J, Qu YB (2013) Enzymatic saccharification and L-lactic acid fermentation of corn stover pretreated with liquid hot water by rhizopus oryzae. Bioresources 8: 4899–4911.
31. Overend R, Chornet E (1994) Severity parameters: an update. Abstracts of Papers of the American Chemical Society 207 Cell Part 1.
32. Sanchez B, Bautista J (1988) Effects of furfural and 5-hydroxymethylfurfural on the fermentation of Saccharomyces cerevisiae and biomass production from Candida guilliermondii. Enzyme Microb Technol 5: 315–318.
33. Taherzadeh MJ, Gustafsson L, Niklasson C, Liden G (2000) Physiological effects of 5-hydroxymethylfurfural on Saccharomyces cerevisiae. Appl Microbiol Biotechnol 6: 701–708.
34. Tengborg C, Galbe M, Zacchi G (2001) Reduced inhibition of enzymatic hydrolysis of steam-pretreated softwood. Enzyme Microb Technol 9–10: 835–844.
35. Peng LC, Chen YC (2011) Conversion of paper sludge to ethanol by separate hydrolysis and fermentation (SHF) using Saccharomyces cerevisiae. Biomass Bioenerg 35: 1600–1606.
36. Tomás-Pejó E, Negro MJ, Sáez F, Ballesteros M (2012) Effect of nutrient addition on preinoculum growth of *S. cerevisiae* for application in SSF processes. Biomass Bioenerg 45: 168–174.
37. Hahn-Hägerdal B, Karhumaa K, Larsson CU, Gorwa-Grauslund M, Gorgens J, et al. (2005) Role of cultivation media in the development of yeast strains for large scale industrial use. Microb Cell Fact 4: 31.
38. Pereira FB, Guimarães PMR, Teixeira JA, Domingues L (2010) Optimization of low-cost medium for very high gravity ethanol fermentations by Saccharomyces cerevisiae using statistical experimental designs. Bioresour Technol 101: 7856–7863.
39. Jiménez-Martí E, Zuzuarregui A, Gomar-Alba M, Gutiérrez D, Gil C, et al. (2011) Molecular response of Saccharomyces cerevisiae wine and laboratory strains to high sugar stress conditions. Int J Food Microbiol 145: 211–220.
40. Bafrncová P, Smogrovicová D, Sláviková I, Pátková J, Dömény Z (1999) Improvement of very high gravity ethanol fermentation by media supplementation using Saccharomyces cerevisiae. Biotechnol Lett 21: 337–341.
41. Palukurty MA, Telgana NK, Bora HSR, Mulampaka SN (2008) Screening and optimization of metal ions to enhance ethanol production using statistical experimental designs. Afr Microbiol Res 2: 87–94.
42. Xue C, Zhao XQ, Yuan WJ, Bai FW (2008) Improving ethanol tolerance of a self-flocculating yeast by optimization of medium composition. World Microbiol Biotechnol 24: 2257–2261.
43. Torija MJ, Rozès N, Poblet M, Guillamón JM, Mas A (2003) Effects of fermentation temperature on the strain population of Saccharomyces cerevisiae. Int Food Microbiol 80: 47–53.
44. Wang L, Templer R, Murphy RJ (2012) High-solids loading enzymatic hydrolysis of waste papers for biofuel production. Appl Energy 99: 23–31.
45. Li CL, Cheng G, Balan V, Kent MS, Ong M, et al. (2011) Influence of physicochemical changes on enzymatic digestibility of ionic liquid and AFEX pretreated corn stover. Bioresour Technol 102: 6928–6936.

46. Yoon HH, Wu ZW, Lee YY (1995) Ammonia-recycled percolation process for pretreatment of biomass feedstock. Appl. Biochem Biotechnol 51–52: 5–19.
47. Yao L, Zhao J, Xie Y, Yang H, Yang W, et al. (2012) Mechanism of diluted acid pretreatment to improve enzymatic hydrolysis of corn stover,. Chemistry and Industry of forest products 32(4): 87–92.
48. Tanaka M, Ikesaka M, Matsuno R (1988) Effect of pore size in substrate and diffusion of enzyme on hydrolysis of cellulosic materials with cellulases. Biotechnol Bioeng 32: 698–706.
49. Wong KK, Deverell KF, Mackie KL (1988) The relationship between fiber-porosity and cellulose digestibility in steam-exploded Pinus radiata. Biotechnol Bioeng 31: 447–456.
50. Weimer PJ, Weston WM (1985) Relationship between the fine structure of native cellulose and cellulose degradability by the cellulase complexes of Trichoderma reesei and Clostridium thermocellum. Biotechnol Bioeng 27: 1540–1547.
51. Mooney CA, Mansfield SD, Touhy MG, Saddler JN (1998) The effect of initial pore volume and lignin content on the enzymatic hydrolysis of softwoods. Bioresour Technol 64: 113–119.
52. Stone J, Scallan A, Donefer E, Ahlgren E (1969) Digestibility as a simple function of a molecule of similar size to a cellulase enzyme. Adv Chem 95: 219–241.
53. Zeng MJ, Mosier NS, Huang CP, Sherman DM, Ladisch MR (2007) Microscopic examination of changes of plant cell structure in corn stover due to hot water pretreatment and enzymatic hydrolysis. Biotechnol Bioeng 97: 265–278.
54. Buranov AU, Mazza G (2010) Extraction and characterization of hemicelluloses from flax shives by different methods. Carbohyd Polym 9: 17–25.
55. Kumar R, Mago G, Balan V, Wyman CE (2009) Physical and chemical characterizations of corn stover and poplar solids resulting from leading pretreatment technologies. Bioresour Technol 100: 3948–3962.
56. Liu L, Sun JS, Li M, Wang SH, Pei HS, et al. (2009) Enhanced enzymatic hydrolysis and structural features of corn stover by FeCl3 pretreatment. Bioresour Technol 100: 5853–5858.
57. Ibarra D, del Río JC, Gutiérrez A, Rodríguez IM, Romero J, et al. (2004) Isolation of high-purity residual lignins from eucalypt paper pulps by cellulase and proteinase treatments followed by solvent extraction. Enzyme Microb Technol 35: 173–181.
58. Zhao J, Xia LM (2009) Simultaneous saccharification and fermentation of alkaline-pretreated corn stover to ethanol using a recombinant yeast strain. Fuel Process Technol 90: 1193–1197.
59. Avci A, Saha BC, Kennedy GJ, Cotta MA (2013) Dilute sulfuric acid pretreatment of corn stover for enzymatic hydrolysis and efficient ethanol production by recombinant Escherichia coli FBR5 without detoxification. Bioresour Technol 142: 312–319.
60. Avci A, Saha BC, Dien BS, Kennedy GJ, Cotta MA (2013) Response surface optimization of corn stover pretreatment using dilute phosphoric acid for enzymatic hydrolysis and ethanol production. Bioresour Technol 130: 603–612.
61. Wan CX, Li YB (2010) Microbial pretreatment of corn stover with Ceriporiopsis subvermispora for enzymatic hydrolysis and ethanol production. Bioresour Technol 130: 6398–6403.

CHAPTER 3

# HELICALLY AGITATED MIXING IN DRY DILUTE ACID PRETREATMENT ENHANCES THE BIOCONVERSION OF CORN STOVER INTO ETHANOL

YANQING HE, LONGPING ZHANG, JIAN ZHANG, AND JIE BAO

## 3.1 BACKGROUND

Pretreatment is the crucial step to overcome the biorecalcitrance of lignocellulose to achieve efficient bioconversion of cellulose into fermentable sugars and then to fermentation products such as ethanol [1-6]. Among various pretreatment methods, dilute sulfuric acid pretreatment is considered to be the one with potential commercial applications [7-12]. The major disadvantages of dilute acid pretreatment include relatively massive acidic waste water generation caused by low solids (lignocellulosic feedstock) content, loss of fermentable sugars during the solids/liquid separation after pretreatment, and relatively high inhibitor compounds generation [13]. To overcome these disadvantages, recent studies on dilute acid pretreatment have tried to increase the feedstock content of lignocellulose solids as high as possible [14-16]. One example in our previous study was a dry dilute acid pretreatment of corn stover, in which the solids content in thc pretreatment was fed to an extreme high of up to 70% of the total feedstock [17] and successfully applied to production of ethanol, lipid, and lactic acid from corn stover [18-21]. This dilute acid pretreatment was called a 'dry' method, because both the corn stover feedstock and the pretreated corn stover product were 'dry' with no free water generation during the pretreatment, while the inhibitor generation was kept at a low level. In this way, the three major disadvantages of dilute acid pretreatment could be overcome: dry-in and dry-out thus no waste water was generated, dry pretreated product thus no solids/liquid separation was needed, and low inhibitor generation maintained a high pretreatment efficiency.

The practice of dry dilute acid pretreatment operation revealed that well mixing of the majority of lignocellulose solids with minimum steam input in the pretreatment reactor was the major challenge. When the dilute acid pretreatment was operated under a low solids/liquid ratio, the mixing of hot steam with corn stover feedstock was relatively easy, because the steam heated the continuous water phase, then the hot liquid heated the solids particles impregnated in the liquid. However, when dilute acid pretreatment was operated under a high solids/liquid ratio, such as the 'dry' condition described above [17], the mixing of the hot steam with the dry solids particles, and the heat transfer from the hot steam to the solids feedstock became very difficult for three reasons: no aqueous phase existed as a continuous phase covering the solids bulk body (heat transfer directly occurred from the hot steam to the solid corn stover), lignocellulose biomass was typically a good insulator to reduce the heat transfer from the surface to the inside part (the surface of a paced pile of biomass was at target temperature but the core of the packed bed was below the desired temperature causing uneven heating), and the steam at a low usage (less than half of the solids used according to Zhang et al. [17]) had to reach the scattered solids particles directly.

On the other hand, since lignocellulose materials possessed high water or steam absorption capacity, the steam entering the pretreatment reactor was quickly absorbed by the lignocellulose materials close to the feeding nozzle regions and could not be dispersed onto the materials in the upper region of the reactor. In a small bench scale reactor, the reactor volume was small and the steam injection could penetrate through the relatively thin packing materials in both the height and diameter of the reactor. However, with the increased pretreatment scale of industrial reactors, enforced mixing is inevitably required because it is not possible to distribute the steam jetting uniformly into the large pretreatment reactors through the thick packing in meters of height and diameter.

The required agitation system should work well with a completely dry solids system at reasonable energy consumption. In our previous studies, the helical ribbon stirred agitation and was found to achieve a well mixing condition of the solids majority with the liquid (enzyme) minority in the simultaneous saccharification and fermentation (SSF) of various pretreated lignocellulose materials [20-22]. However, the difference of mixing scenarios between these fermentation bioreactors and the pretreatment reactor was that in bioreactors, mixing started with the solids feedstock but these solids quickly changed into the liquid slurry, while in the pretreatment reactor, the mixing apparatus had to face a completely solids phase throughout the whole operation time when the dry pretreatment system was applied.

In this work, the mixing performance of helically agitated mixing in the dry dilute acid pretreatment was investigated. The mixing effect by the helical ribbon stirrer was first tested in a mock-up experiment using three reactors of different sizes. In the base of the results of the mock-up experiment, a new pretreatment reactor equipped with a helical ribbon stirrer was designed and the mixing performance of the majority of corn stover solids with minimum steam input was tested. The pretreatment efficiency was evaluated by enzymatic hydrolysis and ethanol fermentation. These results all indicate that the helically agitated mixing worked well for the dry dilute acid pretreatment with the enhanced bioconversion efficiency of corn stover to ethanol. This study may provide a suitable prototype of a pretreatment reactor at high solids loading for future large-scale or industrial-scale pretreatment operations.

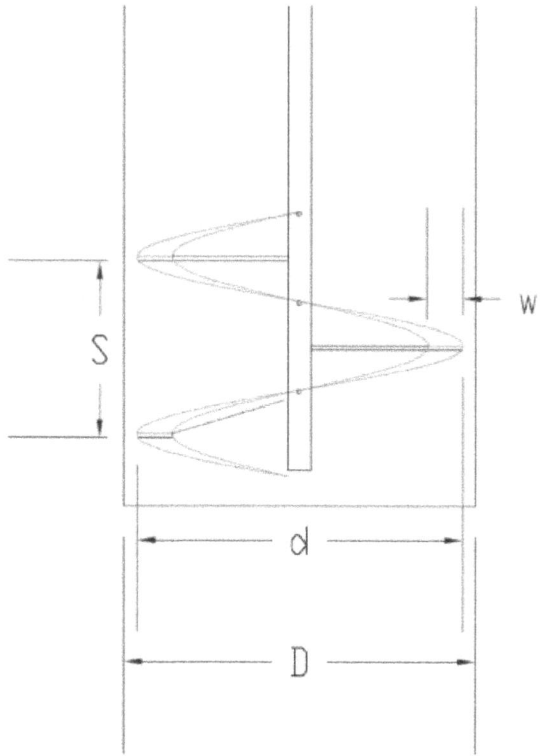


| Reactor volume (L) | D (mm) | Solids content (%, w/w) | Mixing time (s) |
|---|---|---|---|
| 5.0 | 170 | 50 | 180 |
| 50.0 | 384 | 50 | 168 |
| 500.0 | 786 | 50 | 120 |

**FIGURE 1:** Schematic diagram of the fluid dynamic reactors equipped with helical ribbon stirrer. The reactor was a cylinder equipped with a helical ribbon stirrer. The parameters of the helical ribbon stirrer were matched with the change in the diameter of the reactor to keep the character of the helical ribbon the same in the three reactors. The mixing time was calculated by the time to reach constant moisture of corn stover in the reactors. d, diameter of impeller (mm); D, diameter of the reactor (mm); S, pitch size of the helical ribbon (mm); w, ribbon width (mm).

## 3.2 RESULTS AND DISCUSSION

### 3.2.1 PRETREATMENT PERFORMANCE IN THE REACTORS WITH AND WITHOUT MIXING

Helical screw feeders or conveyors are frequently used in lignocellulose processing plants as described in the National Renewable Energy Laboratory (NREL) technical report [12]. However, when the solids content gets very high, the compact screw device does not work well. Authors have tested several screw devices for mixing in the dry pretreatment reactor at corn stover solids of 70% (w/w), but the screw devices were damaged due to the high resistance to the screw movement (data not shown). Therefore, the loosely structured helical ribbon stirrer used in the previous SSF studies of pretreated corn stover at 30% (w/w) solids was tested. The fluid dynamic mock-up experiments at solids content of 50% (w/w) were carried out in several reactors of different sizes to test the mixing efficiency of corn stover with water. Figure 1 shows that the mixing of corn stover solids with water at solids content of 50% (w/w) was completed within 2 to 3 minutes. The positive fluid dynamic results plus the successful applications in high solids loading of SSF suggest that the helical ribbon stirrer might be a suitable agitation apparatus for dry pretreatment with very high solids content of up to 70% (w/w).

The dry dilute acid pretreatment of corn stover in both the helically agitated reactor and the static reactor (without mixing apparatus) was carried out and analyzed. In the previous study, the pretreatment reactor was a 10 L stainless cylinder of 180 mm in diameter and 400 mm in height [17]. Three steam injection nozzles in the bottom of the reactor were relatively sufficient to mix the steam with the lignocellulose materials packed inside the reactor. In this study, the pretreatment reactor was enlarged in diameter from 180 mm to 260 mm and the volume was increased from 10 L to 20 L (the height was kept unchanged at 400 mm), but the number and size of the steam injection nozzles were still the same. The purpose of the new enlarged reactor design was to demonstrate a scale-down example of the industrial pretreatment reactor in which the steam jetting was not efficiently mixed with the lignocellulose materials. The focus of this study is to find a solution for the problem in large-scale pretreatment reactors.

The original research plan on the enlarged pretreatment reactor was to operate the dry dilute acid pretreatment at exactly the same optimal conditions as that in the previous 10 L reactor: 190°C, 3 minutes, and 2.5% sulfide acid. Unfortunately, the new reactor (20 L) was twice as large as the previous one (10 L) and the agitation apparatus provided more heat dissipation. Thus, the steam supply from the same steam generator DZFZ4.5C was not sufficient for the new enlarged reactor and the maximum temperature was only 185°C, 5°C below the planned temperature (190°C).

**TABLE 1:** Pretreatment performance in the static reactor and in the helically agitated reactor

| Pretreatment conditions | Cellulose conversion (%) | Inhibitors in the pretreated CS (g/100 g DM) | | | Sugars in the pretreated CS (g/100 g DM) | | | |
|---|---|---|---|---|---|---|---|---|
| | | Furfural | 5-HMF | Acetate | Glucose | Xylose | O-Glu | O-Xyl |
| In helically agitated reactor | | | | | | | | |
| Case 1: 185°C, 2.0%, 3 minutes | 77.55 | 0.18 | 0.09 | 0.58 | 0.48 | 5.34 | 0.72 | 8.45 |
| Case 2: 185°C, 2.5%, 3 minutes | 87.11 | 0.63 | 0.17 | 0.81 | 1.01 | 10.20 | 1.10 | 2.84 |
| In no agitation reactor | | | | | | | | |
| Case 3: 190°C, 2.0%, 3 minutes | 72.10 | 0.50 | 0.25 | 0.77 | 0.88 | 5.58 | 2.27 | 7.55 |
| Case 4: 190°C, 2.5%, 3 minutes | 85.10 | 0.90 | 0.21 | 1.20 | 1.58 | 8.02 | 1.57 | 4.29 |

*The data in the pretreatment conditions column indicate the pretreatment temperature, acid usage, and residual time, respectively. Experiments were carried out in duplicate and averaged to give the mean values. Enzymatic hydrolysis was carried out at 50°C, 15 FPU enzyme dosage, and 150 rpm agitation for 72 hours. 5-HMF, 5-hydroxymethylfurfural; CS, corn stover; DM, dry matter; FPU, filter paper unit; O-Glu, glucan oligomer; O-Xyl, xylan oligomer.*

On the other hand, the pretreatment efficiency in the new reactor at zero agitation was very poor, the inhomogeneity could be observed even with basic examination: large bulks of dark over-pretreated portions were at the bottom and yellow un-pretreated fresh portions were at the top. The main reasons were not only due to the enlarged diameter, which worsened the mixing of steam and lignocellulose materials, but also (and a major factor) due to the existence of the helical ribbon impeller parts, which severely disturbed the steam flow and created many dead zones inside the reactor. Unless the agitation apparatus was removed from the new enlarged reactor and left the new reactor as an empty cylinder, as the previous 10 L reactor for the static pretreatment operation (which was not possible to operate on the new reactor), the direct comparison of the pretreated materials obtained at the agitated condition and static condition did not accurately reflect the true situation.

Therefore, the authors decided not to compare the pretreated materials obtained from the same enlarged reactor at the agitated and static conditions. Instead, the pretreated materials for evaluation were chosen from the static pretreatment operation in the previous 10 L reactor at the optimal condition (190°C, 3 minutes, and 2.5% sulfide acid), and from the agitated pretreatment operation at the optimal conditions of the enlarged 20 L reactor (185°C, 3 minutes, and 2.5% sulfide acid), although there was 5°C difference in the pretreatment temperature.

Four operation cases are shown in Table 1. Case 1 and Case 2 were operated on the helically agitated reactor (Figure 2a) and lasted for 3 minutes at 185°C at different sulfuric acid usage: 2.0% (2.0 g per 100 g dry solids) for Case 1 and 2.5% for Case 2, respectively. Case 3 and Case 4 were operated on the static reactor without mixing apparatus (Figure 2b) at 190°C for 3 minutes at the same sulfuric acid usage of 2.0% and 2.5%, respectively. There was a temperature difference of 5°C between the two reactors, 185°C at Case 1 and Case 2, and 190°C for Case 3 and Case 4, because of the limitation of steam supply with the increased reactor size. However, the comparison among the operation cases of the two reactors still revealed sufficient information of the impact of helically agitated mixing on the dry pretreatment processing.

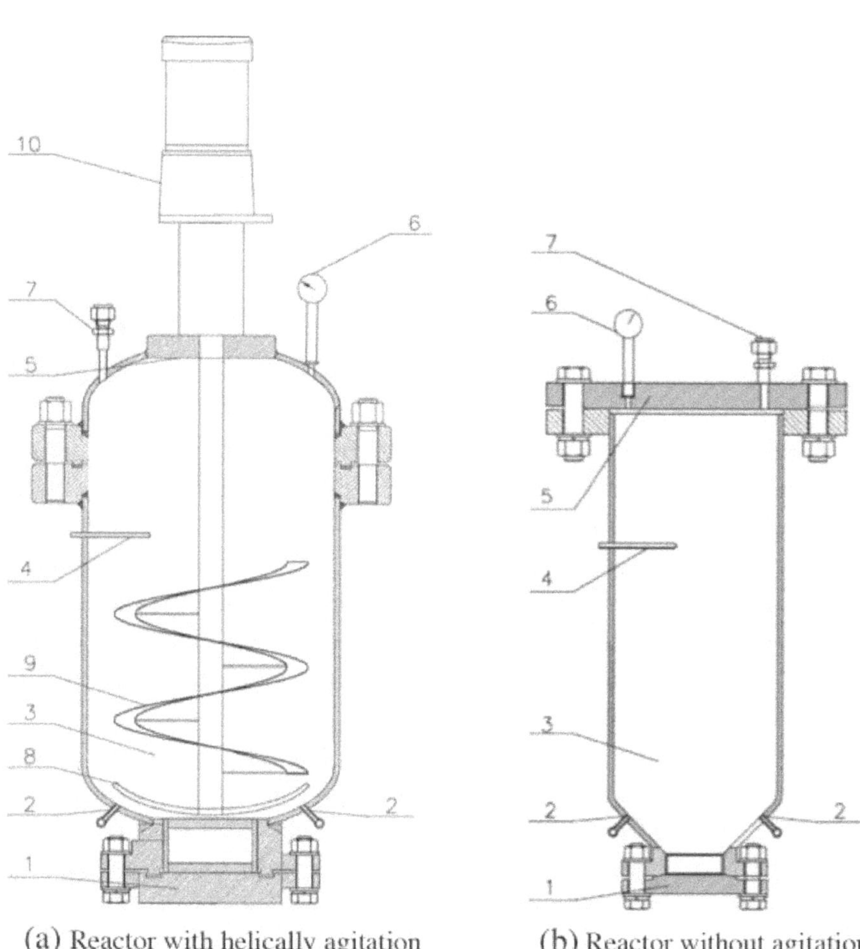


**FIGURE 2:** Schematic diagram of the dry pretreatment reactors with and without helical agitation mechanism. (a) Reactor equipped with helical ribbon impeller with the inner volume of 20 L; (b) Reactor without agitation apparatus with the inner volume of 10 L. 1, product outlet; 2, steam inlet; 3, pretreatment vessel; 4, thermocouple; 5, cap of the reactor; 6, pressure gauge; 7, inert air outlet; 8, anchor stirrer; 9, helical ribbon stirrer; 10, electric motor for driving the helical ribbon impeller.

Table 1 indicates that the helically agitated mixing in the dry pretreatment played a crucial role in promoting the pretreatment efficiency and reducing the inhibitor generation. At the sulfuric acid usage of 2.0% for 3 minutes, the cellulose conversion increased from 72.10% (Case 3) to 77.55% (Case 1), while at the sulfuric acid usage of 2.5% for 3 minutes, the cellulose conversion increased from 85.10% (Case 4) to 87.11% (Case 2). It is worth stressing that the increase of cellulose conversion occurred with a 5°C lower temperature in the helically agitated reactor (185°C) than that in the static reactor (190°C). Generally, the lower temperature in the pretreatment operation leads to lower pretreatment efficiency [9,11,17,23,24]. However, the opposite results were obtained, in which pretreatment efficiency at the lower temperature under the agitated condition was elevated, compared to that under the static pretreatment condition. The result clearly confirmed the advantage of helically agitated mixing on thepretreatment efficiency.

Furthermore, Table 1 also indicates that the inhibitor concentration in the pretreated corn stover dramatically decreased in the helically agitated pretreatment, although generally the inhibitor concentration increased with increasing pretreatment efficiency [14,15,17,25]. The comparison of Case 1 and Case 3 indicates that the three major inhibitors, furfural, 5- hydroxymethylfurfural (5-HMF), and acetic acid, decreased from 0.50, 0.25, and 0.77 g/100 g dry matter (DM) in Case 3 (without mixing) to 0.18, 0.09, and 0.58 g/100 g DM in Case 1 (with helically mixing), respectively. Similarly, the comparison of Case 2 and Case 4 indicates that these inhibitors decreased from 0.90, 0.21, and 1.20 g/100 g DM in Case 4 (without mixing) to 0.63, 0.17, and 0.81 g/100 g DM in Case 2 (with helically mixing), respectively. Corresponding to the inhibitor generation, the glucose and its oligomer concentrations in the pretreated corn stover decreased in the helically agitated reactor, and the xylose and oligomer concentrations were kept approximately the same.

### *3.2.2 CFD MODELING OF THE HELICALLY AGITATED MIXING IN THE DRY PRETREATMENT*

The well mixing of corn stover solids with steam by helical agitation could lead to a uniform distribution of temperature and sulfuric acid concentration, thus overheating at the bottom or underheating near the top of the reactor could be avoided. To illustrate the effect of helical agitation on the dry pretreatment efficiency, the computational fluid dynamics (CFD) method was used to simulate the steam flow with the corn stover solids in the helical ribbon stirrer agitated reactor. A simplified CFD model was established under several assumptions and the fluid dynamic state of the helically agitated reactor was simulated.

Figure 3c indicates that the steam holdup (represented by the conservative gas volume fraction) was significantly improved by helical agitation. At the static state or the low agitation rate (0, 10, 30 rpm), the steam was accumulated to a very limited region near the jetting nozzles and the reactor walls, then quickly diffused upwardly without sufficient contact with solids. When the rotation rate was increased to 50 rpm, the steam holdup increased everywhere in the solids bulk, indicating that a well mass and heat transfer state was established.

Figure 3d also indicates that solids mixing (represented by the liquid velocity distribution) did not occur at no agitation or a low agitation rate in the reactor; with the increasing agitation rate, solids mixing was quickly improved and a reasonable fluid flow regime was established inside the reactor. On the other hand, Figure 3c and d reveal that the improvement of both steam holdup and solids mixing did not require a high agitation rate, thus an agitation rate of 50 rpm should be a suitable value for the present reactor.

The CFD modeling illustrated a relatively broad but clear picture of the improved mixing performance by helical agitation in the pretreatment reactor. The modeling results confirmed the estimation proposed at the beginning of this section.

### *3.2.3 IMPROVING PRETREATMENT EFFICIENCY AND REDUCING INHIBITOR GENERATION BY HELICALLY AGITATED MIXING*

The dry pretreatment performance of corn stover in the helically agitated reactor was optimized by changing pretreatment parameters for maximum hydrolysis yield and minimum inhibitor generation. The results are shown in Tables 2 and 3 and Figure 4.

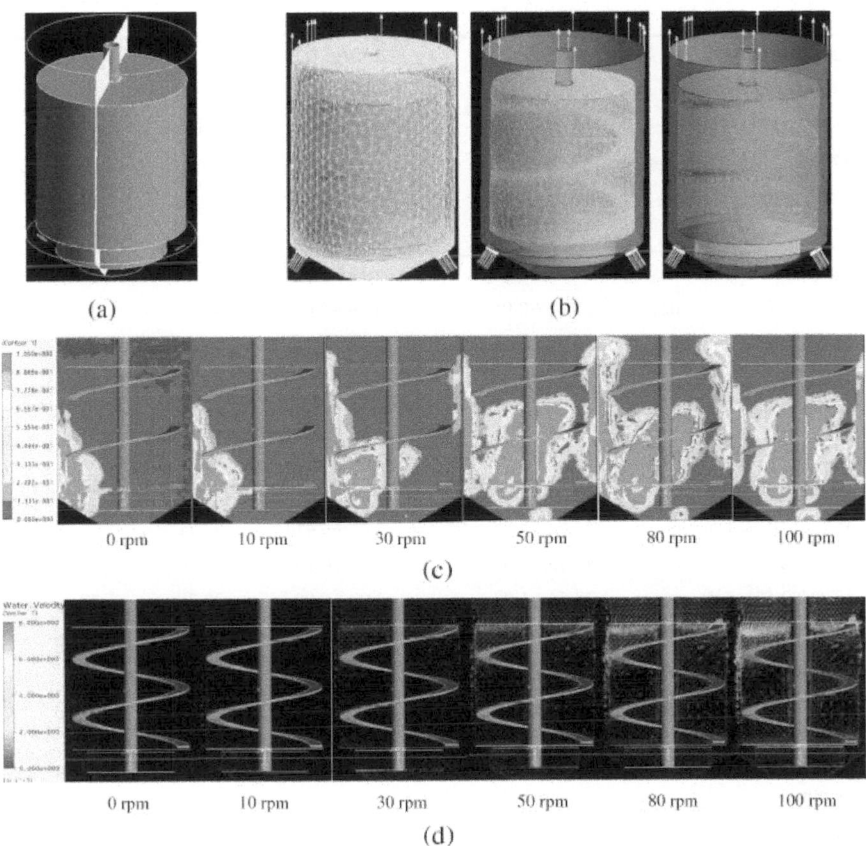


**FIGURE 3:** CFD modeling of steam holdup and solids flow in the helically agitated pretreatment reactor. (a) The reactor diagram in which the flow regime of the cross-section was simulated in the CFD calculation. (b) Geometric structure of the reactor in the CFD model. Left, mesh cells and structure; middle, motion region of the helical ribbon impeller; and right, motion region of the bottom anchor stirrer. (c) Conservative gas volume fraction under different agitation rates. (d) Fluid velocity distribution at different agitation rates. In this modeling, pretreated CS materials were assumed to the high viscose liquid with high apparent viscosity of 2.31 Pa · s; the hot steam stream was assumed to be inert air with a flow rate of 1.75 m/s. CFD, computational fluid dynamics; CS, corn stover.

**TABLE 2:** Glucan and xylan recovery of the dry pretreatment in the helically agitated reactor

| Pretreatment conditions | Cellulose content before pretreatment (%) | Xylan content before pretreatment (%) | Glucan recovery (%) | Xylan recovery (%) |
|---|---|---|---|---|
| Raw corn stover | 37.15±0.22 | 19.86±0.56 | - | - |
| Changing temperature | | | | |
| 165°C , 2.5%, 3 minutes | 38.04±2.04 | 8.57±2.26 | 100.09 | 94.24 |
| 175°C , 2.5%, 3 minutes | 37.23±0.75 | 5.28±0.02 | 96.69 | 79.32 |
| 185°C , 2.5%, 3 minutes | 40.93±0.06 | 2.93±0.16 | 96.50 | 62.00 |
| Changing acid usage | | | | |
| 185°C, 1.5%, 3 minutes | 37.71±1.07 | 6.48±0.18 | 92.05 | 78.89 |
| 185°C, 2.0%, 3 minutes | 37.41±0.06 | 4.16±0.08 | 96.63 | 74.91 |
| 185°C, 2.5%, 3 minutes | 40.93±0.06 | 2.93±0.16 | 96.50 | 62.00 |
| 185°C, 3.0%, 3 minutes | 38.48±0.40 | 2.52±0.17 | 73.97 | 43.18 |
| 185°C, 3.5%, 3 minutes | 36.85±1.25 | 1.79±0.21 | 75.12 | 36.37 |
| 185°C, 4.0%, 3 minutes | 35.98±1.10 | 1.05±0.10 | 65.53 | 29.75 |
| Changing residue time | | | | |
| 185°C, 2.5%, 1 minute | 41.52±5.88 | 4.06±0.41 | 92.71 | 64.08 |
| 185°C, 2.5%, 3 minutes | 40.93±0.06 | 2.93±0.16 | 96.50 | 62.00 |
| 185°C, 2.5%, 5 minutes | 38.54±1.88 | 2.93±0.08 | 82.78 | 57.29 |
| 185°C, 2.5%, 10 minutes | 39.29±0.75 | 2.21±0.22 | 76.44 | 42.96 |
| Changing agitation rate | | | | |
| 185°C, 2.5%, 3 minutes, 10 rpm | 40.23±2.59 | 3.00±0.19 | 84.81 | 54.03 |
| 185°C, 2.5%, 3 minutes, 30 rpm | 38.41±0.58 | 2.86±0.87 | 93.20 | 64.19 |
| 185°C, 2.5%, 3 minutes, 50 rpm | 40.93±0.06 | 2.93±0.16 | 96.50 | 62.00 |

The data in the pretreatment conditions column indicate the pretreatment temperature, acid concentration, and residual time, respectively. Cellulose and xylan content were determined by two-step acid hydrolysis methods described in the Methods section. The recovery of cellulose and xylan was calculated by the ratio of cellulose and xylan content in the dry materials before and after pretreatment. Cellulose and xylan after pretreatment consisted of monosaccharides, oligosaccharides, glucan and xylan components, and furfural in the xylan recovery. All the experiments were carried out in duplicate and error was calculated as standard deviation except the recovery of cellulose and xylan, which was calculated with the total materials from two batches of pretreatment at the same condition.

Table 2 indicates that the cellulose content after the pretreatment was almost the same with the virgin corn stover, but the xylan content decreased sharply with increasing temperature and acid usage. Glucan recovery was almost constant with the increasing temperature in the experimental range (165 to 185°C), but suddenly decreased when the sulfuric acid usage was above 2.5% and the residue time was longer than 3 minutes. This result gave a strong indication that in the present helically agitated reactor, the acid concentration and long pretreatment time were not preferred because the cellulose was easily converted into other degradation compounds such as 5-HMF at such a condition. On the other hand, the xylan recovery was relatively low compared to the cellulose recovery and decreased steadily with the increasing intensity of temperature, acid usage, and residue time. As illustrated in Table 3, the concentrations of the typical inhibitors such as furfural, 5-HMF, and acetic acid increased with increasing pretreatment intensity; the glucose increased with the increasing intensity but the oligomer showed the opposite tendency; xylose increased with pretreatment intensity but when it was too strong, xylose and its oligomer decreased possibly due to the generation of its downstream products such as furfural. The results indicate that at the present pretreatment, xylan degradation was still strong and the pretreatment condition should be compromised by considering cellulose/hemicellulose loss, inhibitor generation, and cellulose conversion, instead of glucose yield only.

Figure 4a indicated the direct cellulose conversion of the pretreated corn stover, while Figure 4b indicated the overall cellulose conversion of the virgin corn stover with the consideration of solid weight loss in the pretreatment. Figure 4a shows that the direct cellulose conversion increased with increasing pretreatment temperature, acid usage, and residue time, which was also in agreement with the tendency of general dilute acid pretreatments. The direct cellulose conversion at different acid usage increased until the acid usage reached 3.0% and remained almost unchanged as the acid usage increased further. The same trend could be observed when the residue time was prolonged. On the other hand, the overall cellulose conversion increased with increasing temperature and also increased with increasing acid usage till 2.5%, then decreased with further increase of acid usage, because of the weight loss mentioned above. The residue time also showed the same tendency on the overall cellulose conversion. The direct cellulose conversion and the overall cellulose conversion of the pretreated corn stover gradually increased with the increasing agitation rate. The pretreated corn stover materials in the reactor were driven out by the constant agitation. Therefore a minimum agitation rate (50 rpm for the 20 L reactor) was maintained because a very low agitation rate was not sufficient to drive the corn stover materials completely out of the reactor, and then led to the loss of cellulose and xylan. The maximum overall cellulose conversion of 83.09% was observed at 185°C, 2.5% acid usage, and 3 minutes of residue time.

### 3.2.4 SIMULTANEOUS SACCHARIFICATION AND ETHANOL FERMENTATION (SSF) OF PRETREATED CORN STOVER

The efficiency of the helically agitated pretreatment was tested by SSF using the dry dilute acid pretreated corn stover as feedstock. The corn stover was pretreated in the helically agitated reactor at 185°C for 3 minutes with 2.5% of sulfuric acid usage (Case 2 in Table 1); in the static reactor, the pretreatment was operated at 190°C for 3 minutes with 2.5% of sulfuric acid usage (Case 4 in Table 1). The pretreated corn stover was biodetoxified to remove the inhibitors until furfural and 5-HMF could not be detected. The SSF of the pretreated and detoxified corn stover was conducted under 30% solids loading (dry materials), 15 filter paper units (FPU)/g dry matter (DM) of cellulase dosage, and the results are shown in Figure 5.

Figure 5 shows that 12 hours' prehydrolysis of corn stover after the helically agitated dry pretreatment released 81.92 g/L of glucose, and increased almost 47% compared to the glucose released from the hydrolysis of corn stover from the static dry pretreatment (55.87 g/L). The prehydrolysis results indicate that the pretreatment efficiency of corn stover from the helically agitated dry pretreatment was significantly improved.

The SSF stage was started after 12 hours' prehydrolysis and the significant improvement of ethanol yield was also observed: the ethanol titer reached 56.20 g/L after 48 hours' SSF using the corn stover from the helically agitated pretreatment, while the ethanol titer was only 44.44 g/L under the same SSF conditions using corn stover from the static pretreatment. The ethanol yield from cellulose using the helically agitated pretreated corn stover was 69.34%, and the yield using statically pretreated corn stover was only 59.14%. The 26.5% and 17.2% increases of ethanol titer and yield were observed, respectively. The results indicated the advantage of the helically agitated well mixing in the dry pretreatment reactor.

The present ethanol titer of 56.20 g/L and yield of 69.43% were still not high enough because of the very high solids loading (30% w/w) and relatively short SSF time (48 hours). However, there is certainly sufficient space for improvement in ethanol titer, yield, and productivity of ethanol product. The helically agitated reactor in this study provided a prototype of dry dilute acid pretreatment processing under the output of no waste water generation, less sugar loss, low inhibitor generation, and low steam consumption.

## 3.3 CONCLUSIONS

The helically agitated mixing significantly improved the efficiency of dry dilute acid pretreatment and reduced inhibitor generation compared to the dry pretreatment without agitation. For the dry dilute acid pretreatment at 70% solids loading of corn stover (dry base), an optimal pretreatment condition was obtained at 185°C, 2.5% of sulfuric acid usage, and lasted for 3 minutes. The ethanol titer and yield from cellulose in the SSF reached 56.20 g/L and 69.43% at 30% solids loading and 15 FPU/g DM cellulase, respectively, corresponding to 26.5% increase in the ethanol titer and 17.2% increase of ethanol yield at the same conditions. The advantage of helically agitated mixing in the dry pretreatment provided a prototype of dry dilute acid pretreatment for future commercial-scale production of cellulosic ethanol.

**TABLE 3:** Impact of the operation parameters on the inhibitor and sugar level of the dry pretreatment in the helically agitated reactor

| Pretreatment conditions | Inhibitors in the pretreated CS (g/100 g DM) | | | Sugars in the pretreated CS (g/100 g DM) | | | |
|---|---|---|---|---|---|---|---|
| | Furfural | 5-HMF | Acetic acid | Glucose | Xylose | O-Glu | O-Xyl |
| Changing temperature | | | | | | | |
| 165°C, 2.5%, 3 minutes, 50 rpm | 0.13±0.02 | 0.03±0.00 | 0.45±0.15 | 0.43±0.01 | 7.58±0.14 | 0.94±0.06 | 9.28±1.11 |
| 175°C, 2.5%, 3 minutes, 50 rpm | 0.24±0.02 | 0.06±0.00 | 0.55±0.01 | 0.60±0.01 | 10.05±0.21 | 0.78±0.03 | 6.42±0.09 |
| 185°C, 2.5%, 3 minutes, 50 rpm | 0.63±0.01 | 0.17±0.01 | 0.81±0.15 | 1.01±0.02 | 10.20±0.02 | 1.10±0.44 | 2.84±0.45 |
| Changing acid concentration | | | | | | | |
| 185°C, 1.5%, 3 minutes, 50 rpm | 0.08±0.01 | 0.04±0.02 | 0.59±0.06 | 0.45±0.05 | 2.39±0.00 | 0.72±0.04 | 11.02 ±0.02 |
| 185°C, 2.0%, 3 minutes, 50 rpm | 0.18±0.06 | 0.09±0.07 | 0.58±0.02 | 0.48±0.00 | 5.34±0.01 | 0.77±0.00 | 8.45±0.38 |
| 185°C, 2.5%, 3 minutes, 50 rpm | 0.63±0.01 | 0.17±0.01 | 0.81±0.15 | 1.01±0.02 | 10.20±0.02 | 1.10±0.44 | 2.84±0.45 |
| 185°C, 3.0%, 3 minutes, 50 rpm | 0.89±0.03 | 0.30±0.03 | 1.29±0.03 | 2.41±0.02 | 9.99±0.11 | 0.39±0.01 | 0.81±0.15 |
| 185°C, 3.5%, 3 minutes, 50 rpm | 0.78±0.02 | 0.36±0.07 | 1.24±0.05 | 2.07±0.02 | 7.85±0.08 | 0.45±0.01 | 0.80±0.10 |
| 185°C, 4.0%, 3 minutes, 50 rpm | 0.84±0.07 | 0.41±0.17 | 1.22±0.16 | 3.50±0.02 | 8.34±0.04 | 0.43±0.07 | 0.45±0.13 |
| Changing residue time | | | | | | | |
| 185°C, 2.5%, 1 minute, 50 rpm | 0.39±0.03 | 0.12±0.01 | 0.73±0.04 | 0.79±0.00 | 9.89±0.03 | 0.90±0.08 | 2.56±0.35 |
| 185°C, 2.5%, 3 minutes, 50 rpm | 0.63±0.01 | 0.17±0.01 | 0.81±0.15 | 1.01±0.02 | 10.20±0.02 | 1.10±0.44 | 2.84±0.45 |
| 185°C, 2.5%, 5 minutes, 50 rpm | 0.80±0.15 | 0.27±0.03 | 1.09±0.07 | 1.21±0.02 | 7.91±0.06 | 0.96±0.11 | 1.35±0.04 |
| 185°C, 2.5%, 10 minutes, 50 rpm | 0.99±0.10 | 0.30±0.02 | 1.13±0.07 | 1.17±0.01 | 6.27±0.07 | 0.69±0.04 | 0.37±0.08 |
| Changing agitation rate | | | | | | | |
| 185°C, 2.5%, 3 minutes, 10 rpm | 0.59±0.04 | 0.18±0.02 | 0.80±0.10 | 0.99±0.20 | 10.65±0.35 | 0.49±0.07 | 1.53±0.00 |
| 185°C, 2.5%, 3 minutes, 30 rpm | 0.52±0.02 | 0.19±0.00 | 0.79±0.11 | 0.97±0.27 | 10.58±0.23 | 0.78±0.05 | 2.04±0.04 |
| 185°C, 2.5%, 3 minutes, 50 rpm | 0.63±0.01 | 0.17±0.01 | 0.81±0.15 | 1.01±0.02 | 10.20±0.02 | 1.10±0.44 | 2.84±0.45 |

*The data in the pretreatment conditions column indicate the pretreatment temperature, acid concentration, and residual time, respectively. All the pretreatment experiments were carried out at 50 rpm. All the experiments were carried out in duplicate. Error was calculated as standard deviation. 5-HMF, 5-hydroxymethylfurfural; CS, corn stover; DM, dry matter; O-Glu, glucan oligomer; O-Xyl, xylan oligomer.*

## 3.4 METHODS

### 3.4.1 RAW MATERIALS

Corn stover was grown in Henan, China, and harvested in fall 2011. The corn stover materials were washed and then dried at 105°C until the weight was constant at which point the moisture was approximately 7% (w/w). The corn stover was then milled coarsely using a beater pulverizer (SF-300; Ketai Milling Equipment, Shanghai, China) and screened through a mesh with the circle diameter of 10 mm, then stored in sealed plastic bags until use.

### 3.4.2 STRAINS AND ENZYME

*Amorphotheca resinae* ZN1 (stored at Chinese General Microorganisms Collection Center, Beijing, China; registration number: CGMCC 7452) was used as the biodetoxification strain for the removal of inhibitors from the pretreated corn stover [18]. *A. resinae* ZN1 was inoculated on the solids of pretreated corn stover which was neutralized with 20% $Ca(OH)_2$ solution to pH 5.5. Biodetoxification started in solid state fermentation mode without any nutrients added and ended when the inhibitors were not detected on HPLC.

*Saccharomyces cerevisiae* DQ1 (stored at Chinese General Microorganisms Collection Center; registration number: CGMCC 2528) was used as the ethanol fermenting strain [22,26]. *S. cerevisiae* DQ1 was first cultured in the synthetic medium (20 g/L glucose, 2 g/L $KH_2PO_4$, 1 g/L $(NH_4)_2SO_4$, 1 g/L $MgSO_4 \cdot 7H_2O$, 1 g/L yeast extracts) for activation and transferred to the same medium without glucose containing the corn stover hydrolysate for adaption according to the procedure described by Zhang et al. [22].

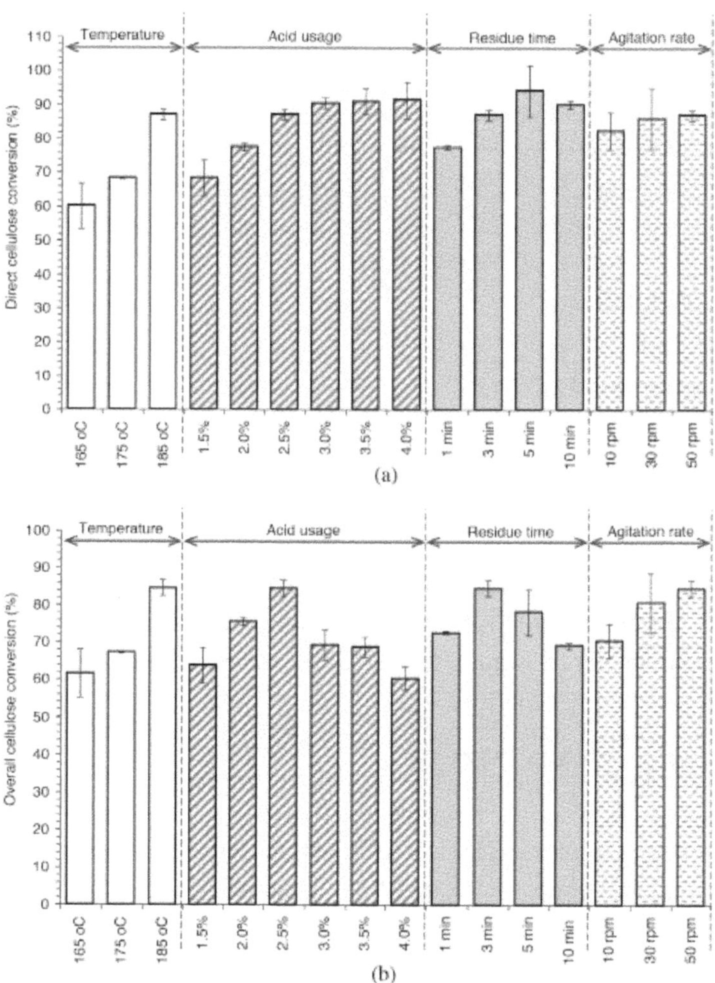


**FIGURE 4:** Enzymatic hydrolysis assay of the pretreatment parameters in the helically agitated reactor. (a) Direct cellulose conversion of the pretreated corn stover. (b) Overall cellulose conversion considering the cellulose recovery after the pretreatment. For detailed pretreatment operation, the experiment of changing temperature was carried out at 2.5% acid usage, 3 minutes of residue time, and 50 rpm agitation rate; the experiment of changing acid usage was carried out at 185°C, 3 minutes of residue time, and 50 rpm agitation rate; the experiment of changing residue time was carried out at 185°C, 2.5% acid usage, and 50 rpm agitation rate; and the experiment of changing agitation rate was carried out at 185°C, 2.5% acid usage, and 3 minutes of residue time. All the enzymatic hydrolysis processes of different pretreatment conditions were carried out at 5% solids loading (dry materials), 15 FPU/g DM cellulase dosage, pH 4.8, and 50°C. DM, dry matter; FPU, filter paper unit.

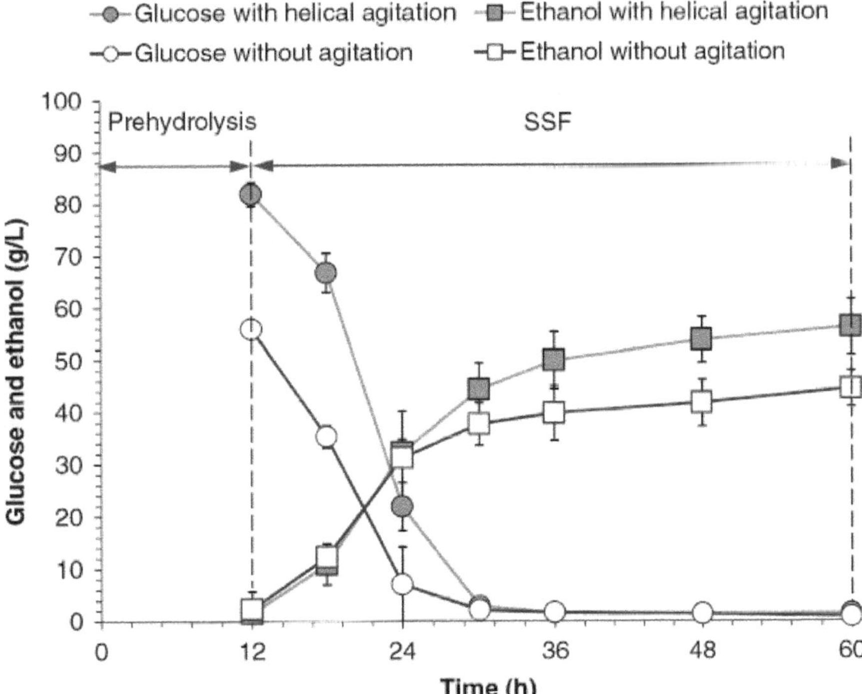


**FIGURE 5:** SSF of the pretreated corn stover. Corn stover was pretreated from the helically agitated reactor (185°C, 2.5%, 3 minutes, 50 rpm) and the no agitation reactor (190°C, 2.5%, 3 minutes), respectively. SSF was carried out under 30% solids loading, with the cellulase dosage of 15 FPU/g DM. The temperature and pH values during prehydrolysis process were controlled at 50°C and 4.8, respectively. These two conditions were set to 37°C and 5.5 during ethanol fermentation process by Saccharomyces cerevisiae DQ1, respectively. DM, dry matter; FPU, filter paper unit; SSF, simultaneous saccharification and fermentation.

The cellulase enzyme Youtell #6 was kindly provided by Hunan Youtell Biochemical Co. (Yueyang, Hunan, China). The filter paper activity of Youtell #6 was 135 FPU/g determined using the NREL Laboratory Analytical Procedure (LAP) LAP-006 [27], and the cellobiase activity was 344 cellobiase units (CBU)/g using the method of Sharma et al. [28].

### 3.4.3 FLUID DYNAMIC MOCK-UP EXPERIMENTS

Mock-up experiments were designed to detect the mixing effect of corn stover and water by helical ribbon stirrer. The experiments were carried out in three reactors with different sizes of 5 L, 50 L, and 500 L. The inner structure of the three scales of reactor was the same, and detailed in Figure 1. Some parameters are also listed in the table below Figure 1. The mixing time of corn stover and water were used to illustrate the mixing effect. First, the corn stover was added into the reactor, and then the water was added from the inlet of the reactor into the corn stover when the agitation was turned on. The mixing time was calculated by the time to reach constant moisture of the corn stover.

### 3.4.4 PRETREATMENT REACTOR

The detailed pretreatment reactor scheme is illustrated in Figure 2. Figure 2a shows the pretreatment reactor equipped with a helical ribbon stirrer. The reactor was a stainless cylinder with the working volume of 20 L (260 mm in diameter and 400 mm in height). The single helical ribbon stirrer was driven by a motor mounted on top of the reactor through an electromagnetic convertor. Figure 2b shows the pretreatment reactor with no mixing apparatus as a comparison. This reactor was previously used in Zhang et al. [17], and a stainless cylinder with a working volume of 10 L (180 mm in diameter and 400 mm in height) was also used. The hot steam was produced from the steam generator (DZFZ4.5C; Zhengyuan Electromechanics, Shanghai, China), then jetted into the reactor from the bottom and dispersed upward through several nozzles at the bottom. Two nozzles (6 mm in diameter) were designed on the distributor to disperse the steam jetted into the reactor at the mean steam flow rate of 0.1 to 1.0 kg per minute.

### 3.4.5 PRETREATMENT OPERATION

The dried corn stover was presoaked with the diluted sulfuric acid solution at the solid to liquid ratio of 2:1 (on the weight basis). The dilute acid solution was poured onto the dry corn stover materials then roughly mixed, and sealed in plastic bags and stayed for 12 hours at the ambient temperature (18 to 25°C). In each operation, 2,100 g of the presoaked corn stover (1,400 g of dry corn stover plus 700 g of dilute acid solution) was fed into the pretreatment reactor, and these corn stover materials roughly occupied the whole space of the reactor to meet the full solids loading condition of the reactor for reduction of steam consumption [17]. All the inlet valves were closed and the helical ribbon stirrer started to operate, then the steam valve was opened to jetting onto the presoaked corn stover. The purge valve was opened twice very briefly (2 to 3 seconds) to release the residual inert air inside the reactor. When the temperature reached the required value, the condition was maintained for a few minutes. To close the pretreatment operation, the steam supply was switched off and the steam inside the reactor was quickly released from the outlet of the reactor. The pretreated corn stover solids were taken out directly from the bottom of the reactor and no free water was released. Two batches of pretreatment at the same conditions were carried out, and the analysis of the pretreated corn stover was averaged from the two batches of pretreated corn stover.

### 3.4.6 PRETREATMENT EFFICIENCY ASSAY BY ENZYMATIC HYDROLYSIS

The pretreated corn stover was assayed following the NREL LAP-009 [29]. One gram of the freshly pretreated corn stover (dry base) was added into 0.1 M citrate buffer (pH 4.8) to prepare the 5% (w/w) solids slurry in the flask. The cellulase dosage was 15 FPU/g DM (dry pretreated corn stover mass) and the hydrolysis lasted for 72 hours at 50°C and 150 rpm of shaking.

Cellulose and xylan recovery was calculated based on the dry weight of corn stover before and after pretreatment. Cellulose components after pretreatment included cellulose, glucose, and glucan oligomers in the dry materials; and xylan components included hemicellulose, xylose, and xylan oligomers in the dry materials. The recovery was defined as the ratio of cellulose and xylan content after pretreatment to those before pretreatment. The direct cellulose conversion of the pretreated corn stover was indicated by the ratio of the glucose produced after the 72 hours' enzymatic hydrolysis (subtracting the initial glucose and glucan oligomers in the pretreated corn stover) to the theoretical glucose released from the cellulose in the pretreated corn stover. The overall cellulose conversion of corn stover was indicated by the ratio of the total glucose produced to the total theoretical glucose released from the original corn stover before pretreatment, in which the cellulose loss in the pretreatment was taken into account. The original cellulose content was calculated by the cellulose content of the pretreated corn stover divided by the cellulose recovery.

### *3.4.7 PRETREATMENT ASSAY BY SSF*

In the SSF process using the dry pretreated corn stover, the higher inhibitor concentration, which was caused by the high solids loading in the pretreatment, would greatly decrease the performance of the fermentative strains. Thus prior to the SSF step, the pretreated corn stover materials were detoxified biologically using the fungus *A. resinae* ZN1 according to the strictly uniformed procedure described in our reports [18,20,21] for all the pretreated SSF cases at the fixed time, temperature, and the operation. The SSF operation of the pretreated and biodetoxified corn stover was carried out in a 5 L helical ribbon stirrer agitated bioreactor as described in Zhang et al. [22]. The SSF operation was carried out at 30% solids (dry pretreated corn stover) concentration, 15 FPU/g DM cellulase dosage. The detoxified corn stover feedstock was sterilized at 115°C for 20 minutes. The operation started with 12 hours' prehydrolysis at 50°C and pH 4.8, then the temperature was reduced to 37°C and the adapted *S. cerevisiae* DQ1 cells were inoculated into the bioreactor at 10% inoculum ratio (v/v) to start the SSF. Samples were taken periodically for analysis of ethanol and glucose. These experiments were carried out using the two separate batches of pretreated corn stover at the same pretreatment condition, and then averaged for the final data and error.

### 3.4.8 COMPUTATIONAL FLUID DYNAMICS (CFD) MODELING OF THE PRETREATMENT REACTOR

The commercial grid generation tool, ICEM CFD 11.0 (Ansys Inc., Canonsburg, PA, USA) was used to generate the three-dimensional grids of the reactor model created in SolidWorks 2010 (Dassault Inc., Vélizy-Villacoublay, France). The impeller agitation was characterized with the multiple reference frame (MRF) model. The mathematical model was solved in CFX 11.0 (Ansys Inc.). The initial and boundary conditions were specified as: 1) the impeller and shaft regions were stationary relative to the fluid domain; 2) no slip wall; 3) the residual error was set as $1 \times 10^{-4}$; and 4) the Eulerian-Eulerian and the k-ϵ turbulence model were applied.

Figure 3a and b show the CFD mesh cells and geometric structure, respectively, in which the gas inlets were identical and the gas outlet was assumed to be released to the completely open cap on top of the reactor. A gas–liquid two-phase flow was assumed under the assumptions of the pseudo-liquid phase of corn stover materials, and the inert and incompressible gas phase of steam vapor. The apparent viscosity of the assumed pseudo-liquid was set to 2.31 Pa · s according to the determination using the torque measurement method. The density of the assumed gas flow was set to 14.18 g/L, which equaled the steam density at 3.0 MPa, 250°C used in the pretreatment. The velocity of the gas phase was calculated to be 1.75 m/s by modeling the typical pretreatment process, which equaled to jetting 700 g of hot steam within 3 minutes into the reactor through two nozzles. The conservative gas volume fraction and the liquid velocity distribution were used to characterize the mixing at different agitation conditions.

### 3.4.9 SUGAR AND INHIBITOR ANALYSIS

Sugars and inhibitors were measured by HPLC (LC-20 AD, RID-10A refractive index detector; Shimadzu, Kyoto, Japan) equipped with an Aminex HPX-87H column (Bio-Rad, Hercules, CA, USA) at the column temperature 65°C. The mobile phase was 5 mM $H_2SO_4$ at the rate of 0.6 mL/min. Samples were filtered through a 0.22 μm membrane before analysis.

### 3.4.10 CELLULOSE AND XYLAN COMPOSITION DETERMINATION

The cellulose and xylan content of corn stover were measured by two-step acid hydrolysis according to NREL LAPs [30,31]. The pretreated corn stover was washed thoroughly with deionized water and oven dried at 105°C overnight to determine the content of water insoluble solids. One hundred milligrams of dried corn stover was added to 1 mL 72% (w/w) $H_2SO_4$ and then incubated at 30°C in a water bath for 1 hour with stirring by glass rod. The mixture was diluted to 29 mL in volume and hydrolyzed at 121°C for 1 hour. The hydrolyzed mixture was neutralized using $CaCO_3$ powder and centrifuged. The supernatant was used for HPLC analysis to measure the glucose and xylose to calculate the cellulose and xylan content.

Oligomers of cellulose and xylan were measured according to NREL LAP [31]. The mixture of 5 g wet pretreated corn stover and 50 mL deionized water was shaken at 180 rpm for 2 hours at 30°C. Then 5 mL filtrate after solids/liquid separation was used for determining the concentration of glucose and xylose, and mixed with 1 mL 72% (w/w) sulfuric acid. The mixture was then diluted to 29 mL in volume and the subsequent process was the same as the two-step hydrolysis. The difference of the sugar concentration before and after acid hydrolysis was calculated as the oligosaccharide content. These experiments were carried out using the two separate batches of pretreated corn stover at the same pretreatment condition, and then averaged for the final data and error.

## REFERENCES


1. Wyman CE, Dale BE, Elander RT, Holtzapple M, Ladisch MR, Lee YY: Coordinated development of leading biomass pretreatment technologies. Bioresour Technol 2005, 96:1959-1966.
2. Yang B, Wyman CE: Pretreatment: the key to unlocking low-cost cellulosic ethanol. Biofuel Bioprod Bioref 2008, 2:26-40.
3. Zhu JY, Pan XJ, Zalesny RS Jr: Pretreatment of woody biomass for biofuel production: energy efficiency, technologies, and recalcitrance. Appl Microbiol Biotechnol 2010, 87:847-857.
4. Alvira P, Tomas-Pejo E, Ballesteros M, Negro MJ: Pretreatment technologies for an efficient bioethanol production process based on enzymatic hydrolysis: a review. Bioresour Technol 2010, 101:4851-4861.
5. Chiaramonti D, Prussi M, Ferrero S, Oriani L, Ottonello P, Torre P, Cherchi F: Review of pretreatment processes for lignocellulosic ethanol production, and development of an innovative method. Biomass Bioenergy 2012, 46:25-35.
6. Galbe M, Zacchi G: Pretreatment: the key to efficient utilization of lignocellulosic materials. Biomass Bioenergy 2012, 46:70-78.
7. Torget R, Werdene P, Himmel M, Grohmann K: Dilute acid pretreatment of short rotation woody and herbaceous crops. Appl Biochem Biotechnol 1990, 24:115-126.
8. Torget R, Werdene P, Himmel M, Grohmann K: Dilute-acid pretreatment of corn residues and short-rotation woody crops. Appl Biochem Biotechnol 1991, 28:75-86.
9. Saha BC, Iten LB, Cotta MA, Wu YV: Dilute acid pretreatment, enzymatic saccharification and fermentation of wheat straw to ethanol. Process Biochem 2005, 40:3693-3700.
10. Lloyd TA, Wyman CE: Combined sugar yields for dilute sulfuric acid pretreatment of corn stover followed by enzymatic hydrolysis of the remaining solids. Bioresour Technol 2005, 96:1967-1977.
11. Hsu TC, Guo GL, Chen WH, Hwang WS: Effect of dilute acid pretreatment of rice straw on structural properties and enzymatic hydrolysis. Bioresour Technol 2010, 101:4907-4913.
12. Humbird D, Davis R, Tao L, Kinchin C, Hsu D, Aden A, Schoen P, Lukas J, Olthof B, Worley M, Sexton D, Dudgeon D: Process Design and Economics for Biochemical Conversion of Lignocellulosic Biomass to Ethanol. NREL: Golden, CO; 2011. [Technical Report NREL/TP-5100-47764]
13. Dong HW, Bao J: Biofuel via biodetoxification. Nat Chem Biol 2010, 6:316-318.
14. Linde M, Jakobsson EL, Galbe M, Zacchi G: Steam pretreatment of dilute H2SO4-impregnated wheat straw and SSF with low yeast and enzyme loadings for bioethanol production. Biomass Bioenergy 2008, 32:326-332.
15. Sassner P, Martensson GG, Galbe M, Zacchi G: Steam pretreatment of H2SO4-impregnated Salix for the production of bioethanol. Bioresour Technol 2008, 99:137-145.
16. Modenbach AA, Nokes SE: The use of high-solids loadings in biomass pretreatment—a review. Biotechnol Bioeng 2012, 109:1430-1442.
17. Zhang J, Wang XS, Chu DQ, He YQ, Bao J: Dry pretreatment of lignocellulose with extremely low steam and water usage for bioethanol production. Bioresour Technol 2011, 102:4480-4488.

18. Zhang J, Zhu ZN, Wang XF, Wang N, Wang W, Bao J: Biodetoxification of toxins generated from lignocellulose pretreatment using a newly isolated fungus Amorphotheca resinae ZN1 and the consequent ethanol fermentation. Biotechnol Biofuels 2010, 3:26.
19. Huang X, Wang YM, Liu W, Bao J: Biological removal of inhibitors leads to the improved lipid production in the lipid fermentation of corn stover hydrolysate by Trichosporon cutaneum. Bioresour Technol 2011, 102:9705-9709.
20. Liu W, Wang YM, Yu ZC, Bao J: Simultaneous saccharification and microbial lipid fermentation of corn stover by oleaginous yeast Trichosporon cutaneum. Bioresour Technol 2012, 118:13-18.
21. Zhao K, Qiao QA, Chu DQ, Gu HQ, Dao TH, Zhang J, Bao J: Simultaneous saccharification and high titer lactic acid fermentation of corn stover using a newly isolated lactic acid bacterium Pediococcus acidilactici DQ2. Bioresour Technol 2013, 135:481-489.
22. Zhang J, Chu DQ, Huang J, Yu ZC, Dai GC, Bao J: Simultaneous saccharification and ethanol fermentation at high corn stover solids loading in a helical stirring bioreactor. Biotechnol Bioeng 2010, 105:718-728.
23. Jensen JR, Morinelly JE, Gossen KR, Brodeur-Campbell MJ, Shonnard DR: Effects of dilute acid pretreatment conditions on enzymatic hydrolysis monomer and oligomer sugar yields for aspen, balsam, and switchgrass. Bioresour Technol 2010, 101:2317-2325.
24. Jung YH, Kim IJ, Kim HK, Kim KH: Dilute acid pretreatment of lignocellulose for whole slurry ethanol fermentation. Bioresour Technol 2013, 132:109-114.
25. Redding AP, Wang ZY, Keshwani RD, Cheng JJ: High temperature dilute acid pretreatment of coastal Bermuda grass for enzymatic hydrolysis. Bioresour Technol 2011, 102:1415-1424.
26. Chu DQ, Zhang J, Bao J: Simultaneous saccharification and ethanol fermentation of corn stover at high temperature and high solids loading by a thermotolerant strain Saccharomyces cerevisiae DQ1. Bioenerg Res 2012, 5:1020-1026.
27. Adney B, Baker J: Measurement of Cellulase Activities. NREL: Golden, CO; 1996. [Laboratory Analytical Procedure (LAP). LAP-006]
28. Sharma S, Sandhu DK, Bagga PS: Physical characterization of isozymes of endo-beta-1,4-glucanase and beta-1,4-glucosidase from Aspergillus species. FEMS Microbiol Lett 1991, 63:99-104.
29. Brown L, Torget R: Enzymatic Saccharification of Lignocellulosic Biomass. Laboratory Analytical Procedure (LAP). LAP-009. NREL: Golden, CO; 1996.
30. Sluiter A, Hames B, Ruiz R, Scarlata C, Sluiter J, Templeton D, Crocker D: Determination of Structural Carbohydrates and Lignin in Biomass. NREL: Golden, CO; 2008. [Laboratory Analytical Procedure (LAP). Technical Report NREL/TP-510-42618]
31. Sluiter A, Hames B, Ruiz R, Scarlata C, Sluiter J, Templeton D: Determination of Sugars, Byproducts, and Degradation Products in Liquid Fraction Process Samples. NREL: Golden, CO; 2008. [Laboratory Analytical Procedure (LAP). Technical Report NREL/TP-510-42623]

## CHAPTER 4

# TWEEN 40 PRETREATMENT OF UNWASHED WATER-INSOLUBLE SOLIDS OF REED STRAW AND CORN STOVER PRETREATED WITH LIQUID HOT WATER TO OBTAIN HIGH CONCENTRATIONS OF BIOETHANOL

JIE LU, XUEZHI LI, RUIFENG YANG, JIAN ZHAO, AND YINBO QU

## 4.1 BACKGROUND

Bioethanol has been widely used as a substitute for fossil fuels [1]. The use of bioethanol produced from lignocellulosic material can reduce the dependence on fossil fuels [2]. In general, lignocellulosic materials are subjected to bioethanol conversion performed in three steps: pretreatment; enzymatic hydrolysis; and fermentation [1]. Pretreatment is crucial to determine conversion efficiency. Studies have investigated and proposed many pretreatment materials/methods, such as alkaline [3], steam explosion [4], ammonia fiber expansion [5], organic solvent [6], and diluted acid [7]. One of the most promising pretreatment processes of lignocelluloses material is liquid hot water (LHW) pretreatment. LHW pretreatment has been considered as an environmentally friendly technology. LHW has been shown to remove most of the hemicelluloses, but a large amount of lignin is retained in water-insoluble solids (WIS). Cellulase can be adsorbed on lignin surfaces during enzymatic hydrolysis, thereby deactivating cellulase. Studies have indicated that surfactant additives can improve enzymatic hydrolysis and bioethanol fermentation of lignocellulosic biomass [8-12]. Tween additives can effectively improve cellulase efficiency during enzymatic hydrolysis and fermentation of lignocellulosic materials [13,14]. For example, the simultaneous saccharification and fermentation (SSF) of steam-pretreated softwood was improved by the addition of Tween 20 due to a combination of increased hydrolysis rate and improved yeast fermentation [15]. Research about the effects of surfactant on SSF of steam-exploded poplar has also shown that the ethanol yield could be increased by 6% by the addition of Tween 80 [16]. Ooshima et al. reported that the rate of SSF of pure cellulose (Avicel) was slightly enhanced by adding Tween 20 [17]. Studies about the mechanism of Tween additives on enzymatic hydrolysis and fermentation of lignocellulosic materials have also been proposed [18-21]. The mechanism mainly focuses on three aspects: protecting free cellulase from deactivation; decreasing cellulase protein adsorption on the substrate; and reducing unproductive binding of enzymes to lignin. Tween additives contain hydrophilic ethylene glycol head groups and a hydrophobic alkyl tail. Absorption of the hydrophobic alkyl group to a hydrophobic surface exposes the hydrophilic ethylene glycol chains, thus making the surface resistant to non-specific protein adsorption [22]. The commonly used Tween additives include Tween 20, Tween 40, Tween 60, and Tween 80. These additives contain the same hydrophilic head group with different lengths of hydrophobic alkyl tail. Different methods of applying Tween additives produce different results of bioethanol fermentation. Tween additives can be applied mainly in three stages: during the pretreatment process [23]; in the enzymatic hydrolysis stage [24]; and in the fermentation stage [24]. However, few reports are available on Tween 40 pretreatment prior to the fed-batch semi-simultaneous saccharification and fermentation (S-SSF) of lignocellulosic biomass. The present study aimed to confirm the effect of Tween 40 pretreatment on the fermentation digestibility of unwashed WIS of reed straw and corn stover pretreated with LHW. Several pretreatment methods were compared. The positive effect of Tween 40 pretreatment with unwashed WIS was observed. The effect of Tween 40 was possibly related to process conditions, so we also investigated the process conditions of Tween 40 pretreatment, such as temperature, time, concentration, and ratio of WIS-to-Tween 40. The residual liquid of Tween 40 pretreatment was recycled to save wash water. Cellulase dosage and feeding methods in fed-batch S-SSF process were also researched to obtain high ethanol concentration. This article presents the results.

## 4.2 RESULTS AND DISCUSSION

### *4.2.1 EFFECT OF DIFFERENT PRETREATMENT METHODS OF TWEEN 40 ON ETHANOL CONCENTRATION USING UNWASHED AND WASHED WIS AS LIGNOCELLULOSIC SUBSTRATES*

Tween 40 was used as an additive in different processes, for example, LHW pretreatment, S-SSF, or as a single pretreatment stage to treat WIS in this study, and the effect of different pretreatment methods of Tween 40 on ethanol concentration using unwashed and washed WIS as substrates is shown in Figure 1. Using untreated raw materials as substrates, the ethanol concentration after fed-batch S-SSF was very low (0.47 g/L for reed straw and 0.48 g/L for corn stover). Therefore, the raw materials without any pretreatment were not suitable for the production of ethanol. The ethanol concentrations of unwashed WIS without any pretreatment were 1.36 g/L (reed straw) and 6.09 g/L (corn stover) (Figure 1). Ethanol concentration was also very low, indicating that unwashed WIS was not suitable for direct fermentation and should undergo pretreatment for follow-up fermentation. The unwashed WIS was washed with water until a neutral condition was obtained. Ethanol concentration significantly increased and reached 34.17 g/L (reed straw) and 32.16 g/L (corn stover) when the washed WIS was used for fermentation. However, a large amount of wash water was used in the washing process, thereby increasing production cost. The ethanol concentration was approximately equal to that of the washed WIS when the unwashed WIS was pretreated with Tween 40, but the washed WIS was treated with Tween 40 to slightly increase ethanol concentration compared with washed WIS without Tween 40, indicating that Tween 40 pretreatment is more suitable for unwashed WIS. To confirm the effect of Tween 40, we used water as a control treatment instead of Tween 40 in the experiment. Ethanol concentrations obtained from water-pretreated unwashed WIS were 22.11 g/L (reed straw) and 28.14 g/L (corn stover), respectively. This result was lower than that obtained by Tween 40 pretreated unwashed WIS (36.18 g/L for reed straw and 38.19 g/L for corn stover, respectively), indicating that Tween 40 pretreatment positively affects the S-SSF ability of the unwashed WIS. Some studies have indicated that calcium hydroxide and sodium bisulfite can eliminate the negative effect of inhibitors because unwashed WIS contains a large number of inhibitors of yeast [25]. Calcium hydroxide and sodium bisulfite added to Tween 40 pretreatment stage was performed. Figure 1 also shows that the ethanol concentrations were retained (calcium hydroxide) or reduced (sodium bisulfite) when the two chemicals were added in the S-SSF of unwashed WIS. Therefore, Tween 40 is possibly the most suitable pretreatment method for unwashed WIS, and detoxication treatment of unwashed WIS for eliminating the negative effect of inhibitors produced by LHW pretreatment is not required.

The Tween 40 was also added into the LHW and S-SSF process to evaluate its function on improving ethanol production, since they could shorten the whole product process and save energy. However, the results show that both ethanol concentrations did not increase (Figure 1). This indicated that adding Tween 40 to either the LHW or S-SSF process was not suitable for unwashed WIS.

Many studies have reported the effect of the surfactant Tween series on enzymatic hydrolysis and SSF and its mechanism [12,15,18-21]. The studies showed that the addition of surfactant Tween improved enzymatic hydrolysis yields and ethanol production. Surfactants can enhance enzymatic digestibility by: 1) changing the substrate structure to make it more accessible to enzymes; 2) stabilizing enzymes to prevent denaturation; 3) increasing positive interactions between substrates and enzymes; and 4) reducing enzyme non-productive binding to lignin and other molecules involved in cellulase activity [24]. Tween contains hydrophilic ethylene glycol head groups and a hydrophobic alkyl tail. Hydrophilic surfactants have been reported to be useful in extracting hydrophobic degradation products from lignin and hemicellulose. Based on the mechanisms, pretreatment with Tween 40 possibly removed some degradation products from lignin and hemicellulose, which were contained in unwashed WIS and had a negative effect on enzymatic hydrolysis and fermentation. Besides, the Tween 40 acted as the above surfactants to enhance enzymatic digestibility of unwashed WIS.

### *4.2.2 OPTIMIZATION OF PROCESS CONDITIONS OF TWEEN 40 PRETREATMENT*

#### *4.2.2.1 EFFECT OF TWEEN 40 PRETREATMENT TEMPERATURE ON ETHANOL CONCENTRATION*

The effect of Tween 40 pretreatment temperature on ethanol concentration is shown in Figure 2. The ethanol concentrations from reed straw and corn stover varied. The ethanol concentration obtained from reed straw remained almost unchanged as Tween 40 pretreatment temperature increased. By contrast, ethanol concentration obtained from corn stover decreased slightly as Tween 40 pretreatment temperature increased before 100°C. Taking supplied water temperature in winter into consideration, we conducted the Tween 40 pretreatment at the lowest temperature of 9°C, and showed the pretreatment temperature had little effect on ethanol yield. Thus, the proposed Tween 40 pretreatment may be conducted at room temperature.

#### *4.2.2.2 EFFECT OF TWEEN 40 PRETREATMENT TIME ON ETHANOL CONCENTRATION*

The effect of Tween 40 pretreatment time on ethanol concentration is shown in Figure 3. Tween 40 pretreatment time ranged from 0 to 90 minutes. At 0 minutes, the unwashed WIS was washed directly with Tween 40 solution at 25°C. Figure 3 shows that the change rule of the ethanol concentrations obtained from reed straw and corn stover were similar. Ethanol concentration increased as Tween 40 pretreatment time was prolonged from 0 to 60 minutes. However, for reed straw, ethanol concentration neither increased nor decreased when pretreatment time was further prolonged, and for corn stover, ethanol concentration decreased slightly. Therefore, the suitable Tween 40 pretreatment time was 60 minutes.

#### *4.2.2.3 EFFECT OF TWEEN 40 CONCENTRATION ON ETHANOL CONCENTRATION*

The effect of Tween 40 concentration on ethanol concentration is shown in Figure 4, in which the ethanol concentration was almost the same for the reed straw and corn stover pretreated at different Tween 40 concentrations. Ethanol concentration initially increased and then decreased as Tween 40 concentration increased. The highest ethanol concentration of 42.21 g/L for the reed straw and corn stover was observed at Tween 40 concentration of 1.5%. Thus, Tween 40 concentration of 1.5% was used for the subsequent experiments.

#### *4.2.2.4 EFFECT OF WIS-TO-TWEEN 40 RATIO ON ETHANOL CONCENTRATION*

The effect of WIS-to-Tween 40 ratio (w/v) on ethanol concentration is shown in Figure 5, wherein the ethanol concentration did not increase when the WIS-to-Tween 40 ratio ranged from 1:4 to 1:8. However, the ethanol concentration increased when the WIS-to-Tween 40 ratio increased to 1:10. More Tween 40 will be consumed when the ratio of WIS to Tween increased. Thus, the WIS-to-Tween 40 of 1:10 was appropriate for ethanol production.

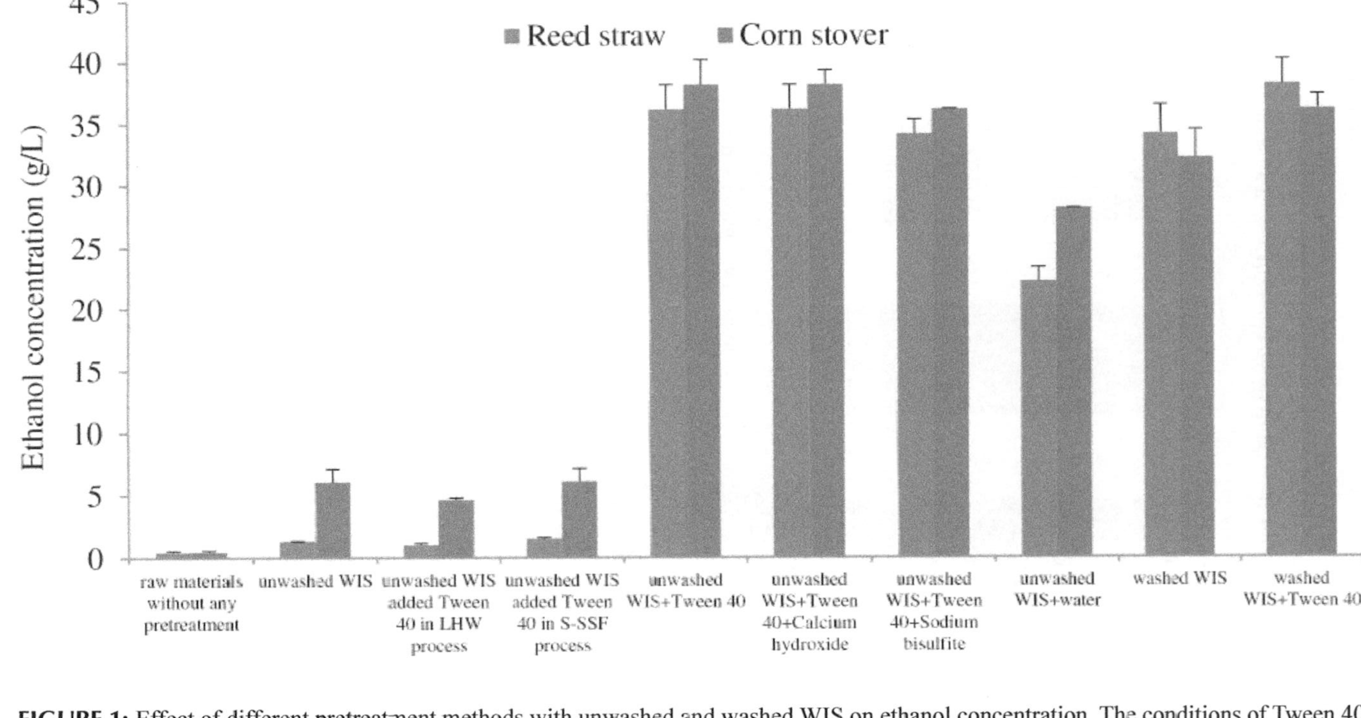


**FIGURE 1:** Effect of different pretreatment methods with unwashed and washed WIS on ethanol concentration. The conditions of Tween 40 pretreatment were as follows: Tween 40 concentration, 1%; the ratio of WIS-to-Tween 40, 1:10 (w/v); pretreatment temperature, 50°C; and pretreatment time, 1 hour. The fed-batch S-SSF conditions were as follows: 1 g dry weight biomass; cellulase loading, 20 FPU/g oven-dried WIS; pre-hydrolysis temperature, 50°C; pre-hydrolysis time, 18 hours; fermentation temperature, 36°C; and fermentation time, 72 hours, and after 6 hours of pre-hydrolysis, 1 g dry weight biomass was supplemented into the flask.

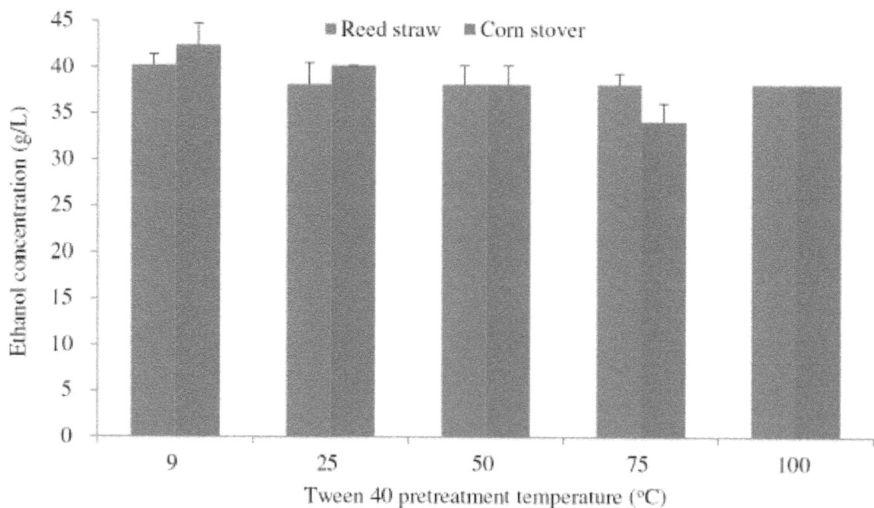


**FIGURE 2:** Effect of Tween 40 pretreatment temperature on ethanol concentration. The conditions of Tween 40 pretreatment and fed-batch S-SSF were the same as that in Figure 1, except pretreatment temperature. S-SSF, semi-simultaneous saccharification and fermentation.

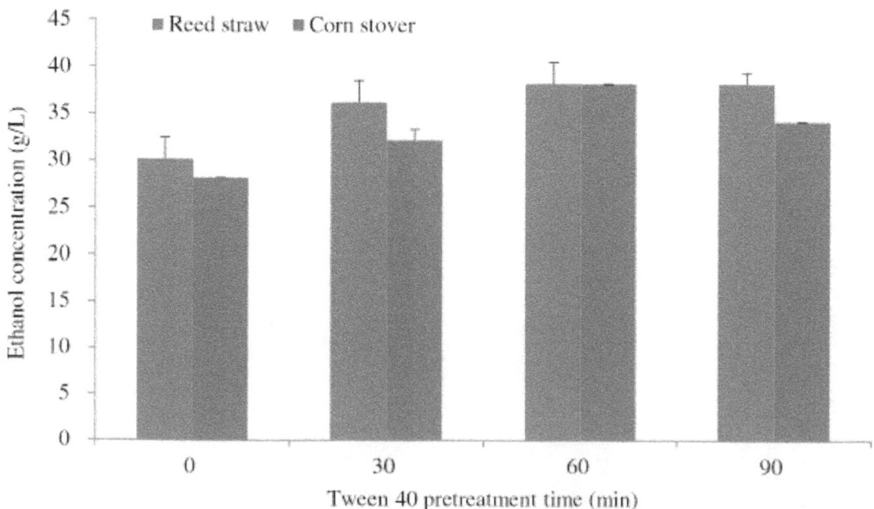


**FIGURE 3:** Effect of Tween 40 pretreatment time on ethanol concentration. The conditions of Tween 40 pretreatment and fed-batch S-SSF were the same as that in Figure 1, except Tween 40 pretreatment temperature of 25°C. S-SSF, semi-simultaneous saccharification and fermentation.

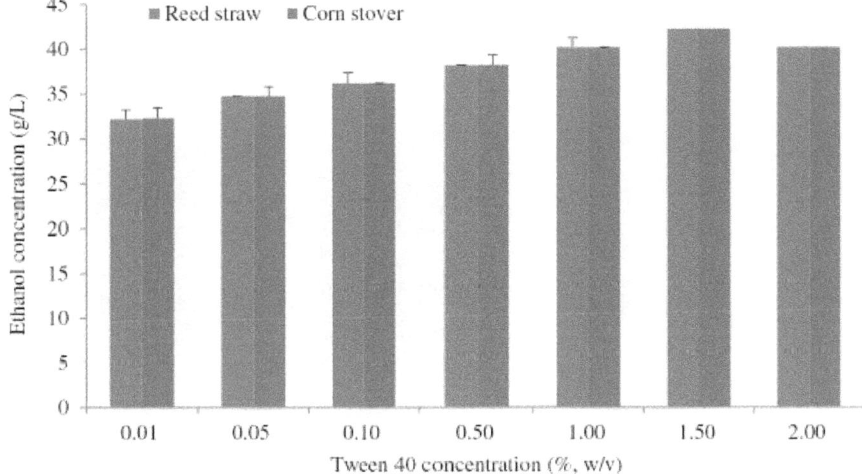


**FIGURE 4:** Effect of Tween 40 concentration on ethanol concentration. The conditions of Tween 40 pretreatment and fed-batch S-SSF were the same as that in Figure 3. S-SSF, semi-simultaneous saccharification and fermentation.

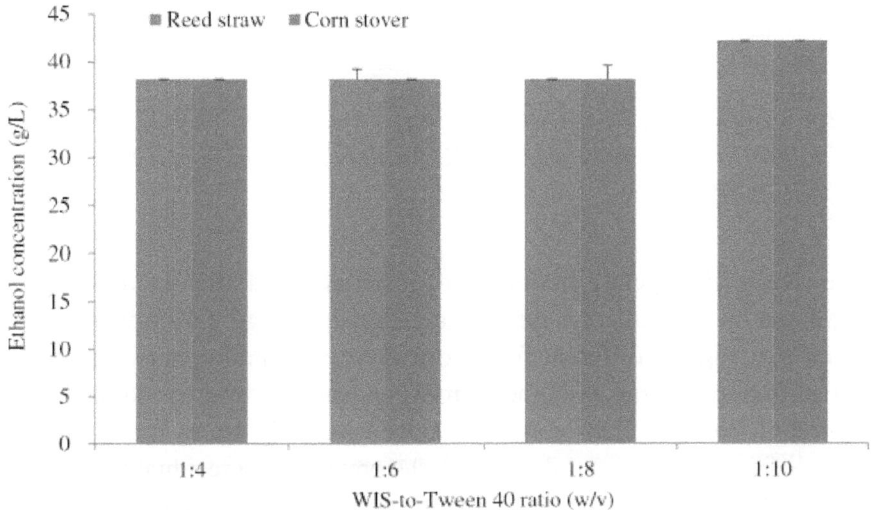


**FIGURE 5:** Effect of WIS-to-Tween 40 ratio on ethanol concentration. Tween 40 concentration of 1.5% was used in Tween 40 pretreatment, and other conditions in Tween 40 pretreatment and fed-batch S-SSF were the same as that in Figure 4. S-SSF, semi-simultaneous saccharification and fermentation.

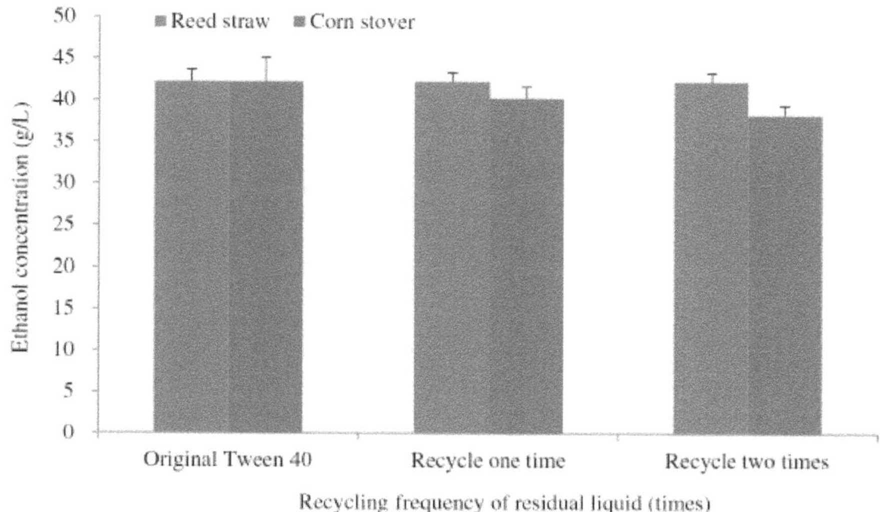


**FIGURE 6:** Effects of recycling frequency of residual liquid after Tween 40 pretreatment on ethanol concentration. Cellulase loading of 25 FPU/g oven-dried WIS was used in fed-batch S-SSF process, and other conditions were the same as that in Figure 5. S-SSF, semi-simultaneous saccharification and fermentation; WIS, water-insoluble solids.

### *4.2.2.5 EFFECTS OF THE RECYCLING FREQUENCY OF RESIDUAL LIQUID AFTER TWEEN 40 PRETREATMENT ON ETHANOL CONCENTRATION*

The effects of recycling frequency of residual liquid after Tween 40 pretreatment on ethanol concentration are shown in Figure 6, in which reed straw was slightly different from corn stover. Ethanol concentration obtained from reed straw remained almost unchanged, whereas ethanol concentration obtained from corn stover decreased slightly as the Tween 40 recycle frequency increased. Overall, the pretreatment residual liquid may be recycled to reduce Tween 40 consumption and decrease the cost of the Tween 40 pretreatment. An optimum number of recycling of the residual liquid will be investigated in further work.

In summary, to obtain a high ethanol yield, WIS obtained from the LHW pretreatment process generally needs to be washed with water or detoxicated with chemicals such as calcium hydroxide to remove some inhibitors that have a negative effect on yeast. A large amount of wash water will be consumed in the washing process, which could lead to an increase in process cost and environmental pollution. For example, approximately 200 mL of water was used to wash 3 g WIS until the pH reached 7, which means that approximately 67 $m^3$ of wash water per ton dry weight of unwashed WIS needs to be consumed. However, as a substitution for washing with water, the Tween 40 pretreatment was used in ethanol production from biomass to save large amounts of wash water. The Tween 40 pretreatment could be performed at room temperature, and the residual liquid after Tween 40 pretreatment could also be recycled to save the surfactant dosage and reduce the risk of environmental pollution caused by secondary wastewater generated after Tween 40 pretreatment. Thus, the Tween 40 pretreatment may develop into an energy-saving and environmentally friendly method that could be used in ethanol production from biomass.

### 4.2.3 OPTIMIZATION OF PROCESS CONDITIONS OF FED-BATCH S-SSF

#### 4.2.3.1 EFFECT OF CELLULASE LOADING IN FED-BATCH S-SSF ON ETHANOL CONCENTRATION

High production cost is the main obstacle hindering the commercialization of bioethanol. An important and expensive input into the biomass conversion system is enzyme loading, which can amount to approximately 60% of the whole cost [26]. Hence, enzyme dosage should be as low as possible. In this study, the effects of cellulase loadings (presented as FPU/g oven-dried WIS) on the fed-batch S-SSF of unwashed WIS pretreated with Tween 40 were investigated. The ethanol concentrations produced with different cellulase loadings are shown in Figure 7. The ethanol concentration increased when the cellulase loading increased from 15 FPU/g to 25 FPU/g of oven-dried WIS in fed-batch S-SSF (Figure 7). The increase in ethanol concentration was not evident for a higher dosage of cellulase. Therefore, cellulase loading of 25 FPU/g oven-dried WIS was sufficient. In a previous study [27], cellulase loading reached 30 FPU/g to 40 FPU/g of oven-dried WIS (reed straw) and 50 FPU/g of oven-dried WIS (corn stover) when the raw materials were not pretreated with Tween 40 prior to S-SSF. Cellulase loading could be reduced by approximately 40% to 50%. This result occurred possibly because Tween 40 pretreatment decreases adsorption of cellulase to the WIS and cellulase deactivation due to lignin. This phenomenon has potential economic implications because the cost of cellulase is a major contributor to process expenses, considering that the price of Tween 40 is lower than that of cellulase. This work will be continued, with the aim of finding cheaper surfactants with the same positive effects as Tween 40.

#### *4.2.3.2 EFFECT OF FED-BATCH METHODS ON ETHANOL CONCENTRATION AND ETHANOL YIELD*

To enhance ethanol concentration and yield, we performed different fed-batch methods. In a previous study, the fed-batch methods of feeding

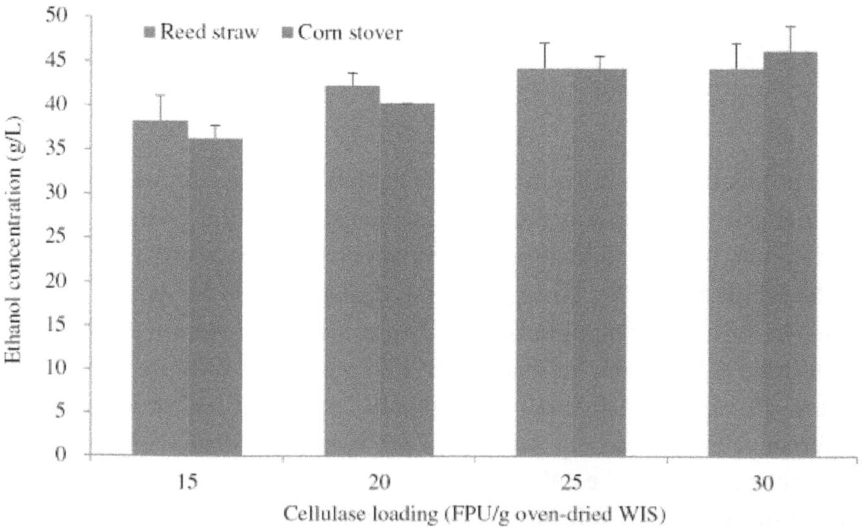


**FIGURE 7:** Effect of cellulase loading on ethanol concentration. The conditions of Tween 40 pretreatment and fed-batch S-SSF were the same as that in Figure 6. S-SSF, semi-simultaneous saccharification and fermentation.

one time at the pre-hydrolysis time of 6 hours obtained high ethanol concentration and yield [27]. On this basis, WIS was fed twice at the pre-hydrolysis time of 16 hours and fermentation time of 6 hours after feeding time twice, respectively. The ethanol concentration and yield at different feeding methods of fed-batch S-SSF are shown in Table 1. Five fed-batch methods (numbered 1 to 5) were evaluated, and the two materials were then compared (Table 1). The results showed that all of the fed-batch methods were beneficial to ethanol production from reed straw and corn stover. Number 5 fed-batch method could obtain a maximum of 56.28 g/L of ethanol concentration and 69.1% ethanol yield (based on glucan of unwashed WIS from LHW process) with reed straw. Number 3 fed-batch method could obtain a maximum of 52.26 g/L ethanol concentration and 71.1% ethanol yield with corn stover. Considering the yield of WIS in LHW pretreatment, the ethanol yield was also calculated on the basis of dry weight of the original feedstock that enters the process, which was 197 g ethanol per kg of reed straw and 233 g ethanol per kg of corn stover, respectively. In future studies, ethanol concentration and yield could be further improved if the fed-batch methods are systematically optimized.

### 4.2.4 COMPARISON OF SEVERAL ETHANOL PRODUCTIONS USING TWEEN AS AN ADDITIVE FOUND IN THE LITERATURE AND IN THE CURRENT STUDY

The data obtained in this study highlighted the importance of the use of the Tween 40 pretreatment method prior to fed-batch S-SSF to increase ethanol concentration and decrease cellulase loading. Several studies have produced ethanol by using Tween as an additive. Table 2 shows a comparison of several ethanol productions in which Tween was used as an additive. In contrast to the results of other studies, the ethanol concentration in the current study reached 56.28 g/L (reed straw) and 52.26 g/L (corn stover), a significantly higher ethanol concentration compared with other studies. Ethanol yield reached 69.1% (reed straw) and 71.1% (corn stover) at a relatively low dosage of cellulase. One of the advantages of Tween

40 pretreatment prior to S-SSF was that the process was beneficial to unwashed WIS, which contains a large number of yeast inhibitors. The usual approach is to use large amounts of water to wash inhibitors until neutral conditions are achieved for obtaining high ethanol yield from unwashed WIS, which could lead to large amounts of wash water consumption. The Tween 40 pretreatment could be performed at room temperature and atmosphere pressure, and the residue liquid of Tween 40 pretreatment could be recycled, leading to a decrease in process cost. Thus, Tween 40 pretreatment is an environmentally friendly, energy-saving, and low-cost method for ethanol production with unwashed WIS. In our further work, Tween 40 pretreatment techniques used to obtain consistently high ethanol results will be developed to improve process efficiency and decrease pretreatment cost and cellulase loadings in subsequent enzymatic hydrolysis of WIS. The mechanism of Tween 40 pretreatment that could improve the fermentable digestibility of unwashed WIS will also be investigated in detail.

## 4.3 CONCLUSIONS

Tween 40 pretreatment prior to bioethanol fermentation of unwashed WIS is a very effective and less costly method of ethanol production with unwashed WIS obtained from LHW pretreatment of corn stover and reed straw. This pretreatment could greatly reduce cellulase loading and save wash water. Higher ethanol concentration was obtained almost without decreasing ethanol yield. The optimum conditions of the Tween 40 pretreatment were as follows: Tween 40 concentration of 1.5%, WIS-to-Tween 40 ratio of 1:10 (w/v), and Tween 40 pretreatment time of 1 hour at ambient temperature. After Tween 40 pretreatment, cellulase loading could be greatly reduced. Residual liquid obtained after Tween 40 pretreatment could be recycled. Unwashed WIS could obtain high ethanol concentrations of 56.28 g/L for reed straw and 52.26 g/L for corn stover by conducting Tween 40 pretreatment prior to bioethanol fermentation with fed-batch S-SSF. Ethanol yields were 69.1% for reed straw and 71.1% for corn stover.

**TABLE 1:** Ethanol concentration and yield at different feed methods using fed-batch S-SSF

| Trial number | Dry weight (g) of WIS fed at different times[b] | | | Volume of fermentation broth (mL) after feeding[a] | | Reed straw | | | Corn stover | | |
|---|---|---|---|---|---|---|---|---|---|---|---|
| | At 6 hours of pre-hydrolysis | At 16 hours of pre-hydrolysis | At 6 hours of fermentation | Reed straw | Corn stover | Ethanol concentration (g/L) | Ethanol yield (%) | Ethanol yield (g/kg feedstock) | Ethanol concentration (g/L) | Ethanol yield (%) | Ethanol yield (g/kg feedstock) |
| 1 | 1 | - | - | 11.44 | 11.60 | 41.21±1.42 | 74.4 | 248 | 39.20±2.84 | 70.3 | 230 |
| 2 | 1 | 0.5 | - | 12.16 | 12.40 | 46.23±2.84 | 71.0 | 237 | 51.26±1.42 | 78.6 | 257 |
| 3 | 1 | 1 | - | 12.88 | 13.20 | 41.21±1.42 | 55.9 | 186 | 52.26±2.84 | 71.1 | 233 |
| 4 | 1 | 0.5 | 0.5 | 12.88 | 13.20 | 53.27 ±1.42 | 72.3 | 241 | 47.24±1.42 | 64.3 | 211 |
| 5 | 1 | 1 | 0.5 | 13.60 | 14.00 | 56.28±0 | 69.1 | 197 | 52.26±0 | 64.6 | 212 |

[a]*Calculated value based on water content of fed WIS. The moisture of fed WIS was 59.02% for reed straw and 61.60% for corn stover;* [b]*weight of initial WIS was 1 g in the S-SSF. The Tween 40 pretreatment conditions were as follows: Tween 40 concentration, 1.5%; ratio of WIS-to-Tween 40, 1:10 (w/v); pretreatment temperature, 25°C; and pretreatment time, 1 hour. The fed-batch S-SSF conditions were as follows: cellulase loading, 25 FPU/g oven-dried WIS; pre-hydrolysis temperature, 50°C; pre-hydrolysis time, 18 hours; fermentation temperature, 36°C; and fermentation time, 72 hours. S-SSF, semi-simultaneous saccharification and fermentation.*

**Table 2.** Comparison of several ethanol productions using Tween as an additive found in the literature and in the current study

| Raw material | Pre-treated method | Tween additives | Adding method of Tween | Fermentation method | Particle size of raw material | Washed or unwashed with water | Prehydrolysis time (hours) + fermentation time (hours) | Enzyme loadings | Ethanol concentration (g/L) | Ethanol yield (%) | Reference |
|---|---|---|---|---|---|---|---|---|---|---|---|
| Reed straw | LHW | Tween 40 | Prior to fermentation | Fed-batch S-SSF | 3 to 5 cm | Unwashed | 18+72 | 25 FPU cellulase/g oven-dried WIS | 56.28 | 69.1 | This study |
| Corn stover | LHW | Tween 40 | Prior to fermentation | Fed-batch S-SSF | 4 to 7 cm | Unwashed | 18+72 | 25 FPU cellulase/g oven-dried WIS | 52.26 | 71.1 | This study |
| Reed straw | LHW | - | - | Fed-batch S-SSF | 20 to 80 mesh | Washed | 18+72 | 40 FPU cellulase/g oven-dried WIS | 39.40 | 75.1 | [27] |
| Corn stover | LHW | - | - | Fed-batch S-SSF | 20 to 80 mesh | Washed | 18+72 | 40 FPU cellulase/g oven-dried WIS | 39.40 | 74.4 | In press |
| Wheat straw | Sulfuric acid | Tween 20 | In the fermentation stage | SSF | 40 mesh | Washed | 72 | 20 FPU cellulase and 40 CBU β-glucosidase/g glucan | 12.4 | 73.9 | [24] |
| Spruce | Steam pretreatment | Tween 20 | In the fermentation stage | SSF | 2 to 10 mm | Washed | 48 | 44 FPU/g cellulose | 20 to 25 | 92 | [15] |

**Table 2.** *Cont.*

| Raw material | Pretreated method | Tween additives | Adding method of Tween | Fermentation method | Particle size of raw material | Washed or unwashed with water | Prehydrolysis time (hours) + fermentation time (hours) | Enzyme loadings | Ethanol concentration (g/L) | Ethanol yield (%) | Reference |
|---|---|---|---|---|---|---|---|---|---|---|---|
| Sugarcane bagasse | Dilute ammonia | Tween 80 | In the pretreatment stage | SSF | 0.05 to 1.5 cm | Washed | 72 | 30 FPU Spezyme CP and 30 CBU Novozyme 188/g glucan | 18 g/100 g of dry biomass | 69 | [23] |
| Sugarcane bagasse | Dilute ammonia | Tween 20 | In the pretreatment stage | SSF | 0.05 to 1.5 cm | Washed | 72 | 30 FPU Spezyme CP and 30 CBU Novozyme 188/g glucan | 15 g/100 g of dry biomass | 59 | [23] |

*CBU, cellobiase unit; LHW, liquid hot water; SSF, simultaneous saccharification and fermentation; S-SSF, semi-simultaneous saccharification and fermentation; WIS, water-insoluble solids.*

## 4.4 MATERIALS AND METHODS

### 4.4.1 MATERIALS

The reed species *Panjin 101* and *Panjin 6* were provided by Yingkou Papermaking Mill, Liaoning Province, China. The reed straw was cut into 3 cm to 5 cm lengths in the mill. Corn stover was collected from a field near Jinzhou New District (Dalian, China). Corn stover was manually cut into pieces of 4 cm to 7 cm in the laboratory. Samples were then homogenized and stored in a plastic bag for subsequent experiments. The chemical compositions of reed straw and corn stover are shown in Table 3. The commercial cellulase used for the fermentation was purchased from Imperial Jade Biotechnology Co, Ltd, Ningxia, China. Saccharomyces cerevisiae was purchased from Angel Yeast Co, Ltd, Hubei, China. The trade name of the yeast was Angel Super Alcohol Active Dry Yeast (molasses base). The yeast was activated prior to fermentation. Approximately 1 g of dry yeast was added to 20 mL of 5% sterilized glucose solution, activated at 38°C for 1 hour, cooled to 28°C to 30°C, and used in the fermentation experiment. The yeast features tolerance with acid (pH 2.5) (http://en.angelyeast.com/contents/1193/16721.html).

**TABLE 3:** Chemical composition (dry weight basis (%)) of raw materials and unwashed WIS

| Chemical composition | Reed straw | | Corn stover | |
|---|---|---|---|---|
| | Raw material | Unwashed WIS | Raw material | Unwashed WIS |
| Benzene-alcohol (2:1) extractive | 8.39±0.10 | 12.14±0.10 | 10.95±0.07 | 19.97±0.19 |
| Glucan | 40.52±0.03 | 55.87±0.03 | 38.75±0.04 | 57.07±0.03 |
| Xylan | 25.86±0.19 | 3.05±0.03 | 23.51±0.18 | 1.80±0.02 |
| Acid-insoluble lignin | 16.22±0.02 | 17.67±0.03 | 15.62±0.16 | 16.89±0.03 |
| Acid-soluble lignin | 2.0±0.1 | 0.6±0 | 2.4±0 | 0.4±0.1 |
| Ash | 3.59±0.14 | 8.67±0.10 | 3.65±0.08 | 0.40±0.04 |

### 4.4.2 LHW PRETREATMENT

LHW pretreatment was conducted in a 15 L digester (machine making factory of Shanxi University of Science and Technology, Shanxi, China). The digester was a cylinder, with an axis passed through its middle portion. The digester could rotate around the motor-driven axis to ensure material uniformity. The digester was electrically heated. Approximately 700 g of raw materials and 7,000 mL of deionized water were loaded in the digester. The pretreatment temperature was controlled at 210°C, the heating time to maximum temperature was 100 minutes, and pretreatment time at the maximum temperature was set to 20 minutes. The cooling down time was approximately 15 minutes. After LHW pretreatment, the WIS and the prehydrolysates were separated by filtration using a cloth bag. The WIS was divided into two fractions. In one fraction, the residual prehydrolysate was removed by a hydraulic machine and named as unwashed WIS, whereas the other fraction was washed with water until the pH reached 7 and was named as washed WIS. The unwashed and washed WIS were stored in a refrigerator at 4°C and used for the subsequent experiments. The moisture content of the unwashed WIS of reed straw and corn stover was 56.82% and 56.06%, respectively. The moisture content of the washed WIS of reed straw and corn stover was 59.39% and 62.17%, respectively. The yields of WIS after LHW pretreatment were 59.64% for reed straw and 57.34% for corn stover, respectively. The chemical compositions of unwashed WIS pretreated with LHW are also shown in Table 3.

### 4.4.3 TWEEN 40 PRETREATMENT PRIOR TO FED-BATCH S-SSF

Approximately 12 g of washed or unwashed WIS and Tween 40 according to a certain ratio of substrate to Tween 40 were added to 250 mL Erlenmeyer flasks. For comparison, sometimes the calcium hydroxide and sodium bisulfite of 0.5 g per 100 g oven-dried WIS were added. The pretreatment temperature was controlled at 25°C to 100°C, and the pretreatment time was set to 0 to 90 minutes. When the Tween 40 pretreatment time was 0 minutes, the substrate was directly washed using Tween 40.

After the pretreatment, the mixtures were divided into two fractions with a cloth bag. The solid fraction was used for the follow-up fed-batch S-SSF. The liquid fraction was used for the Tween 40 recycle utilization experiment.

### 4.4.4 FED-BATCH S-SSF

The fermentation experiment was conducted in 100 mL Erlenmeyer flasks. A specific calculated mass of solid cellulase was first dissolved in HAc-NaAc buffer of pH 4.8 in the flask. The dosage of cellulase was 15 FPU/g to 30 FPU/g of oven-dried weight WIS, and the ratio of WIS:buffer was 1:10 (w/v). Then, the pre-weighted WIS of 1 g (on oven-dried weight) was added into the flask, and sealed with rubber stoppers equipped with syringe needles to remove the generated carbon dioxide. The flasks were placed in the water shaker. In the pre-hydrolysis phase, the medium temperature was maintained at 50°C, and the pre-hydrolysis time was fixed at 18 hours according to our previous study [27]. The initial pH in S-SSF with calcium hydroxide was 4.89 (reed straw) and 4.91 (corn stover), and with sodium bisulfite was 4.81 (reed straw) and 4.80 (corn stover), respectively. For optimization of process conditions of Tween 40 pretreatment and fed-batch S-SSF of unwashed WIS, the pre-weighted WIS of 1 g (on oven-dried weight) was fed into the Erlenmeyer flask at 6 hours of prehydrolysis time. For evaluation of different feeding methods, WIS was fed at different prehydrolysis times and fermentation times according to different fed-batch strategies given in Table 1. After the pre-hydrolysis, the medium temperature was adjusted to the fermentation temperature of 36°C and maintained during the subsequent SSF. Approximately 0.2 mL of activated yeast was added into the medium. The fermentation experiments were performed for 72 hours. The ethanol and glucose concentrations were determined using the SBA-40D Biological Sensing Analyzer (Biology Institute of the Shandong Academy of Sciences, Jinan, China). Each experiment was performed using three parallel samples and the standard error was calculated using Microsoft Excel (Redmond, WA, USA) software.

### 4.4.5 CHEMICAL COMPOSITION ANALYSIS

The contents of xylan, acid-insoluble lignin, ash, acid-soluble lignin, and benzene-alcohol (2:1) extractives were determined according to the Chinese National Standard methods, namely, GB/T2677.9–1994, GB/T2677.8–1994, GB/T2677.3–1993, GB/T10337–1989, and GB/T2677.6–1994, respectively. The glucan content was determined according to National Renewable Energy Laboratory (NREL) methods.

The glucan content was calculated using formula (1):

$$\text{Glucan content (\%)} = \frac{[\text{glucose}]\times 0.087\times 0.9}{m}\times 100\% \quad (1)$$

where [glucose] is glucose concentration (g/L), m is mass of oven-dried solid residues (g), 0.087 is volume of acid hydrolysis liquid (L), and 0.9 is conversion factor for glucose to glucan.

The ethanol yield was calculated using formula (2):

$$\text{Ethanol yield (\%)} = \frac{[EtOH]}{f\times biomass\times 1.111\times 0.51}\times 100 \quad (2)$$

where [EtOH] is the ethanol concentration at the end of the fermentation minus any ethanol produced from the enzyme and medium (g/L), f is the glucan fraction of dry biomass (g/g), biomass is the dry biomass concentration at the beginning of the fermentation (g/L), 0.51 is the conversion factor for glucose to ethanol based on the stoichiometric biochemistry of yeast, and 1.111 is the conversion factor of cellulose to equivalent glucose.

## REFERENCES

1. Limayem A, Ricke SC: Lignocellulosic biomass for bioethanol production: current perspectives, potential issues and future prospects. Progr Energ Combust Sci 2012, 38:449-467.
2. Balat M: Production of bioethanol from lignocellulosic materials via the biochemical pathway: a review. Energ Convers Manag 2011, 52:858-875.
3. Kim I, Han JI: Optimization of alkaline pretreatment conditions for enhancing glucose yield of rice straw by response surface methodology. Biomass Bioenerg 2012, 46:210-217.
4. Yu ZD, Zhang BL, Yu FQ, Xu GZ, Song AD: A real explosion: The requirement of steam explosion pretreatment. Bioresour Technol 2012, 121:335-341.
5. Li BZ, Balan V, Yuan YJ, Dale BE: Process optimization to convert forage and sweet sorghum bagasse to ethanol based on ammonia fiber expansion (AFEX) pretreatment. Bioresour Technol 2010, 101:1285-1292.
6. Sathitsuksanoh N, Zhu ZG, Zhang YHP: Cellulose solvent- and organic solvent-based lignocellulose fractionation enabled efficient sugar release from a variety of lignocellulosic feedstocks. Bioresour Technol 2012, 117:228-233.
7. Hsu TC, Guo GL, Chen WH, Hwang WS: Effect of dilute acid pretreatment of rice straw on structural properties and enzymatic hydrolysis. Bioresour Technol 2010, 101:4907-4913.
8. Helle SS, Duff SJB, Cooper DG: Effect of surfactants on cellulose hydrolysis. Biotechnol Bioeng 1993, 42:611-617.
9. Kaya F, Heitmann JA, Joyce TW: Influence of surfactants on the enzymatic hydrolysis of xylan and cellulose. Tappi J 1995, 78:150-157.
10. Kristensen JB, Börjesson J, Bruun MH, Tjerneld F, Jørgensen H: Use of surface active additives in enzymatic hydrolysis of wheat straw lignocelluloses. Enzyme Microb Technol 2007, 40:888-895.
11. Sindhu R, Kuttiraja M, Preeti VE, Vani S, Sukumaran RK, Binod P: A novel surfactant-assisted ultrasound pretreatment of sugarcane tops for improved enzymatic release of sugars. Bioresour Technol 2013, 135:67-72.
12. Castanon M, Wilke CR: Effects of the surfactant tween 80 on enzymatic hydrolysis of newspaper. Biotechnol Bioeng 1981, 23:1365-1372.
13. Kumar R, Wyman CE: Effect of additives on the digestibility of corn stover solids following pretreatment by leading technologies. Biotechnol Bioeng 2009, 102:1544-1557.
14. Eckard AD, Muthukumarappan K, Gibbons W: Enzyme recycling in a simultaneous and separate saccharification and fermentation of corn stover: a comparison between the effect of polymeric micelles of surfactants and polypeptides. Bioresour Technol 2013, 132:202-209.
15. Alkasrawi M, Eriksson T, Börjesson J, Wingren A, Galbe M, Tjerneld F, Zacchi G: The effect of Tween-20 on simultaneous saccharification and fermentation of softwood to ethanol. Enzyme Microb Technol 2003, 33:71-78.
16. Ballesteros I, Olivia JM, Carrasco J, Cabanas A, Navarro AA, Ballesteros M: Effect of surfactant and zeolites on simultaneous saccharification and fermentation

of steam-exploded poplar biomass to ethanol. Appl Biochem Biotechnol 1998, 70–72:369-381.

17. Ooshima H, Sakata M, Harano Y: Enhancement of enzymatic-hydrolysis of cellulose by surfactant. Biotechnol Bioeng 1986, 28:1727-1734.
18. Eriksson T, Börjesson J, Tjerneld F: Mechanism of surfactant effect in enzymatic hydrolysis of lignocelluloses. Enzyme Microb Technol 2002, 31:353-364.
19. Qing Q, Yang B, Wyman CE: Impact of surfactants on pretreatment of corn stover. Bioresour Technol 2010, 101:5941-5951.
20. Yang MH, Zhang AM, Liu BB, Li WL, Xing JM: Improvement of cellulose conversion caused by the protection of Tween-80 on the adsorbed cellulose. Biochem Eng J 2011, 56:125-129.
21. Seo DJ, Fujita H, Sakoda A: Structural changes of lignocelluloses by a nonionic surfactant, Tween 20, and their effects on cellulase adsorption and saccharification. Bioresour Technol 2011, 102:9605-9612.
22. Shen L, Guo A, Zhu XY: Tween surfactants: adsorption, self-organization, and protein resistance. Surf Sci 2011, 605:494-499.
23. Cao S, Aita GM: Enzymatic hydrolysis and ethanol yields of combined surfactant and dilute ammonia treated sugarcane bagasse. Bioresour Technol 2013, 131:357-364.
24. Qi BK, Chen XR, Wan YH: Pretreatment of wheat straw by nonionic surfactant-assisted dilute acid for enhancing enzymatic hydrolysis and ethanol production. Bioresour Technol 2010, 101:4875-4883.
25. Cantarella M, Cantarella L, Gallifuoco A, Spera A, Alfani F: Comparison of different detoxification methods for steam-exploded poplar wood as a substrate for the bioproduction of ethanol in SHF and SSF. Process Biochem 2004, 39:1533-1542.
26. Nguyen QA, Saddler JN: An integrated model for the technical and economic evaluation of an enzymatic biomass conversion process. Bioresour Technol 1991, 35:275-282.
27. Lu J, Li XZ, Yang RF, Yang L, Zhao J, Liu YJ, Qu YB: Fed-batch semi-simultaneous saccharification and fermentation of reed pretreated with liquid hot water for bio-ethanol production using Saccharomyces cerevisiae. Bioresour Technol 2013, 144:539-547.

# CHAPTER 5

# THE DEVELOPMENT AND USE OF AN ELISA-BASED METHOD TO FOLLOW THE DISTRIBUTION OF CELLULASE MONOCOMPONENTS DURING THE HYDROLYSIS OF PRETREATED CORN STOVER

AMADEUS Y. PRIBOWO, JINGUANG HU, VALDEIR ARANTES, AND JACK N. SADDLER

## 5.1 INTRODUCTION

One of the key steps in a biomass-to-ethanol process is the enzymatic hydrolysis of the cellulosic component to fermentable sugars. Typically, a mixture of complementary cellulase and other, so-called, accessory enzymes (such as hemicellulases, GH61, etc.) are required to effectively break down the structural cellulose and hemicellulose polysaccharides to their component sugars [1,2]. However, various technoeconomic analyses have indicated that the cost of enzymatic hydrolysis is still unacceptably high, primarily because of the high enzyme loadings required to achieve effective hydrolysis [3]. As a result, a considerable amount of research has focussed on ways to try to improve the efficiency of hydrolysis while using low protein/enzyme loadings. Various strategies have been assessed, such as increasing substrate digestibility through biomass pretreatments [4,5], improving the efficiency of enzyme cocktails [6,7], and reusing the enzymes for multiple rounds of hydrolysis [8,9]. The last two strategies, in particular, have benefitted from better characterization of the specific roles and actions of individual enzymes and their synergistic interaction during cellulose hydrolysis.

However, getting a better understanding of the individual enzyme's interaction with the substrate during hydrolysis of lignocellulosic substrates has been challenging, primarily because of the lack of specific techniques that can overcome both the complexity of the enzyme mixture and the interference caused by the heterogeneous lignocellulosic substrates. Many of the biochemical techniques that might be used lack the resolution to specifically probe individual enzymes and proteins. For example, the enzyme Cel7A from *T. reesei* has a very similar molecular weight to that of Cel6A and Cel7B and, consequently, these three enzymes typically show up as a single band after gel electrophoresis [1]. Another commonly used technique is to characterize and evaluate distribution of enzymes based on their activities on model substrates such as carboxymethyl cellulose (CMC), filter paper, or a number of chromophoric substrates such as p-Nitrophenyl-based substrates [10]. Unfortunately, many of these model substrates are not specific enough to distinguish individual enzymes. Protein chromatography techniques have also been utilized to fractionate the enzyme mixture down to its individual components [11,12]. However, this approach is laborious and, depending on the purification protocols used, the enzyme mixture may not always completely separate into its individual components [13]. In addition, interference caused by substrate materials such as lignin auto-fluorescence limits the use of traditional protein chromatography techniques and protein labelling techniques using fluorescent dyes [14].

Primarily due to the limitations of the assay methods that have been employed, most of the previous enzyme-cellulosic substrate interaction studies have used purified enzymes or reconstituted mixtures of purified enzymes [15,16] and/ or model substrates such as pure cellulose or substrates with a very low lignin content [17,18] to simplify the subsequent enzyme assays and analyses. While these studies have advanced our understanding of enzymes-substrate interaction, they have not looked at the interactions occurring during the hydrolysis of an industrially relevant lignocellulosic substrate using a complete enzyme mixture.

In recent work, the distribution of individual enzymes present in a commercial cellulase mixture (Accellerase 1000) was assessed during the hydrolysis of steam pretreated corn stover (SPCS) [1]. A combination of methods such as, gel electrophoresis, zymograms, activity assays using chromophoric substrates, and mass spectrometry were used to define the general distribution patterns of some of the enzymes during SPCS hydrolysis [1]. However, although we were able to semi-quantitatively assess enzyme distribution using these techniques, we were not able to quantitatively follow the adsorption profiles of individual enzymes.

It is well known that antibodies can bind to specific antigens and this ability has been used as the basis for many assays [19-21]. This specific recognition and binding has been utilized in various techniques including the enzyme-linked immunosorbent assay (ELISA). The ELISA method uses antibodies linked to a reporter enzyme to specifically recognize and bind a target compound in a mixture of compounds. This specific compound or protein can then be quantified by adding a substrate for the reporter enzyme and measuring the concentration of the product [22]. The ELISA method, using monoclonal and/or polyclonal antibodies (MAbs and PAbs, respectively) raised against various cellulase enzymes, has been successfully used to quantify target enzymes both in culture filtrates and commercial enzyme preparations [19].

A double-antibody sandwich ELISA, which is an ELISA-based technique using a pair of antibodies to sandwich the target compound and specifically quantify it among other compounds in the mixture, has been successfully used to quantify the amount of Cel7A in a crude culture broth with minimal interference from other enzymes or other materials present in the broth [23]. Improved specificity of the assay was achieved when MAb was used as the coating antibody and PAb as the second, detecting antibody [23]. In related work, Buhler et al. (1991) optimized a double-antibody sandwich ELISA for Cel7B in a culture broth using MAb as the coating antibody and PAb as the detecting antibody. They were able to show that the assay was both sensitive and specific for Cel7B [24].

However, the feasibility of using ELISA to quantify specific proteins present in the supernatant after hydrolysis of a lignocellulosic substrate has not yet been demonstrated.

In the work described here, a double-antibody sandwich ELISA was developed and used to quantify some of the specific cellulase enzymes present in the supernatant during the hydrolysis of SPCS. A double-antibody sandwich ELISA was used to specifically quantify the amount of cellulase monocomponents Cel7A, Cel6A, and Cel7B present in a commercial enzyme mixture. The sensitivity was improved by subjecting the enzyme samples to a pH adjustment treatment and/or a heat treatment. While lignocellulosic substrate derived materials did interfere with the assay, this interference could be minimized by simple dilution.

## 5.2 RESULTS AND DISCUSSIONS

### *5.2.1 DETERMINATION OF THE SPECIFICITY OF THE DIFFERENT MONOCLONAL ANTIBODIES (MABS) AND POLYCLONAL ANTIBODIES (PABS)*

We initially wanted to ensure that the MAb and PAb that we had been provided were specific for their target cellulase monocomponents. The specificity of Cel7A, Cel6A, and Cel7B MAbs were initially assessed using Western Blots against Cel7A that had been purified from a commercial Celluclast mixture as well as against the Cel7A component that was known to be present in the 3 commercial enzyme mixtures. The Cel7A MAb Western Blot showed a single band corresponding to the purified Cel7A and a major band at molecular weight (MW) ~ 70 kDa, which is the molecular weight of the Cel7A, present in the 3 commercial enzyme mixtures (Figure 1A). Although the Cel6A MAb also showed a band of protein at MW ~70 kDa when assayed against the 3 commercial enzyme mixtures (Figure 1B), this MAb did not react with the purified Cel7A. In addition to the major bands at MW ~70 kDa, Cel7A and Cel6A MAb Western Blots both showed multiple bands with commercial enzyme mixtures. Although we could not be certain if these bands corresponded to multiple isoforms of the target enzyme or actual unspecific bindings to other proteins without further experiments, we did not expect these apparent multiple bindings to significantly influence the specificity of the ELISA for 2 reasons. Firstly, the intensity of these other bands was significantly less compared to the band intensity of the expected target enzyme. Therefore, given the low protein concentration required for ELISA (< 5 μg/ml), this apparent unspecific binding (if any) would not likely to have any significant influence to the specificity of the assay. Secondly, the differing banding patterns between Cel7A and Cel6A Western Blots seemed to suggest a specific rather than an unspecific binding. The Western Blot that used the Cel7B MAb did not recognize the purified Cel7A but recognized a band of protein at MW ~60 kDa in all of the 3 commercial enzyme mixtures (Figure 1C). Therefore, it appears that all 3 MAbs were reactive and specific for their target enzymes.

The specificity and reactivity of Cel7A and Cel6A PAbs were also determined by Western Blots by using purified Cel7A and Cel6A from Celluclast as well as 2 commercial enzyme mixtures. The PAb against Cel7A was specific for its target enzyme since it reacted only with purified Cel7A and not with the purified Cel6A (Figure 2A). However, the PAb against Cel6A recognized both the purified Cel7A and Cel6A (Figure 2B). Possible contamination by Cel6A in the purified Cel7A fraction did not appear to be an issue as the Cel6A MAb did not react with the purified Cel7A preparation (Figure 1B). The reactivity and specificity of the Cel7B PAb was next determined using Western Blots against purified Cel7A, Cel6A, and Cel7B as well as against 3 commercial cellulase mixtures. It was apparent that the Cel7B PAb recognized the purified Cel7B but also cross-reacted with the purified Cel7A and Cel6A (Figure 2C). However, this cross-reactivity with the Cel6A and Cel7B PAbs was not expected to influence the specificity of the double-antibody sandwich ELISA since both the Cel6A and Cel7B MAbs were shown to be specific to their respective target enzymes (Figure 1B and C).

### *5.2.2 OPTIMIZATION OF THE ASSAY PROTOCOLS TO IMPROVE THE SENSITIVITY OF THE DOUBLE-ANTIBODY SANDWICH ELISA*

Previous work had shown that a double-antibody sandwich ELISA, using a combination of a MAb and a PAb as the capture and detecting antibodies respectively, resulted in improved specificity compared to normal ELISA or to a sandwich ELISA using PAb as the capture and MAb as the detecting antibody [19,23]. Thus, we next used an MAb as the capture antibody and a PAb as the detecting antibody to assay different concentrations of each of the 3 antibodies MAb, PAb, and goat-anti rabbit IgG conjugated to alkaline phosphatase (GAR-AP). In this way, we hoped to assess the sensitivity of the assay in detecting purified Cel7A at concentrations ranging from 0–2.5 μg/ml.

Although two concentrations of Cel7A MAb (10 and 50 μg/ml diluted in 1x Phosphate-Buffered Saline or PBS) were initially assessed, as both concentrations gave similar absorbance values (Figure 3A), a MAb concentration of 10 μg/ml was used in subsequent work. Previous work had also determined that a concentration of 10 μg/ml was sufficient to coat the bottom surface of a well in a typical 96-well ELISA plate [25]. The concentrations of the PAb (detecting antibody) and GAR-AP, the tertiary antibody, were similarly optimized over the same range of Cel7A concentrations. A concentration of 0.14 μg/ml of PAb Cel7A and 1/500 dilution of GAR-AP (corresponding to 1 μg/ml of GAR-AP) were found to improve the sensitivity of the assay for all the three enzymes (Figure 3B and C). These concentrations of antibodies were then also used for the Cel6A and Cel7B based ELISA's.

Despite the increased sensitivity gained by optimizing the concentrations of all 3 antibodies, the improved signal was still quite low when compared to previously reported values [23]. Therefore, to try to further increase the sensitivity of the assay, the enzyme samples were subjected to pH adjustment and heat treatments prior to addition to the well. Although previous work had shown that the antigen-antibody interactions are typically optimum at $pH > 7$ [24], fungal derived enzymes are typically buffered and used at around $pH < 5$. We therefore brought the enzyme samples up to pH 7.5 using PBS buffer prior to their addition to the wells.

Previously, Riske et al. (1990) had reported that a heat-sensitive fungal product caused a signal reduction with Cel7A ELISA and that this interference disappeared after the cellulase preparation was boiled, resulting in increased ELISA sensitivity [23]. Therefore, to see if we could also obtain the same beneficial effect, the enzyme monocomponents were also heated at 100°C for 10 minutes to determine if a heat treatment might also improve sensitivity. When the Cel7A and Cel6A ELISA's were subjected to a heat treatment at 100°C for 10 minutes in a pH 5.0 buffer, followed by dilution in PBS buffer at pH 7.5, the sensitivity of ELISA increased by about 6× and 10× respectively for Cel7A and Cel6A, at an enzyme concentration of 2.5 μg/ml when compared to the untreated samples (Figure 4A and B). However, heat treatment decreased the sensitivity for the Cel7B based ELISA (Figure 4C). Therefore, the enzyme samples for the Cel7B ELISA were not heated but directly diluted in PBS buffer and then added to the wells.

As mentioned earlier, the improved signal achieved by heating the enzymes used for the Cel7A and Cel6A based ELISA's was likely caused by the removal of interfering heat-sensitive materials present in the samples [23] or by protein denaturation which may lead to the opening up of the protein structure, exposing the antigen to the antibody. The ineffectiveness of heating the Cel7B may indicate that the interfering materials may not interfere with the Cel7B based ELISA system. This differential response to the heat treatment highlights the need to optimize the double-antibody sandwich ELISA for each specific enzyme-antibody system.

### 5.2.3 HOW SPECIFIC IS THE ELISA TO THE ENZYME OF INTEREST?

The specificity of each ELISA was next determined by comparing the absorbance values of each enzyme when it was added as a single component and when it was added as a mixture of 4 purified enzymes (Cel7A, Cel6A, Cel7B, and Cel5A). For all of the enzyme based ELISAs (Cel7A, Cel6A, and Cel7B ELISA), the standard curves obtained with the pure enzymes was similar to those obtained with the reconstituted mixture especially when the target enzyme concentration was less than 1 μg/ml (Figure 5A, B, and C). It was apparent that the ELISA double-antibody sandwich assay was able to specifically quantify a target enzyme when it was present in a mixture with 3 other cellulase monocomponents.

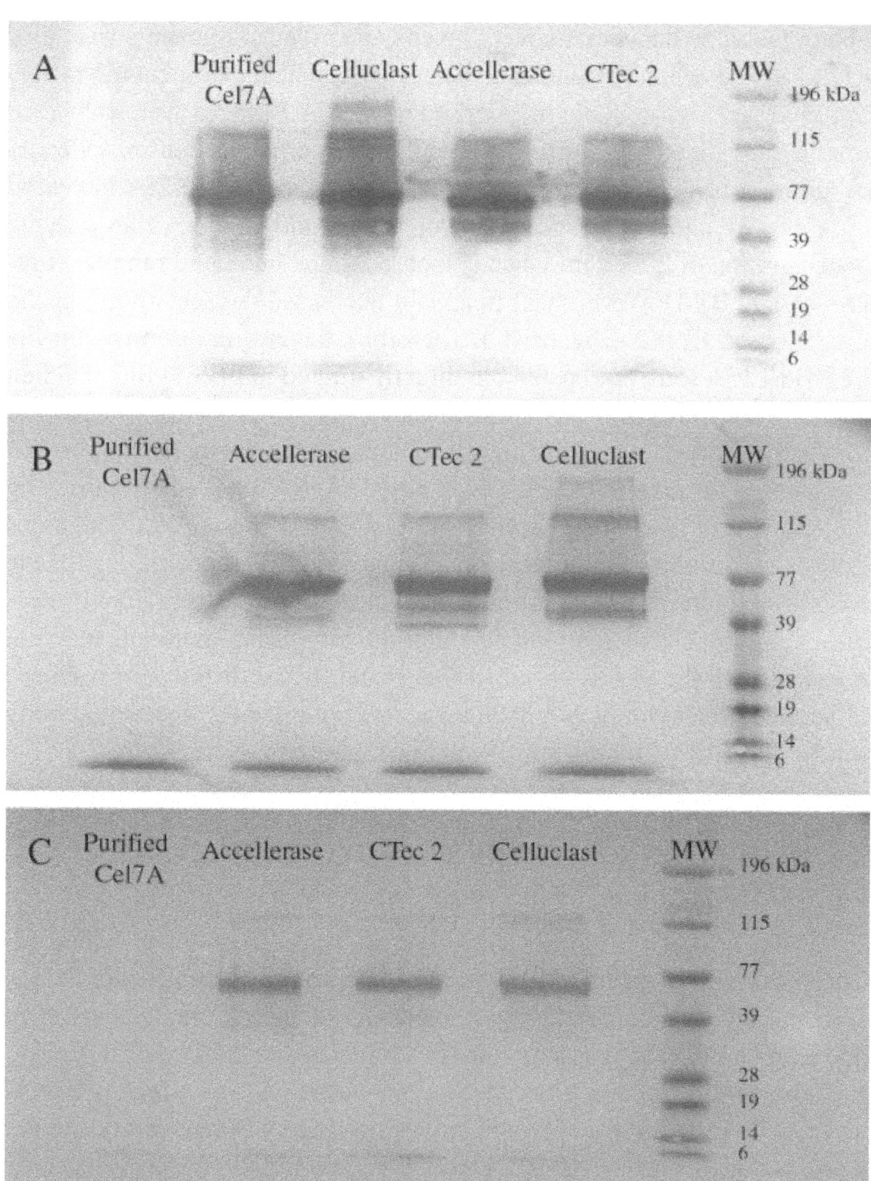


**FIGURE 1:** Reactivity and specificity of MAbs against Cel7A (A), Cel6A (B), and Cel7B (C) as determined using Western Blots.

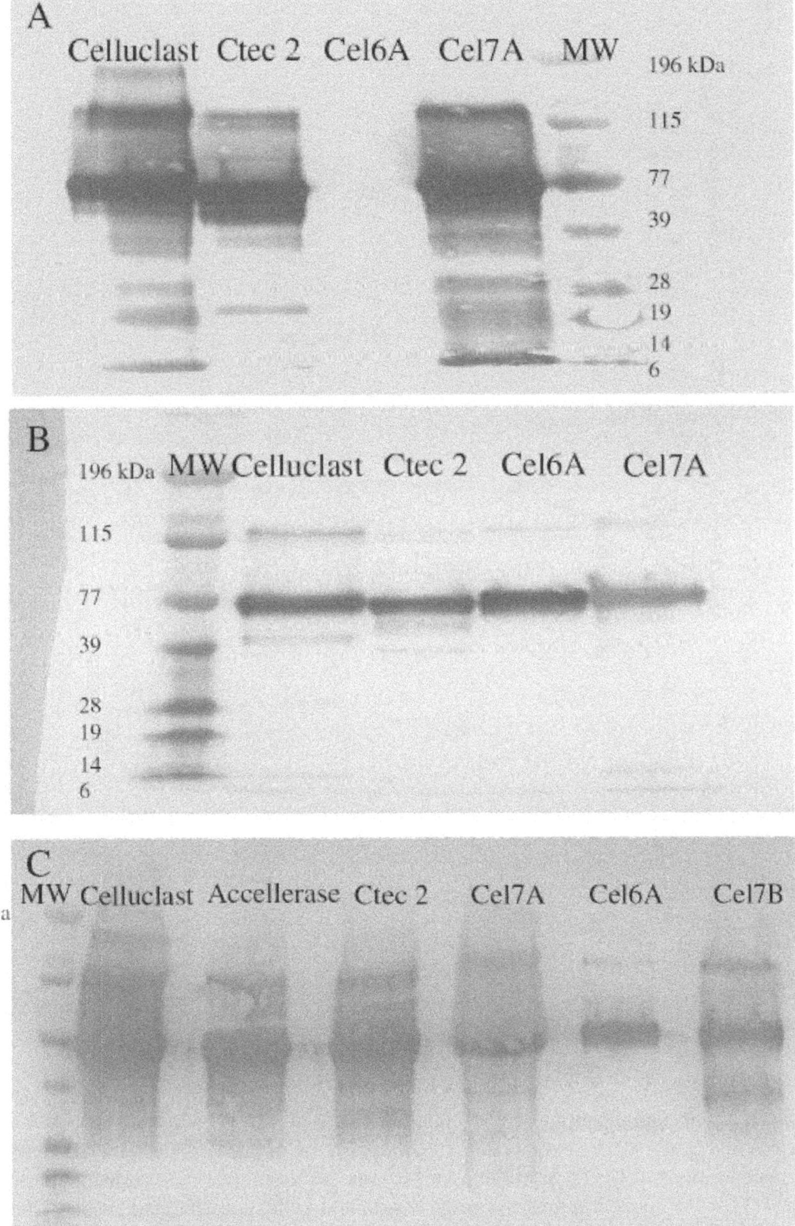


**FIGURE 2:** Reactivity and specificity of PAb against Cel7a (A), Cel6A (B), and Cel7B (C) as determined using Western Blots.

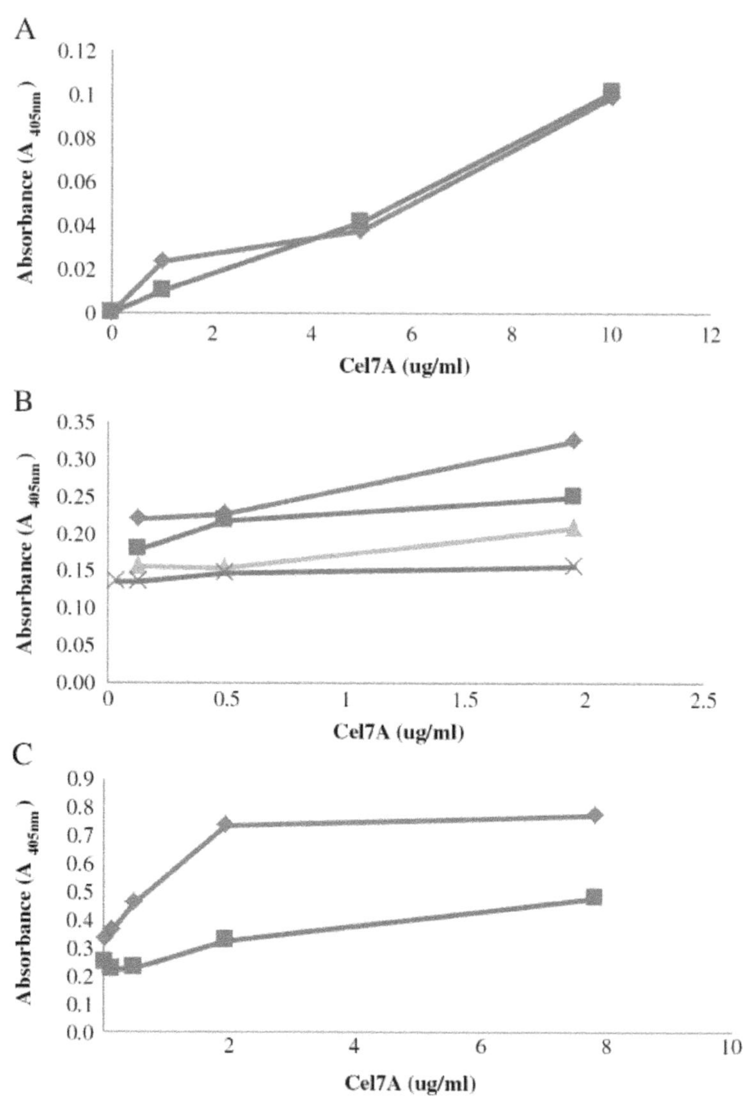


**FIGURE 3:** Optimization of the concentrations of MAb (A), PAb (B), and the third antibody, GAR-AP (C) over a range of concentration of purified Cel7A. (A). Two different concentrations of Cel7A MAb 10 μg/ml (◊) and 50 μg/ml (□) were added to the well. PAb and GAR-AP concentrations were kept constant at 1/400 and 1/1750 dilutions, respectively (B). Using 10 μg/ml Cel7A MAb, the Cel7A PAb was diluted to different degrees: 14(◊), 7(□), 3.5(Δ), and 1.75(X) μg/ml. The GAR-AP was diluted at 0.3 μg/ml. (C). The concentration of GAR-AP was varied by diluting it to 1 (◊) or 0.3 (□)μg/ml in PBS. Cel7A MAb concentration was kept at 10 μg/ml, and Cel7A PAb was diluted to 14 μg/ml in PBS.

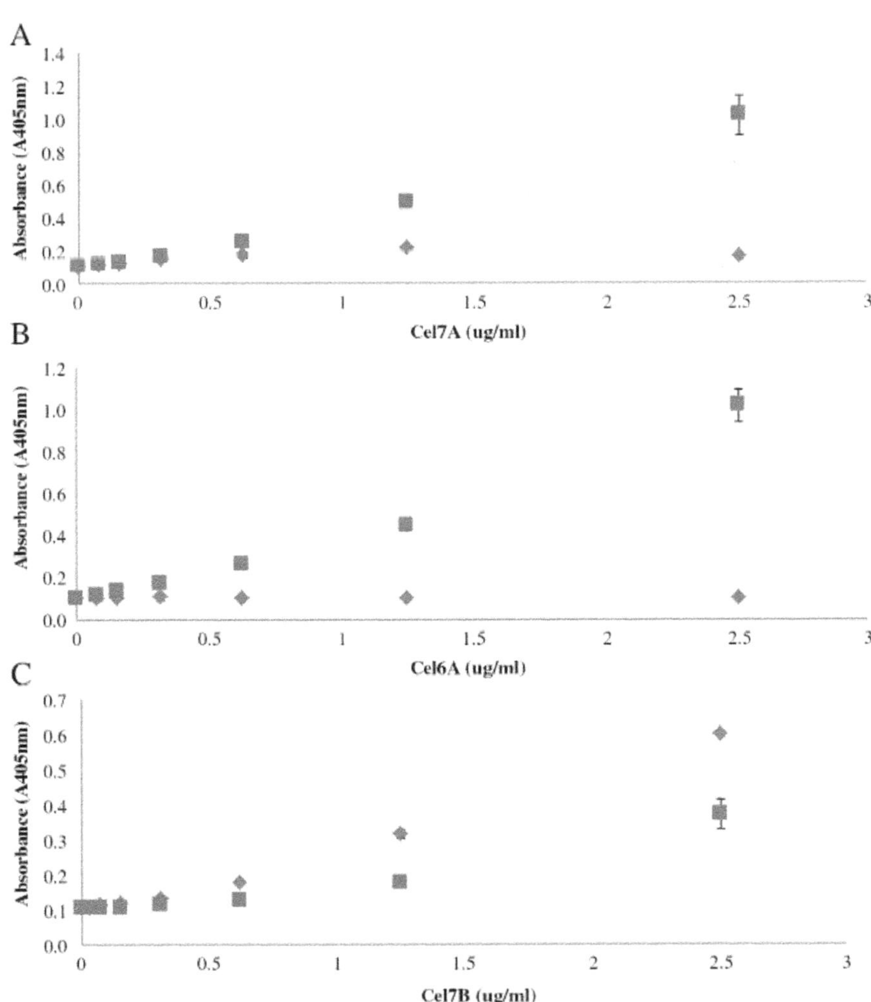


**FIGURE 4:** The effect of a heat treatment on the sensitivity of ELISA for pure Cel7A (A) and Cel6A (B) and Cel7B (C). Heated enzyme samples were boiled in Na-acetate buffer pH 5.0 at 100°C for 10 minutes and then serially diluted in PBS (□). Non-heated samples were directly diluted in PBS (◊).

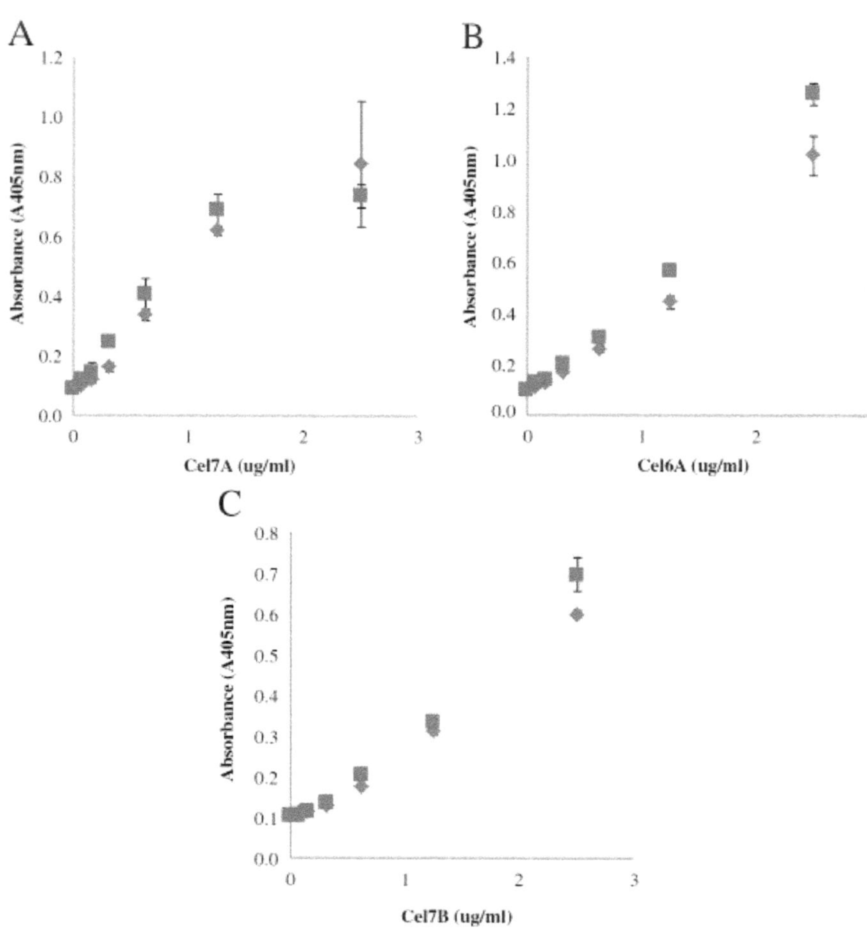


**FIGURE 5:** The specificity of ELISA for Cel7A (A), Cel6A (B), and Cel6A (C) as measured using pure enzymes (◊) and reconstituted mixtures of the 4 purified enzymes Cel7A, Cel6A, Cel7B, and Cel5A (□).

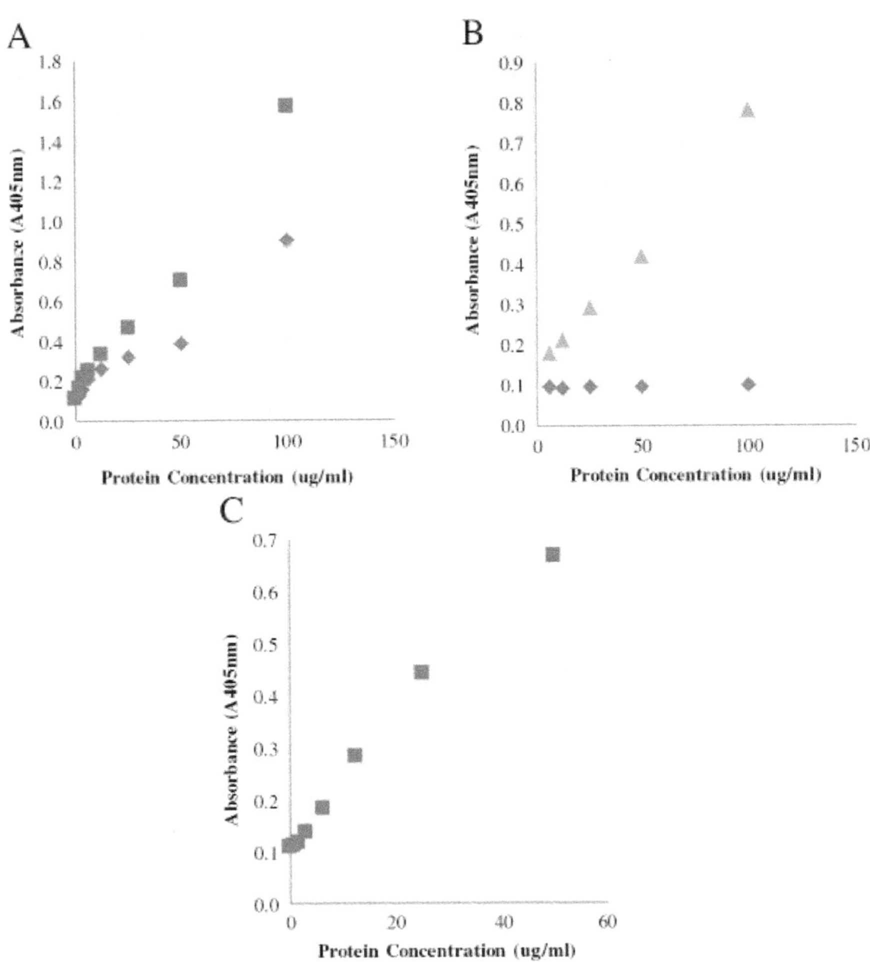


**FIGURE 6:** The construction of a standard curve for Cel7A ELISA (A) using whole commercial enzyme mixtures Accellerase 1000 (◊), CTec 2 (□), Cel6A ELISA (B) using Accellerase 1000 (◊) and Celluclast 1.5L (Δ), and Cel7B ELISA (C) using CTec 2 (□).

We next wanted to determine if a whole commercial enzyme mixture could be used to make a standard curve, thus obviating the need for purified enzymes. A commercial enzyme mixture was diluted to 200 µg protein/ml in Na-acetate buffer (0.05 M, pH 5.0). When using the Cel7A and Cel6A based ELISA's, the commercial enzyme mixtures were heated, serially diluted 2-fold in PBS and then added to the wells. By sufficiently diluting the enzyme mixtures, a relatively linear standard curve could be obtained with whole enzyme mixtures when using the Cel7A and Cel7B based ELISA's (Figure 6A and C). A linear standard curve was also obtained with the Cel6A ELISA. However, this linear standard curve was only obtained with Celluclast and not with Accellerase or CTec 2 (Figure 6B).

The linear standard curve obtained for all of the target enzymes highlighted the ability of the double-antibody sandwich ELISA to detect the target enzyme even when present in complex enzyme mixtures. The high specificity of the MAbs could also be the reason why Cel6A ELISA only worked with Celluclast and not with other commercial enzyme mixtures as the Cel6A MAb was developed by colleagues at the National Renewable Energy Laboratory (NREL) to detect Cel6A in Celluclast whereas the PAb was developed commercially by Alpha Diagnostics using a synthesized peptide. Although both the MAb and PAb's against Cel6A recognized the Cel6A present in Celluclast, Accellerase and CTec 2 (Figure 1B and 2B), the lower ELISA signal observed in the latter two commercial enzyme mixtures might be a result of a slight change in antigen recognition by the MAb. When the concentration of the enzymes and antibodies are high, as in the case of the Western Blot studies (30 µg of enzyme samples and 250 µg MAb or PAb), there is likely enough interaction between the enzymes and antibodies, resulting in a significant band on the membrane. However, when the enzyme concentration is low (< 0.1 µg), as in the case with the ELISA, the lower binding affinity between the antibodies and Cel6A in Accellerase and CTec 2 would result in a lower ELISA signal. It was apparent that a double-antibody sandwich ELISA was specific for target enzymes providing appropriate MAbs and PAbs were available. Given the recent rapid development of enzyme cocktails to which new-and-improved enzymes have been introduced, (i.e. CTec3) the highly specific nature of the antibody-antigen interaction shown in this assay will likely require the development of specific MAbs and PAbs that will recognize individual enzymes present in these new and improved enzyme mixtures.

### *5.2.4 DETERMINING THE POSSIBLE INTERFERENCE OF SUBSTRATE DERIVED MATERIALS ON THE ELISA*

Although various ELISA based methods have been used to quantify cellulase enzymes, these assays have only been applied to commercial enzyme mixtures or to culture filtrates [19,23,24,26]. The use of an ELISA to try to follow the distribution of cellulase enzymes during enzymatic hydrolysis of a realistic, lignocellulosic substrate has not, so far, been described in the literature As a result, there is limited information on the possible influence of interfering materials that will likely be present when attempts are made to use an ELISA in this situation.

Previous work on the use of ELISA's to detect residual agrochemicals in soil samples had shown that humic substances in soil may result in an overestimation of chemical concentrations [20,27,28], and that sample dilution could be used to minimize interference [20]. As a similar type of interference might occur with biomass-derived materials such as soluble lignin fragments, supernatants derived from steam pretreated corn stover (SPCS), steam pretreated poplar (SPP), steam pretreated douglas fir (SPDF), and Avicel were assessed for their possible influence on the double-antibody ELISA. The supernatants were diluted in PBS to varying degrees to determine if a simple dilution could minimize the interference caused by these materials.

It was apparent that the undiluted biomass derived supernatants resulted in considerable interference with all of the Cel7A, Cel6A, and Cel7B based ELISAs (Figure 7). The Cel7A ELISA either over or under estimated the amount of enzyme (Figure 7A) with the supernatants derived from the SPCS (5× higher) and SPP substrates resulting in an overestimation and the SPDF and Avicel supernatants in an underestimation (Figure 7A). In contrast, only the SPP supernatants caused a signal overestimation with Cel6A ELISA while the SPCS, SPDF, and Avicel supernatants gave a signal that was lower than the PBS control (Figure 7B). Interference with Cel7B based ELISA was only assessed with the SPCS supernatant which caused a slight overestimation (Figure 7C). To assess if a simple dilution could minimize interference, each supernatant was diluted 10× or 100× in PBS. It was apparent that the interference caused by the addition of the undiluted SPCS supernatant could be minimized at both dilution levels (Figure 7A). This dilution strategy was also effective on both the Cel6A and Cel7B based ELISA's and a 100-fold dilution in PBS seemed to consistently give an ELISA signals similar to the PBS control for both Cel6A and Cel7B ELISA (Figure 7B and 7C).

### *5.2.5 CAN AN ELISA BE USED TO FOLLOW ENZYME DISTRIBUTION DURING SPCS HYDROLYSIS?*

We next wanted to assess if the double-antibody sandwich ELISA could be used to quantitatively monitor the time course of individual enzyme adsorption (Cel7A, Cel6A, and Cel7B) during the hydrolysis of SPCS. It was apparent that all 3 enzymes exhibited different adsorption profiles when incubated with SPCS (Figure 8A, B, and C). Most of Cel7A immediately adsorbed to the SPCS after mixing, leaving only about 30% of Cel7A in the supernatant. After 3 hours of hydrolysis, Cel7A started to desorb back to the supernatant with maximum desorption occurring after 6 hours of hydrolysis with about 65% of the initial Cel7A detected in the supernatant. Over prolonged hydrolysis, the concentration of Cel7A in the supernatant decreased progressively (Figure 8A). This partially reversible adsorption of Cel7A confirmed previous work where a combination of techniques, such as zymogram, SDS-PAGE, and enzyme activity assays, were used to semi-quantitatively determine specific Cel7A adsorption/desorption during SPCS hydrolysis [1].

In contrast, Cel6A directly adsorbed onto the SPCS within the first 3 hours and remained tightly bound throughout the course of hydrolysis (Figure 8B). Previous work that looked at Cel6A adsorption used purified Cel6A due to a lack of a specific assay able to monitor Cel6A in the presence of other enzymes. The irreversible adsorption of Cel6A observed in this study using commercial enzyme mixtures was in a good agreement with this previous work [29].

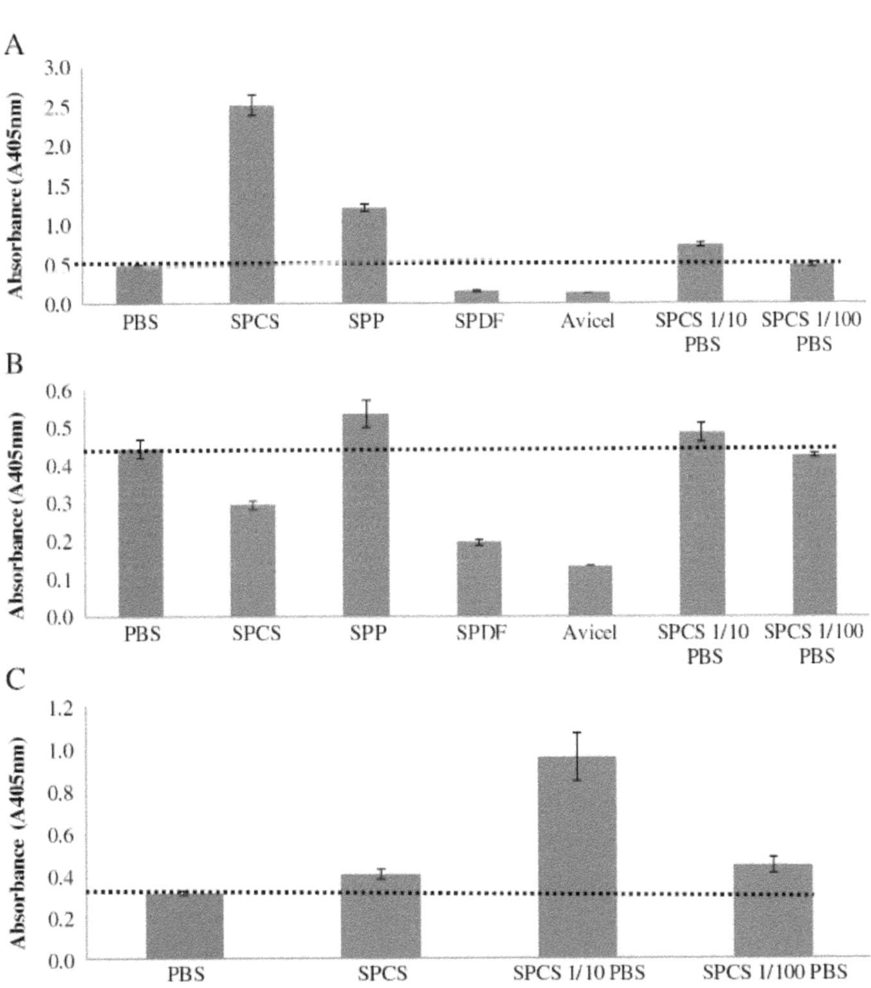


**FIGURE 7:** Effect of substrate supernatants on Cel7A ELISA (A), Cel6A ELISA (B), and Cel7B ELISA (C). Amount of purified enzymes added: 1.25 μg/ml.

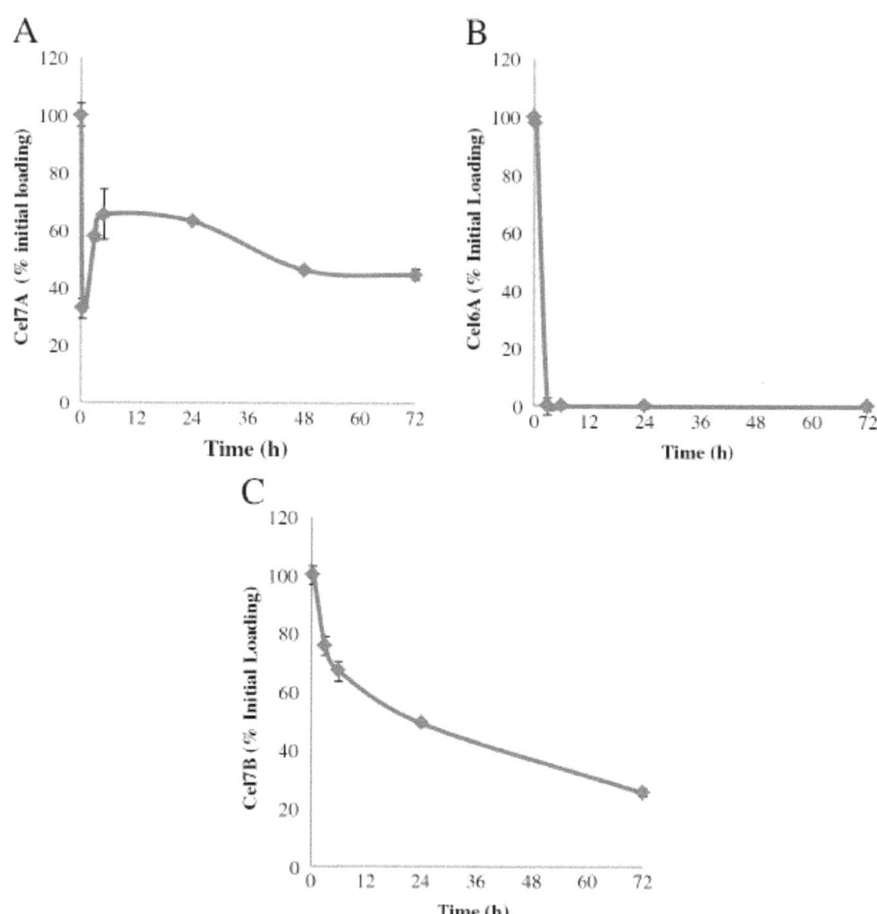


**FIGURE 8:** Adsorption profiles of Cel7A (A), Cel6A (B), and Cel7B (C) during hydrolysis of SPCS as determined by a double-antibody sandwich ELISA. Samples for Cel7A ELISA were obtained by hydrolyzing SPCS at 2% substrate consistency in 0.05 M Na-acetate buffer pH 5.0 with 20 FPU/ g cellulose Accellerase 1000. Samples for Cel6A and Cel7B ELISA were obtained from SPCS hydrolysis using 20 FPU/ g cellulose Celluclast and 40 CBU/ g cellulose β-glucosidase.

Compared to Cel7A and Cel6A, the adsorption of Cel7B was more gradual with the amount of Cel7B detected in the supernatant continuously declining over the 72 h hydrolysis (Figure 8C). Prior to developing the ELISA method, we had tried to follow the specific adsorption profile of Cel7B by monitoring its profile as determined by zymograms using CMC and xylan as substrates [1]. The quantitative adsorption profiles obtained using the ELISA profile were in a good agreement with the qualitative results obtained previously using zymograms during the 72 h hydrolysis [1].

## 5.3 CONCLUSIONS

A simple, high-throughput assay that can specifically follow and quantify individual enzymes present in the complex enzyme mixtures that are used to hydrolyse pretreated lignocellulosic substrates was developed and demonstrated. The protocols for an immunoassay using antibodies against Cel7A, Cel6A, and Cel7B were developed with the hope of using the method to follow the distribution of individual enzymes during hydrolysis. A combination of MAb's and PAb's, as the respective coating and detecting antibodies, was used to develop a double-antibody sandwich ELISA. This method was able to detect and quantify individual enzymes when present in cellulase mixtures. The assay was sensitive over a range of relatively low enzyme concentration (0 – 1 μg/ml), provided the enzymes were first pH adjusted and/or heat treated to increase their antigenicity. Although lignocellulosic hydrolysates resulted in varying degrees of interference with the assay, the interference could be minimized by diluting the samples in PBS buffer. The immunoassay was employed to quantitatively monitor the adsorption of cellulase monocomponents, Cel7A, Cel6A, and Cel7B that are present in both Celluclast and Accellerase 1000, during the hydrolysis of SPCS. All three enzymes exhibited different individual adsorption profiles. The specific and quantitative adsorption profiles observed with the ELISA method was in agreement with earlier work where more laborious enzyme assay techniques were used.

## 5.3 METHODS AND MATERIALS

### 5.3.1 PURIFICATION OF CELLULASE MONOCOMPONENTS, CEL7A, CEL6A, CEL7B, AND CEL5A

The cellulase monocomponents Cel7A, Cel6A, Cel7B, and Cel5A were purified from Celluclast (Novozyme) using previously described methods [12,30-32]. The Ninhydrin assay [33] was then used to determine the concentrations of these purified enzymes as well as the commercial enzyme mixtures. Bovine serum albumin (BSA, Sigma) was used as the protein standard.

### 5.3.2 PREPARATION OF ANTIBODIES AND DETERMINATION OF THEIR SPECIFICITY

MAbs against Cel7A, Cel6A, and Cel7B as well as PAb against Cel7B were a kind gift from Dr. Larry Taylor of the National Renewable Energy Laboratory (NREL). PAbs against Cel7A and Cel6A were prepared commercially by Alpha Diagnostic International, Texas. Briefly, synthetic peptides containing amino acid sequence with high antigenicity from enzymes Cel7A and Cel6A were identified and synthesized The peptide sequence used to raise the Cel7A PAb was R-A-Q-S-A-C-T-L-Q-S-E-T-H-P-P-L-T-W-Q-K, and that for Cel6A PAb was C-D-T-L-D-K-T-P-L-M-E-Q-T-L-A-D-I-R. Following peptide conjugation, antibodies were raised by immunizing rabbits with these peptides. The antibody titers in the rabbit sera and its reactivity to the target peptide were tested using ELISA. Once the test results met the required criteria, the antibody was then purified from the sera by using affinity columns coated with the respective peptide.

The specificity of all MAbs and PAbs were first tested against purified enzymes and enzyme mixtures by using the Western Blot technique following a protocol described by the assay kit producer (Immun-Blot Assay Kit, Bio-Rad). The reactivity and specificity of MAbs against all 3 enzymes (Cel7A, Cel6A, and Cel7B) were tested against purified Cel7A from Celluclast and 3 commercial enzyme mixtures (30 μg each) Accellerase 1000 (Genencor-DuPont), Celluclast, and Cellic CTec 2 (Novozymes). PAbs against Cel7A and Cel6A were tested against purified Cel7A and Cel6A from Celluclast as well as the commercial cellulase mixtures Celluclast and Cellic CTec 2. The specificity and reactivity of PAb against Cel7B were similarly tested against purified Cel7A, Cel6A, and Cel7B from Celluclast as well as the 3 enzyme mixtures Accellerase 1000, Celluclast, and Cellic CTec 2.

Briefly, purified enzymes and enzyme mixtures were separated using sodium dodecyl sulphate polyacrylamide gel electrophoresis (SDS-PAGE) on 4-12% (w/v) Bis-Tris Criterion XT polyacrylamide gels (Bio-Rad). Following electrophoresis, the polyacrylamide gel was equilibrated in the transfer buffer (Towbin buffer containing 25 mM Tris, 192 M glycine, and 20% (v/v) methanol) for 30 minutes. The proteins in the polyacrylamide gel were then transferred to a polyvinylidene difluoride (PVDF) membrane using a Trans-Blot Semi-Dry Electrophoretic Transfer Cell (Bio-Rad) for 60 minutes at 15 V. After washing with Tris-buffered saline containing 0.05% (v/v) Tween 20 (TTBS), the membrane was immersed in Tris-buffered saline (TBS) containing 3% (w/v) gelatin to block any unoccupied sites on the membrane. Antibodies to be tested were then added at a concentration of 5 μg/ml diluted in TTBS containing 1% (w/v) gelatin, and the membrane was incubated for 1 hour. Bound MAbs were detected by immersing the membrane in TTBS-1% (w/v) gelatin containing 1/3000 dilution of goat anti-mouse-IgG antibody conjugated to alkaline phosphatase (GAM-AP, Bio-Rad) for 1 hour whereas bound PAbs were detected by using goat anti-rabbit-IgG antibody conjugated to alkaline phosphatase (GAR-AP, Bio-Rad). After a final wash, the membrane was developed by incubation in the color development/substrate solution containing 5-bromo-4-chloro-3'-indolyphosphate p-toluidine salt (BCIP) and nitro-blue tetrazolium chloride (NBT) for 30 minutes. The reaction was stopped by immersing the membrane in nanopure water for 10 minutes.

### *5.3.3 OPTIMIZATION OF DOUBLE-ANTIBODY SANDWICH ELISA*

A double-antibody sandwich ELISA was developed as it was previously shown to have improved specificity for a target cellulase enzyme present in a cellulase enzyme mixture. MAbs were used as the coating antibodies and PAbs as the detecting antibodies to minimize possible interference from other enzymes, sugars and other materials that may be present in the enzyme mixture [23]. Unless otherwise stated, all reagents were added at a volume of 100 μl, and incubation was carried out at 37°C. Maxisorp plates (Nunc) were coated with MAb diluted in 1× phosphate-buffered saline (PBS) pH 7.5 at 4°C overnight. The wells were then washed with PBS and blocked with 2% (w/v) BSA diluted in 1× PBS for 2 hours. After the wells were washed, enzyme standards and/or samples were added to the wells and incubated for 2 hours. As antibody-antigen interaction is optimum at pH > 7 [24], the enzyme samples were added to the wells after dilution in PBS pH 7.5 to ensure that the enzyme samples were in a solution at greater that pH > 7. Purified Cel7A, Cel6A, and Cel7B were serially diluted (concentrations 0–2.5 μg/ml) in PBS to develop standard curves. After incubation with each of the enzymes, the plate was washed, and the PAb, diluted in PBS with 1% (w/v) BSA, was added to each well. The plate was then incubated for 1 hour. Following another washing step, the third antibody, a commercial GAR-AP (Bio-Rad) diluted in PBS with 1% (w/v) BSA, was added to the wells and incubated for another hour. After a final washing step, 35 mg/ml of p-nitrophenylphosphate (Bio-Rad), a substrate for alkaline phosphatase (AP), was added to the wells and the plate was incubated at room temperature for 30 minutes or until sufficient colour had developed. Colour development was stopped by adding 400 mM glycine-NaOH. The amount of enzymes bound to the sandwich ELISA was quantified by measuring the absorbance of p-nitrophenyl at 405 nm.

### *5.3.4 DETERMINING THE CONCENTRATIONS OF THE MAB, PAB, AND THE ENZYME-ANTIBODY CONJUGATE*

The concentrations of the MAb, PAb, and GAR-AP were optimized for the Cel7A ELISA. Various concentrations of each antibody were tested against a series of concentrations of purified Cel7A. During each antibody optimization, the concentrations of the other two antibodies were kept constant. MAb's against Cel7A was tested at two different concentrations of 10 and 50 μg/ml. Once the concentration of the MAb was optimized, the PAb against Cel7A was assayed at concentrations of 1.75, 3.5, 7, and 14 μg/ml. Similarly, two different dilutions (1/500 and 1/1750 or 1 and 0.3 μg/ml, respectively) of the third antibody, (the GAR-AP conjugate) were assessed.

### 5.3.5 OPTIMIZATION OF SAMPLE TREATMENTS

As heat treatment had previously been used successfully to improve the sensitivity of an ELISA system for Cel7A [23] we investigate the possible influence of heat treatment on the ELISA when 5 μg/ml of each of the purified enzymes were heated at 100°C for 10 minutes. Each enzyme was heated in either Na-acetate buffer (0.05 M pH 5.0) or in PBS pH 7.5. After cooling the samples to room temperature, the enzymes that had been heated in Na-acetate buffer were first diluted with PBS and then added to the ELISA plate. Samples heated in PBS were directly added to the wells at the same final concentration. Unheated samples were added as controls.

### 5.3.6 DETERMINATION OF THE SPECIFICITY OF ELISA

The specificity of each ELISA was determined by comparing the ELISA signal of the target enzyme in the absence and presence of the 3 other cellulase enzymes (Cel7A, Cel6A, Cel7B, and Cel5A). The reconstituted enzyme mixture consisted of 5 μg/ml of the target enzyme and 2.5 μg/ml of each of the other 3 cellulase enzymes in Na-acetate buffer (0.05M, pH 5.0). For Cel7A and Cel6A ELISA, the reconstituted enzyme mixture was heated at 100°C for 10 minutes, serially diluted in PBS to make a standard curve, and then added to the well. Similarly, 5 μg/ml of the pure enzyme sample was subjected to the same treatment. The standard curve obtained from the purified enzyme sample was then compared with that obtained from the reconstituted enzyme mixture. The specificity of Cel7B ELISA was determined in a similar manner except that the enzyme samples were not heated but directly added to the wells after dilution in PBS. The specificity of ELISA was also tested using commercial enzyme mixtures to determine if a dilution of a commercial enzyme mixture can be used to construct a standard curve, obviating the need to use purified enzymes. Commercial enzyme mixtures were diluted in Na-acetate buffer (0.05M, pH 5.0), subjected to the heat treatment when required (i.e. for Cel7A and Cel6A ELISA), serially diluted in PBS, and then added to the wells.

### *5.3.7 LIGNOCELLULOSIC FEEDSTOCKS AND THEIR PRETREATMENT*

An agricultural residue (corn stover), softwood (Douglas-fir) and hardwood (hybrid poplar) chips were used as feedstocks and were pretreated by $SO_2$-catalyzed steam pretreatment. The pretreatments were performed at near optimal conditions that had previously been determined to provide maximum hemicellulose recovery while ensuring effective enzymatic hydrolysis of the cellulose component (steam pretreatment: corn stover [34], Douglas-fir [35], and poplar [36]). After pretreatment, the cellulose rich water insoluble components were washed, filtered and refrigerated for long-term storage. The details of the pretreatment conditions and the chemical compositions of the pretreated substrates have been described earlier [36,37].

### *5.3.8 INFLUENCE OF LIGNOCELLULOSIC DERIVED COMPONENTS PRESENT IN THE HYDROLYSIS SUPERNATANTS ON THE ELISA*

Other than the enzymes, lignocellulosic hydrolyzates can contain various materials derived from the biomass such as soluble phenolic compounds that may interfere with the ELISA. Therefore, to try to determine the possible influence of these substrate materials on the ELISA's, lignocellulosic supernatants obtained from steam pretreated corn stover (SPCS), steam pretreated poplar (SPP), steam pretreated douglas fir (SPDF), and Avicel PH-101 (Sigma), a pure crystalline cellulose substrate, were incubated in 0.05 M Na-acetate buffer pH 5.0 for 24 hours at 50°C with rotational mixing in an incubator (Combi-D24) in the absence of any enzymes. After centrifugation to remove the solid substrate, a known concentration of the target enzyme was added to these supernatants. The same enzyme concentration diluted in

0.05 M Na-acetate buffer pH 5.0 was used as a control. These samples were subjected to heat treatment when required, diluted in PBS and then added to the well. The influence of sugar was not determined as previous work had shown that sugars did not interfere with the ELISA when a MAb was used as the first antibody [23]. As previous work had suggested that "diluting-out" these substrate-derived materials could minimize their interference of the ELISA [20] the supernatants were diluted 10 or 100 times with PBS.

### *5.3.9 ENZYMATIC HYDROLYSIS OF SPCS*

The enzymatic hydrolysis of SPCS was carried out in 15 ml tubes (Corning) in four replicates at 50°C with a rotational mixing at 20 rpm. The SPCS was diluted to 2% (w/v) solid loading with Na-acetate buffer (0.05 M, pH 5.0) to a total volume of 5 ml. Accellerase 1000 was added at 51 mg protein/g glucan, which corresponded to 20 FPU/g glucan. Similarly, SPCS hydrolysis was also carried out using Celluclast at 20 FPU/g glucan or 52 mg protein/g glucan. Concurrently, SPCS was also incubated in Na-acetate buffer (0.05 M, pH 5.0), in the absence of enzymes, to serve as a substrate alone control (SPCS SC).

During hydrolysis, samples were taken at different time points over a period of 72 hours. After centrifugation, the unbound proteins in the supernatant were recovered by transferring the supernatant into 15 ml tubes. One ml of the supernatant was collected and heated at 100°C for 10 minutes for subsequent glucose measurement using the glucose oxidase assay [38]. The remaining supernatant was stored at 4°C for subsequent ELISA assay using the optimized conditions to determine any changes in Cel7A, Cel6A, and Cel7B concentrations during hydrolysis.

### *5.3.10 THE DEVELOPMENT OF A DOUBLE-ANTIBODY SANDWICH ELISA TO QUANTIFY CEL7A, CEL6A, AND CEL7B ADSORPTION DURING SPCS HYDROLYSIS*

ELISA plates were incubated with 10 μg/ml of MAb in PBS at 4°C overnight. The wells were then washed with PBS and blocked with 2% (w/v)

BSA diluted in PBS for 2 hours. After the wells were washed, enzyme standards and/or samples were added to the wells and incubated for 2 hours. For the Cel7A and Cel6A ELISA's, before the addition of samples to the ELISA plate, the purified enzyme samples or the hydrolysate samples were first heated at 100°C for 10 minutes. The heat treatment was always done in Na-acetate buffer (0.05 M pH 4.8). After cooling to room temperature, the samples were diluted in PBS and then added to the ELISA plate. This dilution in PBS not only adjusted the pH of the added samples but also diluted any interfering materials that might be present in lignocellulosic supernatants. After incubation with the enzyme samples for 2 hours, the plate was washed with PBS. A PAb toward the enzyme of interest was added at a concentration of 14 μg/ml diluted in PBS containing 1% (w/v) BSA. The plate was then incubated for 1 hour. Following another washing step, the third antibody, a commercial GAR-AP (Bio-Rad) diluted 1/500 in PBS containing 1% (w/v) BSA, was added and incubated for another hour. After a final washing step, p-nitrophenylphosphate (Bio-Rad) was added, and the plate was incubated until sufficient colour had developed. The colour development was stopped by adding 400 mM glycine-NaOH. The amount of enzymes bound to the sandwich ELISA was quantified by measuring the absorbance of p-nitrophenyl at 405 nm.

By following this protocol, the amount of Cel7A, Cel6A, and Cel7B present in SPCS hydrolysates (unbound proteins) during 72-hour hydrolysis could be quantified. Purified Cel7A, Cel6A, and Cel7B were used to make standard curves. In each of the ELISA assays, the SPCS SC were included and treated in the same way as the hydrolysate samples, to determine the possible influence of any materials in the hydrolysates. The initial enzyme in buffer without any substrate (enzyme control-EC) was also included, to determine the initial concentration of each enzyme. Protein samples for Cel7A ELISA were obtained from SPCS hydrolysis using 20 FPU/ g cellulose of Accellerase 1000. Those for Cel6A and Cel7B ELISA were obtained from SPCS hydrolyzed by 20 FPU/ g Celluclast complemented with 40 CBU/ g cellulose of β-glucosidase.

## REFERENCES

1. Pribowo A, Arantes V, Saddler JN: The adsorption and enzyme activity profiles of specific Trichoderma reesei cellulase/xylanase components when hydrolyzing steam pretreated corn stover. Enzyme Microb Technol 2012, 50(3):195-203.
2. Banerjee G, Scott-Craig JS, Walton JD: Improving enzymes for biomass conversion: a basic research perspective. Bioen Res 2010, 3(1):82-92.
3. Humbird D, Davis R, Tao L, Kinchin C, Hsu D, Aden A, Schoen P, Lukas J, Olthof B, Worley M, Sexton D, Dudgeon D: Process design and economics for biochemical conversion of lignocellulosic biomass to ethanol - dilute-acid pretreatment and enzymatic hydrolysis of corn stover. NREL/TP 2011, 5100-47764.
4. Kumar L, Chandra R, Chung PA, Saddler J: Can the same steam pretreatment conditions be used for most softwoods to achieve good, enzymatic hydrolysis and sugar yields? Bioresour Technol 2010, 101(20):7827-7833.
5. Del Rio LF, Chandra RP, Saddler JN: The effect of varying organosolv pretreatment chemicals on the physicochemical properties and cellulolytic hydrolysis of mountain pine beetle-killed lodgepole pine. Appl Biochem Biotechnol 2010, 161(1–8):1-21.
6. Darias R, Villalonga R: Functional stabilization of cellulase by covalent modification with chitosan. J Chem Technol Biotechnol 2001, 76(5):489-493.
7. Boer H, Koivula A: The relationship between thermal stability and pH optimum studied with wild-type and mutant Trichoderma reesei cellobiohydrolase Cel7A. Eur J Biochem 2003, 270(5):841-848.
8. Tu MB, Zhang X, Kurabi A, Gilkes N, Mabee W, Saddler J: Immobilization of beta-glucosidase on Eupergit C for lignocellulose hydrolysis. Biotechnol Lett 2006, 28(3):151-156.
9. Tu M, Chandra RP, Saddler JN: Recycling cellulases during the hydrolysis of steam exploded and ethanol pretreated lodgepole pine. Biotechnol Prog 2007, 23(5):1130-1137.
10. Sharrock KR: Cellulase assay-methods - a Review. J Biochem Biophys Methods 1988, 17(2):81-105.
11. Medve J, Lee D, Tjerneld F: Ion-exchange chromatographic purification and quantitative analysis of Trichoderma reesei cellulases cellobiohydrolase I, II and endoglucanase II by fast protein liquid chromatography. J Chromatogr 1998, 808(1–2):153-165.
12. Medve J, Karlsson J, Lee D, Tjerneld F: Hydrolysis of microcrystalline cellulose by cellobiohydrolase I and endoglucanase II from Trichoderma reesei: adsorption, sugar production pattern, and synergism of the enzymes. Biotechnol Bioeng 1998, 59(5):621-634.
13. Yu AHC, Lee D, Saddler JN: A quantitative approach to the study of the adsorption-desorption of cellulase components in a crude cellulase mixture. Biotechnol Tech 1993, 7(10):713-718.

14. Gao D, Chundawat SPS, Uppugundla N, Balan V, Dale BE: Binding characteristics of trichoderma reesei cellulases on untreated, ammonia fiber expansion (AFEX), and dilute-acid pretreated lignocellulosic biomass. Biotechnol Bioeng 2011, 108(8):1788-1800.
15. Varnai A, Viikari L, Marjamaa K, Siika-aho M: Adsorption of monocomponent enzymes in enzyme mixture analyzed quantitatively during hydrolysis of lignocellulose substrates. Bioresour Technol 2011, 102(2):1220-1227.
16. Valjamae P, Sild V, Pettersson G, Johansson G: The initial kinetics of hydrolysis by cellobiohydrolases I and II is consistent with a cellulose surface - erosion model. Eur J Biochem 1998, 253(2):469-475.
17. Igarashi K, Uchihashi T, Koivula A, Wada M, Kimura S, Okamoto T, Penttila M, Ando T, Samejima M: Traffic Jams Reduce Hydrolytic Efficiency of Cellulase on Cellulose Surface. Science 2011, 333(6047):1279-1282.
18. Ramos LP, Zandona A, Deschamps FC, Saddler JN: The effect of Trichoderma cellulases on the fine structure of a bleached softwood kraft pulp. Enzyme Microb Technol 1999, 24(7):371-380.
19. Kolbe J, Kubicek C: Quantification and identification of the main components of the trichoderma cellulase complex with monoclonal-antibodies using an Enzyme-Linked-Immunosorbent-Assay (Elisa). Appl Microbiol Biotechnol 1990, 34(1):26-30.
20. Conde S, Suyama K, Itoh K, Yamamoto H: Application of commercially available fenitrothion-ELISA kit for soil residue analysis. J Pestic Sci 2008, 33(1):51-57.
21. Drow D, Manning D: Indirect sandwich enzyme-linked Immunosorbent-assay for rapid detection of streptococcus-pneumoniae type-3 antigen. J Clin Microbiol 1980, 11(6):641-645.
22. Voller A, Bartlett A, Bidwell D: Enzyme immunoassays with special reference to elisa techniques. J Clin Pathol 1978, 31(6):507-520.
23. Riske FJ, Eveleigh DE, Macmillan JD: Double-antibody sandwich enzyme-linked-immunosorbent-assay for cellobiohydrolase-i. Appl Environ Microbiol 1990, 56(11):3261-3265.
24. Buhler R: Double-antibody sandwich enzyme-linked-immunosorbent-assay for quantitation of endoglucanase-i of trichoderma-reesei. Appl Environ Microbiol 1991, 57(11):3317-3321.
25. Crowther JR: The ELISA Guidebook: Totowa. NJ, USA: Humana Press; 2000.
26. Lynd L, Zhang Y: Quantitative determination of cellulase concentration as distinct from cell concentration in studies of microbial cellulose utilization: Analytical framework and methodological approach. Biotechnol Bioeng 2002, 77(4):467-475.
27. Toscano I, Gascon J, Marco M, Rocha J, Barcelo D: Atrazine interaction with tropical humic substances by Enzyme Linked Immunosorbent Assay. Analusis 1998, 26(3):130-134.
28. Krotzky A, Zeeh B: Pesticides report .33. Immunoassays for residue analysis of agrochemicals: Proposed guidelines for precision, standardization and quality control. Pure Appl Chem 1995, 67(12):2065-2088.
29. Palonen H, Tenkanen M, Linder M: Dynamic interaction of Trichoderma reesei cellobiohydrolases Ce16A and Ce17A and cellulose at equilibrium and during hydrolysis. Appl Environ Microbiol 1999, 65(12):5229-5233.

30. Gama F, Vilanova M, Mota M: Exo- and endo-glucanolytic activity of cellulases purified from Trichoderma reesei. Biotechnol Tech 1998, 12(9):677-681.
31. Rosgaard L, Pedersen S, Langston J, Akerhielm D, Cherry JR, Meyer AS: Evaluation of minimal Trichoderma reesei cellulase mixtures on differently pretreated barley straw substrates. Biotechnol Prog 2007, 23(6):1270-1276.
32. Zhou J, Wang Y, Chu J, Zhuang Y, Zhang S, Yin P: Identification and purification of the main components of cellulases from a mutant strain of Trichoderma viride T 100–14. Bioresour Technol 2008, 99(15):6826-6833.
33. Starcher B: A ninhydrin-based assay to quantitate the total protein content of tissue samples. Anal Biochem 2001, 292(1):125-129.
34. Ohgren K, Galbe M, Zacchi G: Optimization of steam pretreatment of SO2-impregnated corn stover for fuel ethanol production. Appl Biochem Biotechnol 2005, 121:1055-1067.
35. Wu M, Chang K, Gregg D, Boussaid A, Beatson R, Saddler J: Optimization of steam explosion to enhance hemicellulose recovery and enzymatic hydrolysis of cellulose in softwoods. Appl Biochem Biotechnol 1999, 77–9:47-54.
36. Bura R, Chandra R, Saddler J: Influence of xylan on the enzymatic hydrolysis of steam-pretreated corn stover and hybrid poplar. Biotechnol Prog 2009, 25(2):315-322.
37. Arantes V, Saddler JN: Cellulose accessibility limits the effectiveness of minimal cellulase loading on the efficient hydrolysis of pretreated lignocellulosic substrates. Biotechnol Biofuels 2011, 4:3.
38. Berlin A, Maximenko V, Bura R, Kang KY, Gilkes N, Saddler J: A rapid microassay to evaluate enzymatic hydrolysis of lignocellulosic substrates. Biotechnol Bioeng 2006, 93(5):880-886.

## CHAPTER 6

# UNDERSTANDING OF ALKALINE PRETREATMENT PARAMETERS FOR CORN STOVER ENZYMATIC SACCHARIFICATION

YE CHEN, MARK A. STEVENS, YONGMING ZHU, JASON HOLMES, AND HUI XU

## 6.1 INTRODUCTION

Alkaline pretreatment is one of several chemical pretreatment technologies that has been intensively investigated. It employs various alkaline reagents including sodium hydroxide [1,2], calcium hydroxide [3,4], potassium hydroxide [5], aqueous ammonia [6,7], ammonia hydroxide [8], and sodium hydroxide in combination with hydrogen peroxide [9,10]. Mechanistically, alkali is believed to cleave hydrolysable linkages in lignin and glycosidic bonds of polysaccharides, which causes a reduction in the degree of polymerization and crystallinity, swelling of the fibers, as well as disruption of the lignin structure [11]. In addition, alkaline saponification of acetyl and uronic ester bonds also improves the enzymatic accessibility of the polysaccharides [12]. The effectiveness of alkaline pretreatment is dependent on the physical structure and chemical composition of the substrate as well as the treatment conditions. In general, alkaline pretreatment is more effective on hardwood, herbaceous crops, and agricultural residues, which have a lower lignin content, than on substrates such as softwood, which contain high amounts of lignin.

Although alkaline pretreatment has been studied on different types of lignocellulosic biomass including switchgrass, corn stover, wheat straw, rice straw, and rice hulls [13], most of the research on alkaline pretreatment has focused on optimization of the process parameters to improve substrate digestibility [13-15]. To achieve this goal, extremely high chemical loading and enzyme dosages were frequently used. Relatively little attention has been paid to process waste management, including chemical recovery and recycle, which has proven to be an indispensable component of the biorefineries [16]. A literature survey also indicates that both alkali concentration in pretreatment solution (g alkali/g pretreatment liquor or g alkali/L pretreatment liquor) and alkali loading based on biomass solids (g alkali/g dry biomass) have been widely used as indicators of alkali strength. The dual approaches make it difficult to compare the chemical consumption in different process scenarios and to evaluate the cost effectiveness of this pretreatment technology. Thus the objectives of this study were to examine the effect of alkaline pretreatment parameters on the digestibility of substrate and to identify whether alkali solution concentration or its dosage on biomass determines hydrolysis yield. The economic feasibility of the alkaline pretreatment process was also evaluated. Corn stover was selected as a model feedstock and a series of alkaline pretreatments were conducted based on a central composite design involving three process variables. Sodium hydroxide was chosen as the pretreatment chemical since it is widely used in the well-established pulp and paper industry. The efficiency of pretreatment was then evaluated by measuring total sugar release from enzymatic hydrolysis of the pretreated substrates.

## 6.2 MATERIALS AND METHODS

### 6.2.1 FEEDSTOCK COLLECTION AND PREPARATION

Corn stover was harvested in the Midwest (United States). Concurrent with the corn (grain) harvest, all residue (leaves stalks and husks) above 12" from the ground was collected. The corn stover residue was then milled to a 6 mm particle size using a Thomas Wiley mill. The moisture content of the corn stover was about 10%. Compositional analysis of the raw corn stover shows that it contains 40.21% glucan, 22.28% xylan, and 19.54% acid insoluble lignin on a dry basis.

### 6.2.2 ENZYMES

Both Cellic® CTec2 and the experimental accessory enzymes were obtained from Novozymes A/S (Bagsvaerd, Denmark). This particular batch of Cellic® CTec2 had a protein concentration of 141.6 mg protein/g as determined by the bicinchoninic acid (BCA) assay (Pierce, Rockford, Ill.). Cellic® CTec2 and the experimental enzyme cocktails were stored at 4°C and −30°C, respectively, until needed for hydrolysis of pretreated corn stover.

### 6.2.3 ALKALINE PRETREATMENT

Two sets of alkaline pretreatment studies were conducted to identify: 1) the pretreatment parameters which have the most influence on substrate digestibility and 2) whether the alkali concentration in the aqueous phase or the alkali loading on a dry corn stover basis determines the pretreatment efficiency.

Alkaline pretreatment of corn stover was conducted in a LABOMAT reactor (Type BFA-12, Mathis, Switzerland) with a digitally controlled infrared heating system that has a temperature range of 20-200°C. The instrument is equipped with a variable speed rotary disk and can be programmed to operate with up to 8 stainless steel cylindrical 1 liter beakers simultaneously. Each beaker was loaded with approximately 500 grams of material including corn stover, 50% (w/w) sodium hydroxide solution, and deionized water. Beaker contents were mixed thoroughly to achieve a total solid loading of 11% and the desired alkali loading. Eight stainless steel balls (Dia 10 mm) were added to the beakers to promote more adequate mixing during rotary movement of the beakers. Pretreatment temperature was monitored with a thermocouple inserted through one of the reactor caps. The heat-up time needed to reach target temperature was approximately 10–25 mins, depending on the setpoint pretreatment temperature. Time zero (for pretreatment) was taken to be the time at which the center of the reactor reached the target temperature. After pretreatment, the beakers were immediately quenched in an ice bath for rapid cooling. Corn stover from two replicate beakers treated under similar conditions was recovered, combined, and washed intensively with deionized water to remove soluble phenolics and other degradation products. The washed corn stover was stored at 4°C.

### *6.2.4 ENZYMATIC HYDROLYSIS*

Batch enzymatic hydrolysis was performed in 50 mL Nalgene polycarbonate centrifuge tubes (Thermo Scientific, Pittsburgh, PA). Alkali pretreated corn stover was mixed with 50 mM sodium acetate buffer (pH 5.0) supplemented with enzymes as well as 2.5 mg/L lactrol to prevent microbial growth. The final total solids concentration was 10% (w/w). The reaction mixtures (20 g) were agitated in a hybridization incubator (Combi-D24, FINEPCR®, Yang-Chung, Seoul, Korea) at 50°C for 120 hrs. To evaluate pretreatment efficiency as well as the effect of accessory enzymes on hydrolysis performance, pretreated corn stover was hydrolyzed with an enzyme blend at 4 mg protein/g glucan dosage. The enzyme mixture contained 90% protein from Cellic® CTec2, 3.33% protein from arabinofuranosidase which has activity on single substituted arabinose side chain, 3.33% arabinofuranosidase which has activity on double substituted arabinose side chain, and 3.33% β-xylosidase was tested. Pretreated corn stover hydrolyzed with 100% Cellic® CTec2 also at 4 mg protein/g glucan was used as a control. At the end of hydrolysis, 600 μL of hydrolysate were transferred to a Costar Spin-X centrifuge filter tube (Cole-Parmer, Vernon Hills, IL) and filtered through a 0.2 μm nylon filter during centrifugation (14,000 rpm, 20 mins). Supernatant was acidified with 5 μL of 40% (w/v) sulfuric acid to deactivate residual enzyme activity and analyzed by HPLC for sugar concentrations.

**TABLE 1:** Central composite design of alkaline pretreatment of corn stover

| Sample ID | NaOH loading (g NaOH/g dry corn stover) | Pretreatment temp (°C) | Pretreatment time (min) |
|---|---|---|---|
| symbol | X1 | X2 | X3 |
| 1 | 0.052 | 74 | 102 |
| 2 | 0.100 | 95 | 75 |
| 3 | 0.070 | 130 | 75 |
| 4 | 0.070 | 95 | 120 |
| 5 | 0.088 | 116 | 102 |
| 6 | 0.052 | 74 | 48 |
| 7 [a] | 0.070 | 95 | 75 |
| 8 | 0.088 | 116 | 48 |
| 9 | 0.040 | 95 | 75 |
| 10 | 0.070 | 60 | 75 |
| 11 | 0.088 | 74 | 102 |
| 12 | 0.052 | 116 | 102 |
| 13 | 0.070 | 95 | 30 |
| 14 | 0.052 | 116 | 48 |
| 15 | 0.088 | 74 | 48 |
| 16 [a] | 0.070 | 95 | 75 |

[a] *Center point of the central composite design.*

### *6.2.5 FEEDSTOCK COMPOSITIONAL ANALYSIS AND SUGAR ANALYSIS*

Total solids content, structural carbohydrate, and lignin content of raw corn stover and alkali pretreated corn stover were analyzed using standard laboratory analytical procedures (LAP) developed by the National Renewable Energy Laboratory (NREL) [17,18]. Sugar samples from compositional analysis were measured using an Agilent 1200 series modular HPLC (Santa Clara, CA) equipped with an Aminex HPX-87P column (Bio-Rad, Richmond, CA), while sugars released from hydrolysis of pretreated corn stover were analyzed using a Rezex ROA-Organic acid $H^+$ column (8%) (7.8 × 300 mm) (Phenomenex Inc., Torrance, CA). The methodology was described in detail in Chen et al. [16]. The overall glucan/xylan conversions from hydrolysis were calculated based on sugar concentrations in the enzyme hydrolysis supernatant and composition of the pretreated feedstock using a method similar to that published by Zhu et al. [19].

### 6.2.6 EXPERIMENTAL DESIGN AND STATISTICAL ANALYSIS

A central composite design was used to reduce the total number of experiments needed to explore the relationship between pretreatment condition and compositional change of pretreated corn stover, as well as its glucan/xylan conversion. The statistical software SAS JMP, version 8 was used for the 3 × 3 central composite design in which 16 pretreatment combinations were derived by altering the three independent variables: alkaline loading, temperature, and time (Table 1) and to analyze the experimental data obtained. The selection of the factorial levels was based on previous studies (data not shown) and the parameters were varied from 60–130°C for temperature, 0.01-0.10 g NaOH/g dry corn stover for chemical dose, and 30–120 mins for pretreatment time. All pretreatment and hydrolysis were performed in duplicate unless otherwise stated. When data have been collected in accordance with the experimental design, the response variable (Y) was fitted to the appropriate empirical equations (second order polynomial regression equations) to identify the key variables:

$$Y = \beta_o + \beta_1 x_1 + \beta_2 x_2 + \beta_3 x_3 + \beta_{11} x_1^2 + \beta_{22} x_2^2 + \beta_{33} x_3^2 + \beta_{12} x_1 x_2 + \beta_{13} x_1 x_3 + \beta_{23} x_{21} x_3$$

where the response variable Y represents compositional change of pretreated corn stover or glucan/xylan conversion and the variables $x_1$, $x_2$, and $x_3$ correspond to alkaline loading, pretreatment temperature, and time, respectively. The predicted response was therefore correlated to the intercept ($\beta_0$), linear ($\beta_1$, $\beta_2$, $\beta_3$), interaction ($\beta_{12}$, $\beta_{13}$, $\beta_{23}$) and quadratic coefficients ($\beta_{11}$, $\beta_{22}$, $\beta_{33}$) which can be calculated from the experimental data. The quality of fit of the polynomial model equation was expressed by the coefficient of determination. An effect is significant if its p-value is less than 0.05.

## 6.3 RESULTS AND DISCUSSION

### *6.3.1 ALKALINE PRETREATMENT OF CORN STOVER*

Table 2 summarizes the compositional change of corn stover following pretreatment. During alkaline pretreatment, the cleavage of hydrolyzable linkages such as α- and β- aryl ethers in lignin and glycosidic bonds in carbohydrates constitute the primary reactions that lead to the dissolution of lignin and carbohydrate with lower alkali stability [20]. More than 95% of the cellulose in corn stover was preserved in alkaline pretreatment, which can be explained by the low reactivity of cellulose with alkali and also its high crystallinity [20,21]. Dissolution of hemicellulose and lignin, however, varied significantly depending on the pretreatment conditions (Table 2). Table 3 shows the effect of the pretreatment parameters on xylan recovery as well as on delignification. The statistical analysis indicates that among the variables that have a statistically significant effect on lignin removal from corn stover (three first-order effects, three second-order effects, and one interaction effect), NaOH loading had the most significant impact (regression coefficient $\beta_1 = 8.73$), indicating the highest sensitivity of lignin content to alkali charge. When alkali loading increased from 0.04 to 0.1 g/g corn stover, the residual lignin decreased from 67.5 to 20.1% (Table 2). Although pretreatment at high alkali loading, temperature, and longer residence time can maximize delignification and therefore improve substrate digestibility, high severity pretreatment conditions may also lead to undesired sugar loss through dissolution and degradation of hemicellulose. Similar to lignin degradation, depolymerization of hemicellulose is also significantly affected by the three parameters with alkali loading having the greatest effect (regression coefficient $\beta_1$=−2.922). Xylan degradation increased by 20% when NaOH loading increased from 0.04 to 0.1 g/g corn stover (Table 2). These reaction mechanisms imply that a balance between extent of delignification and preservation of carbohydrate has to be established in order to achieve maximum overall sugar yield.

**TABLE 2:** Composition of washed pretreated corn stover solids

| Temp (°C) | NaOH loading (g NaOH/g corn stover) | Time (mins) | Composition (%) [a,b] | | | Recovery (%) [b,c] | | | |
|---|---|---|---|---|---|---|---|---|---|
| | | | Glucan | Xylan | AIL | Glucan | Xylan | AIL | Total |
| 74 | 0.052 | 102 | 48.11 | 25.75 | 14.44 | 96.16 | 92.88 | 59.41 | 80.38 |
| 95 | 0.100 | 75 | 61.20 | 28.20 | 6.42 | 93.25 | 77.53 | 20.12 | 61.26 |
| 130 | 0.070 | 75 | 57.85 | 25.98 | 9.74 | 95.75 | 77.61 | 33.17 | 66.55 |
| 95 | 0.070 | 120 | 55.46 | 27.20 | 8.98 | 96.47 | 85.39 | 32.14 | 69.95 |
| 116 | 0.088 | 102 | 59.53 | 26.91 | 5.36 | 95.20 | 77.67 | 17.65 | 64.30 |
| 74 | 0.052 | 48 | 46.63 | 25.31 | 14.84 | 96.66 | 94.69 | 63.28 | 83.35 |
| 95 d | 0.070 | 75 | 54.60 | 26.16 | 8.37 | 97.64 | 84.44 | 30.81 | 71.91 |
| 116 | 0.088 | 48 | 60.24 | 27.53 | 4.74 | 96.89 | 79.92 | 15.69 | 64.68 |
| 95 | 0.040 | 75 | 45.25 | 25.06 | 15.20 | 97.69 | 97.64 | 67.50 | 86.80 |
| 60 | 0.070 | 75 | 48.48 | 25.18 | 12.19 | 97.55 | 91.44 | 50.47 | 80.90 |
| 74 | 0.088 | 102 | 54.03 | 25.64 | 8.06 | 95.83 | 82.07 | 29.40 | 71.31 |
| 116 | 0.052 | 102 | 50.67 | 24.68 | 13.36 | 95.70 | 84.13 | 51.94 | 75.95 |
| 95 | 0.070 | 30 | 51.60 | 26.00 | 11.48 | 96.43 | 87.68 | 44.13 | 75.14 |
| 116 | 0.052 | 48 | 49.43 | 25.03 | 14.07 | 96.46 | 88.13 | 56.50 | 78.46 |
| 74 | 0.088 | 48 | 54.34 | 25.54 | 9.35 | 97.13 | 82.39 | 34.38 | 71.87 |
| 95 d | 0.070 | 75 | 54.22 | 26.33 | 8.21 | 97.81 | 84.29 | 30.26 | 71.59 |

*[a] The composition of insoluble solids are based on oven-dry weight. [b] Values are expressed as averages of three replicate samples. Coefficient of variation (CV) is below 2.1 %. [c] Values were calculated based on the total recovery, composition of alkaline pretreated, as well as that of raw corn stover. [d] Center point of the central composite design.*

### 6.3.2 ENZYMATIC HYDROLYSIS OF PRETREATED CORN STOVER

Glucan and xylan conversions for hydrolysis and for the overall process (pretreatment and hydrolysis) are presented in Table 4. Statistical analysis of the hydrolysis data, which examines the relationship between pretreatment parameters and conversions, is summarized in Table 5. All four models have $R^2$ values between 0.91 and 0.97, indicating that a large fraction of the variation in responses can be accounted for by the independent variables. The analysis of variance also showed that the second order polynomial regression models are highly significant (p value<0.0001) (Table 5).

**TABLE 3:** Statistical analysis of the effects of pretreatment parameters on corn stover xylan recovery and delignification[a]

| | Xylan recovery | | Lignin removal | |
|---|---|---|---|---|
| $R^2$ | 0.942 | | 0.985 | |
| Prob. > F | < 0.0001* | | < 0.0001* | |
| Terms | Estimate | p value | Estimate | p value |
| NaOH loading | −2.922 | < 0.0001* | 8.730 | < 0.0001* |
| Temperature | −0.002 | < 0.0001* | 0.003 | < 0.0001* |
| Time | −0.0003 | 0.015* | 0.001 | < 0.0001* |
| NaOH loading × temperature | 0.028 | 0.027* | 0.053 | 0.003* |
| NaOH loading × time | 0.008 | 0.374 | −0.014 | 0.272 |
| Temperature × time | $-9.09 \times e^{-6}$ | 0.258 | $-1.376 \times e^{-5}$ | 0.208 |
| NaOH loading × NaOH loading | 31.764 | 0.0218* | −137.865 | < 0.0001* |
| Temperature × temperature | $-1.62 \times e^{-6}$ | 0.865 | $-8.50 \times e^{-5}$ | < 0.0001* |
| Time × time | $8.93 \times e^{-6}$ | 0.133 | $-3.325 \times e^{-5}$ | < 0.0003* |

[a] *Numbers with asterisk indicate that the term has a significant effect at 95 % confidence interval.*

Alkali loading and temperature have a significant effect on glucan conversion, which is consistent with previous studies investigating alkaline pretreatment of various lignocellulosic feedstocks [13,22]. Glucan conversion during hydrolysis is positively correlated with NaOH loading. An increase of NaOH loading from 0.04 to 0.1 g/g corn stover improved glucan conversion by 35% during hydrolysis (Table 4). Since more than 95% of the original glucan was preserved in the solid fraction following pretreatment, this increase was also reflected in the overall process yield. To reach 70% overall glucan conversion at 4 mg protein/g glucan enzyme dose, approximately 0.08 g NaOH/g corn stover was required.

**TABLE 4:** Enzymatic hydrolysis of alkaline pretreated corn stover[a]

| Temp (°C) | NaOH loading (g NaOH/g corn stover) | Time (mins) | Hydrolysis yield[b] | | Pretreatment and hydrolysis yield[c] | | Enzyme accessibility[d] | |
|---|---|---|---|---|---|---|---|---|
| | | | Glucan | Xylan | Glucan | Xylan | Glucan | Xylan |
| 74 | 0.052 | 102 | 47.35% | 45.56% | 45.53% | 42.32% | 56.79% | 55.30% |
| 95 | 0.100 | 75 | 82.31% | 75.67% | 76.75% | 58.67% | 88.19% | 79.84% |
| 130 | 0.070 | 75 | 59.83% | 67.21% | 57.28% | 52.16% | 88.42% | 79.87% |
| 95 | 0.070 | 120 | 69.63% | 70.47% | 67.17% | 60.17% | 83.98% | 77.39% |
| 116 | 0.088 | 102 | 82.63% | 76.62% | 78.66% | 59.51% | 94.12% | 84.55% |
| 74 | 0.052 | 48 | 49.29% | 42.53% | 47.65% | 40.27% | 64.21% | 58.07% |
| 95 d | 0.070 | 75 | 68.42% | 72.94% | 66.81% | 61.59% | 89.75% | 80.58% |
| 116 | 0.088 | 48 | 82.32% | 79.22% | 79.76% | 63.32% | 91.42% | 82.03% |
| 95 | 0.040 | 75 | 47.18% | 38.04% | 46.09% | 37.14% | 59.40% | 50.27% |
| 60 | 0.070 | 75 | 55.72% | 53.82% | 54.35% | 49.22% | 66.25% | 62.45% |
| 74 | 0.088 | 102 | 75.71% | 73.24% | 72.55% | 60.11% | 90.89% | 80.43% |
| 116 | 0.052 | 102 | 61.11% | 61.76% | 58.49% | 51.96% | 95.57% | 83.29% |
| 95 | 0.070 | 30 | 68.01% | 64.40% | 65.59% | 56.46% | 86.79% | 76.17% |
| 116 | 0.052 | 48 | 55.42% | 60.37% | 53.45% | 53.20% | 80.88% | 71.78% |
| 74 | 0.088 | 48 | 70.54% | 72.13% | 68.51% | 59.43% | 88.90% | 80.73% |
| 95 [d] | 0.070 | 75 | 68.71% | 73.00% | 66.99% | 61.95% | 89.95% | 81.20% |

[a] *Values are expressed as averages of two replicate samples. Coefficient of variation (CV) is below 2.5%.* [b] *Glucose and xylose yield was calculated based on glucan and xylan content in pretreated and washed corn stover.* [c] *glucose and xylose yield was calculated based on glucan and xylan content in raw corn stover.* [d] *Glucan (or xylan) enzyme accessibility was defined as the fraction of glucan (or xylan) in biomass that can be converted to monomeric sugars by enzyme activities. They were obtained by measuring glucan or xylan conversions at an extremely high enzyme dosage (50 mg Cellic® CTec2/g glucan).*

All the linear and quadratic model terms that include alkali loading and temperature have a significant effect on xylan conversion during hydrolysis and for the overall process (Table 5). Among linear terms, NaOH loading had the greatest effect on the responses while this variable had a significant interaction with temperature ($p\text{-value}<0.05$). Xylan conversion during hydrolysis is positively correlated with NaOH loading. An increase in NaOH loading from 0.04 to 0.1 g/g corn stover improved xylan conversion by 37% during hydrolysis (Table 4). However, alkali delignification processes are usually accompanied by dissolution and degradation of hemicellulose [21]. When NaOH loading exceeded a certain limit (approximately 0.08 g/g corn stover), the substantial loss of carbohydrates during pretreatment can offset increased substrate digestibility.

Temperature is the second most important parameter affecting hydrolysis conversion. The models indicate that the optimal temperature ranges are 103–106°C and 93–97°C for glucose and xylose release, respectively. An increase in temperature accelerates delignification. However, severe pretreatment conditions can lead to lignin condensation reactions that form carbon-carbon bonds between lignin subunits, thereby limiting its removal and consequently reducing glucan/xylan conversion [23]. In addition, higher temperatures also increase carbohydrate loss through random chain cleavage as well as peeling reactions, which can greatly reduce the sugar yield from the overall process [24].

Although alkaline pretreatment and chemical pulping share many similarities in reaction chemistry and substrate physicochemical changes, the desired outcomes from pretreatment and pulping are very different. The purpose of chemical pulping is to remove lignin and improve paper strength. Most of the pulp mills, with the exception of those practicing high yield pulping, delignify biomass extensively to save on bleaching chemical costs. The final kappa number for unbleached pulp is typically between 15–30, which corresponds to 2.5–4.5% lignin content or more than 90% delignification [25]. As a result, there is a significant loss of hemicellulose due to peeling reactions and the overall pulp yield is in the range of 45–50% [26]. In the case of biomass pretreatment, the purpose is to depolymerize cellulose/hemicellulose into fermentable sugars; high carbohydrate yields are essential for economic viability. The conditions used in biomass pretreatment are much milder than pulping including lower alkali charge, lower temperatures and shorter residence times. Consequently, the pulp yield after pretreatment is significantly higher (61–72%) than that of the chemical pulping process.

**TABLE 5:** Statistical analysis of the effects of pretreatment parameters on corn stover hydrolysis[a]

| | Glucose (hydrolysis) | | Xylose (hydrolysis) | | Glucose (overall) | | Xylose (overall) | |
|---|---|---|---|---|---|---|---|---|
| R2 | 0.951 | | 0.966 | | 0.914 | | 0.939 | |
| Prob>F | < 0.0001* | | < 0.0001* | | < 0.0001* | | < 0.0001* | |
| Terms | Estimate | pvalue | estimate | pvalue | Estimate | pvalue | Estimate | pvalue |
| NaOH loading | 6.417 | < 0.0001* | 5.084 | < 0.0001* | 5.962 | < 0.0001* | 2.987 | < 0.0001* |
| Temperature | 0.002 | < 0.0001* | 0.002 | < 0.0001* | 0.001 | < 0.0001* | 0.001 | < 0.0001* |
| Time | 0.0003 | 0.159 | 0.0003 | 0.0592 | 0.0002 | 0.336 | $9.604 \times e^{-5}$ | 0.467 |
| NaOH loading × temperature | −0.004 | 0.852 | −0.052 | 0.0017* | −0.004 | 0.835 | −0.042 | 0.0022* |
| NaOH loading × time | 0.004 | 0.786 | −0.006 | 0.594 | 0.0001 | 0.998 | −0.003 | 0.743 |
| Temperature × time | $6.139 \times e^{-6}$ | 0.663 | $6.798 \times e^{-6}$ | 0.491 | $4.46 \times e^{-6}$ | 0.763 | $3.804 \times e^{-7}$ | 0.948 |
| NaOH loading × NaOH loading | −26.181 | 0.263 | −86.943 | < 0.0001* | −43.260 | 0.086 | −76.638 | < 0.0001* |
| Temperature × temperature | $-7.616 \times e^{-5}$ | 0.0002* | $-6.44 \times e^{-5}$ | < 0.0001* | $-7.753 \times e^{-5}$ | 0.0002* | $-5.918 \times e^{-6}$ | < 0.0001* |
| Time × time | $8.50 \times e^{-6}$ | 0.411 | $-4.701 \times e^{-6}$ | 0.514 | $5.53 \times 10^{-6}$ | 0.628 | $3.898 \times e^{-7}$ | 0.948 |

*a Numbers with asterisk indicate that the term has a significant effect at 95 % confidence interval.*

### 6.3.3 IMPACT OF HEMICELLULASE SUPPLEMENT

Alkaline pretreated corn stover had a xylan content of 25–28%, which implies that hemicellulases should be indispensable components in the hydrolysis of biomass pretreated under alkaline conditions. To efficiently hydrolyze the xylan and xylooligomers remaining after pretreatment, CTec2 was supplemented with a 3% (based on protein dose) replacement of an experimental hemicellulase mixture which contained accessory enzymes such as arabinofuranosidases and β-xylosidase. Hydrolysis was conducted with 4 mg protein/g glucan enzyme dose at 10% total solids loading for 120 hrs. Alkaline pretreated corn stover hydrolyzed with CTec2 only was used as the control. Supplementation of CTec2 with accessory hemicellulases only marginally increased the glucan to glucose conversion by 0–2% (data not shown). The relatively small improvement observed in this study can be attributed to the presence of hemicellulase activities in CTec2. These hemicellulases help remove hemicellulose that physically blocks access to cellulose by cellulase [27]. Hemicellulases also contribute to the decrease in the concentrations of high molecular weight xylooligomers, which have been reported to be highly inhibitory towards cellulase activities [28]. On the other hand, the conversion of xylan to xylose was significantly enhanced by supplementation with accessory enzymes. Depending on the pretreatment conditions, corn stover samples hydrolyzed with CTec2 supplemented with accessory enzymes had xylan conversions that were 6–17% higher than their respective controls (Figure 1). The effect was less pronounced for substrate pretreated with lower NaOH loadings (0.040 and 0.052 g NaOH/g corn stover), possibly due to the fact that their poor accessibility limits enzyme-substrate interaction (Table 4). The higher xylan to xylose yield obtained for hydrolysis with Cellic® CTec2 supplemented with accessory enzymes can be explained as follows. Complete hydrolysis of xylan requires synergistic effect of endo-β-1,4 xylanase, β-xylosidase on xylan backbone and accessory enzymes for hydrolyzing various substituted xylans [29]. However, many xylanases are not capable of cleaving glycosidic bonds between xylose units that are substituted [30]. α-arabinofuranosidase and β-arabinofuranosidase remove the arabinose substituents from the xylan backbone, as indicated by the 1- to 4-fold increase in arabinose concentration in the hydrolysate compared to the controls (data not shown). Removal of side chains allows better access by the xylanase to the linkage between backbone components of the polysaccharide; In addition, β-xylosidase acts synergistically with xylanases and releases xylose monomers from xylobiose and short chain xylooligosaccharides, which contributes to the higher xylan to xylose conversion [28].

### 6.3.4 EFFECT OF ALKALI SOLUTION CONCENTRATION AND BIOMASS ALKALI LOADING

In previous studies on alkaline pretreatment, both alkali solution concentration (g alkali/g pretreatment liquor) and biomass alkali loading (g alkali/g biomass) are used as indicators of chemical strength [14,15]. To distinguish the effects of solution strength vs. biomass alkali loading on biomass digestibility, pretreatment of corn stover was conducted at 0.05 g, 0.10 g, and 0.15 g dry corn stover/g slurry. At each solid loading, three NaOH dosages (0.06, 0.08, and 0.10 g NaOH/g corn stover) were used to compare the pretreatment efficiency. In the experiment, pretreatment temperature and residence time were maintained at 90°C and 120 mins, respectively. Compositional analysis of the corn stover pretreated under the 9 different test conditions was performed; delignification of the pretreated corn stover is shown in Figure 2. The extent of delignification was closely correlated with biomass alkali loading. For a given NaOH loading based on corn stover dry weight, lignin removal was relatively stable regardless of the total solids content during pretreatment.

The pretreated corn stover was washed with deionized water and hydrolyzed with Cellic® CTec2 at 4 mg protein/g glucan for 120 hrs at 8.5% total solids loading. Figure 3 shows the effect of biomass alkali loading on glucose and xylose yields for hydrolysis (Figure 3a) and also for the combined pretreatment and hydrolysis processes (Figure 3b). The effect of alkali solution concentration is shown in Figure 4. Essentially, the enzymatic digestibility of the corn stover correlates better with biomass alkali loading than with alkali solution concentration. Glucose/xylose concentrations in the hydrolysate increased as the alkali charge on dry corn stover increased (Figure 3). On the other hand, no correlation can be established between glucan/xylan conversion and NaOH solution concentration (Figure 4).

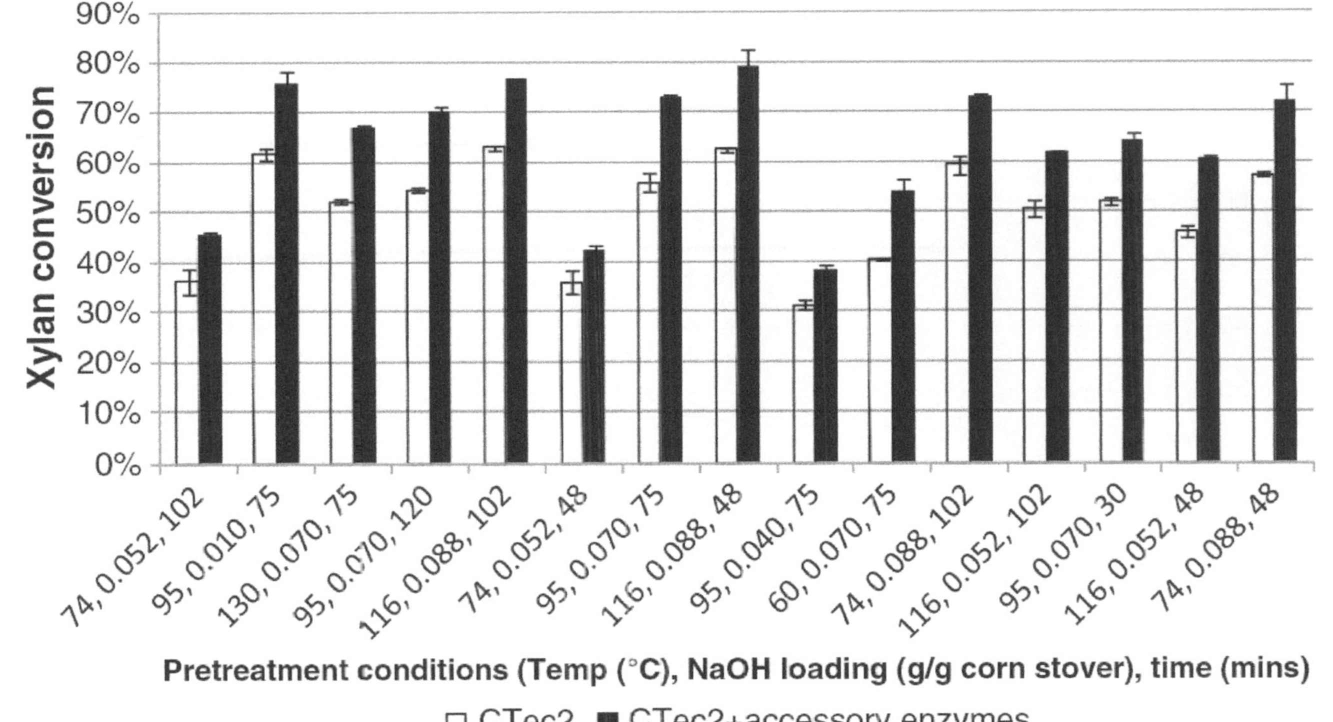


**FIGURE 1:** Improvement of xylan conversion of alkaline pretreated corn stover by accessory hemicellulases.

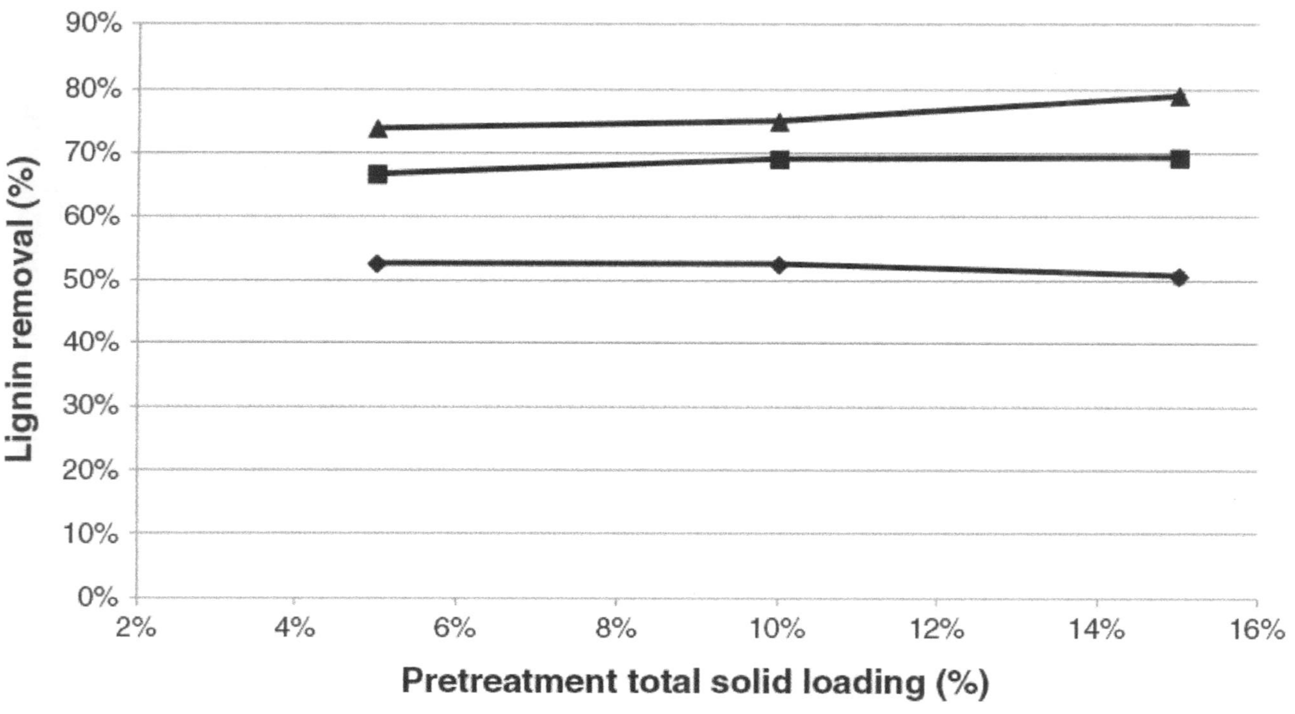


**FIGURE 2:** Effect of alkali solution concentration and biomass alkali loading on delignification of corn stover.

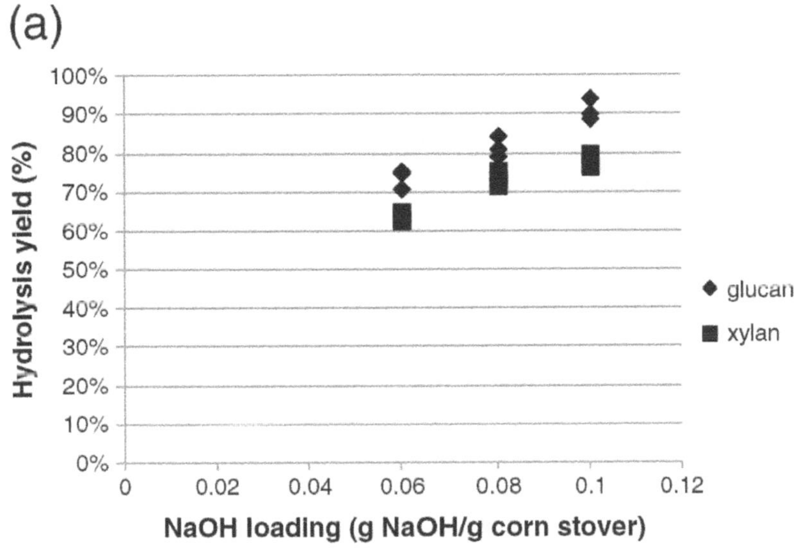


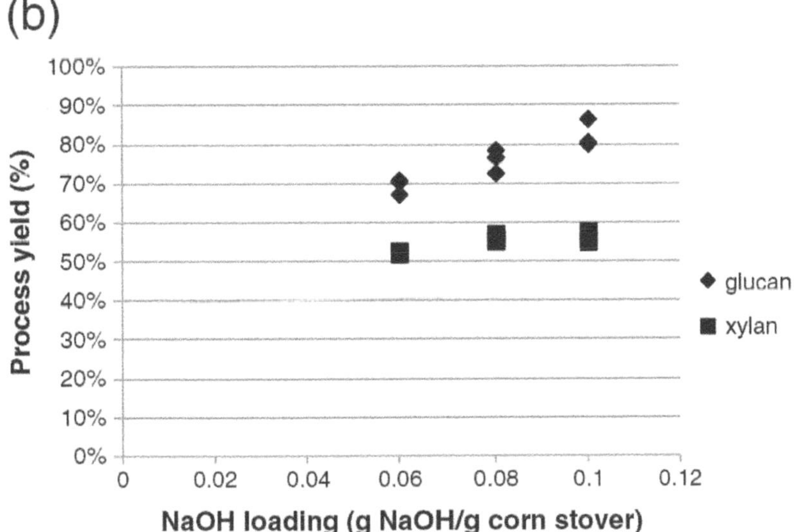


**FIGURE 3:** Relationship between biomass alkali loading and (a) glucan and xylan conversion during hydrolysis and (b) glucan and xylan conversion for the combined processes of pretreatment and hydrolysis. Hydrolysis of pretreated corn stover was conducted at 8.5% total solids level.

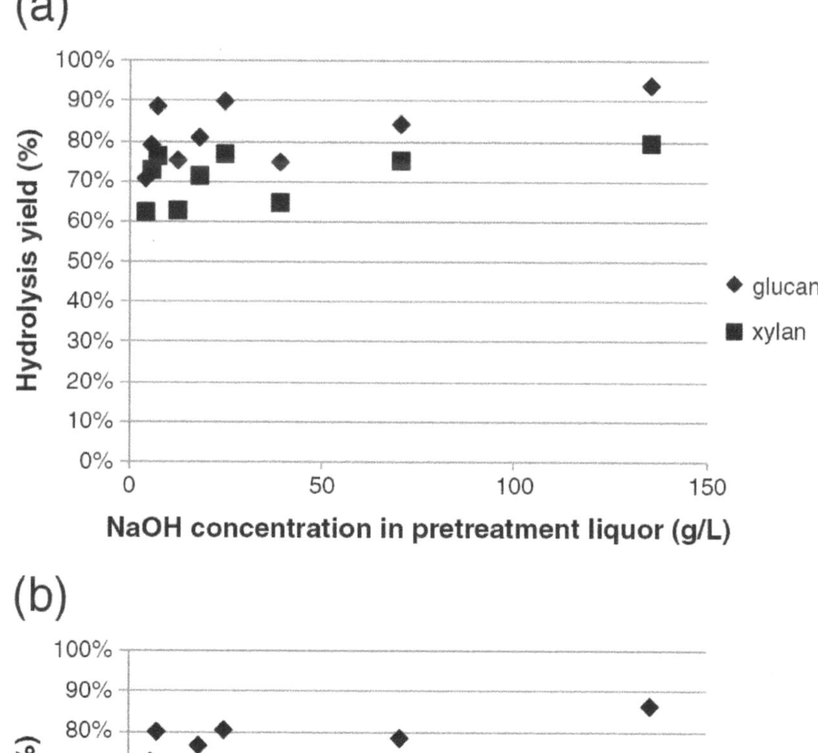


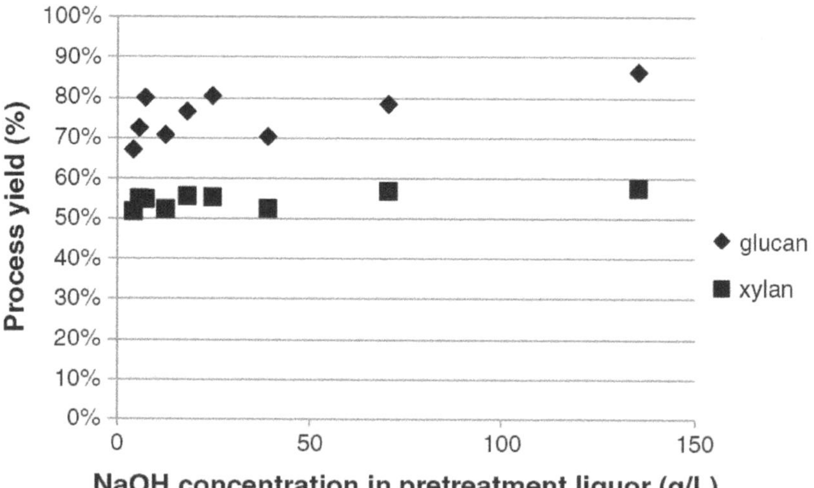


**FIGURE 4:** Relationship between alkali solution concentration and (a) glucan and xylan conversion in hydrolysis and (b) glucan and xylan conversion for the combined processes of pretreatment and hydrolysis. Hydrolysis of pretreated corn stover was conducted at 8.5 % total solids level.

Among the three fractions that constitute lignocellulosic materials, cellulose is relatively stable under alkaline conditions due to its high degree of polymerization and high crystallinity. However, hemicellulose is more labile and a significant loss of hemicellulose can occur as a consequence of the dissolution and peeling of undissolved polysaccharides. In addition, alkaline saponification of acetyl and uronic ester groups in hemicellulose proceeds readily and contributes significantly to their partial crystallization [31]. The major reactions that lead to the removal of lignin are the cleavage of α- and β- ether bonds in phenolic units and of β- ether linkages in non-phenolic units. In these reactions, NaOH participates in the ionization of C1 and/or C2 hydroxyl groups on monosaccharide rings, free phenolic hydroxyl groups, and hydroxyl groups at α- or γ- position in lignin monomers. Because NaOH is consumed as these reactions proceed [20], it stands to reason that NaOH loading on dry corn stover is more crucial in determining substrate digestibility than is alkali solution concentration. Based on this study, it can also be concluded that a reduction in chemical consumption is unlikely to be realized simply by increasing the solids loading in the pretreatment reactor.

### *6.3.5 POTENTIAL OF ALKALINE PRETREATMENT TECHNOLOGY*

Alkaline pretreatment has a unique application in many integrated biorefineries where value added products, other than ethanol, are produced from lignocellulosics. The advantage of this pretreatment technology lies in the fact that it would create a washed clean substrate which is highly digestible and rich in cellulose and xylan. After enzymatic hydrolysis, a relatively clean sugar stream (both xylose and glucose) could be obtained at reasonably high yield and economically relevant enzyme dose. For processes that are highly sensitive to impurities (inhibitors, salts), alkaline pretreatment is certainly a better choice.

Since alkali loading is the most crucial parameter affecting hydrolysis efficiency and alkali loading on dry biomass governs the digestibility of pretreated corn stover, chemical cost becomes one of the major components of the operating cost as well as total capital investment. For a biomass-to-ethanol plant that has a capacity of 50 million gallons of ethanol per year (processing 2,205 dry ton corn stover per day) [32], approximately 176 tons of NaOH is required per day for pretreatment given the fact that 0.08 g NaOH/g corn stover is needed to reach satisfactory glucan and xylan conversions. The black liquor generated during alkaline pretreatment has to be treated before it can be recycled or released to the environment. Spent chemicals from alkaline processes can be separated from biomass by washing and regenerated through well-established lime kiln technology. The black liquor is concentrated in evaporators to form concentrated black liquor (65–80% solids) which can then be combusted in a recovery boiler to generate sodium carbonate from inorganic sodium. The sodium carbonate salt is subsequently dissolved in water and sent to a causticizing plant to regenerate NaOH by contact with slaked lime. The resulting calcium carbonate is filtered off and returned to a lime kiln where burnt lime is produced, slaked and returned to the causticizer [33,34]. The estimated capital cost of such a chemical recovery system is approximately $121.7–242.1 million [35,36]. Depending on the system installed, this cost may exceed the total equipment cost ($232 million) proposed by the National Renewable Energy Laboratory for a lignocellulosic ethanol plant using dilute acid pretreatment [32]. Therefore, from an economic point of view, an alkali-based biorefinery is less economically attractive unless the cost of chemical recovery can be significantly reduced or, alternatively, low cost recovery systems can be identified and commercialized. A great opportunity to implement alkaline pretreatment process while significantly reducing capital investment would be the repurposing of existing Kraft paper mills to bioethanol plants [37]. Repurposing can take advantage of proven manufacturing infrastructure, existing skilled operating personnel, and an established biomass supply chain [38]. Another possibility would be to co-locate bio-ethanol plants with existing pulp mills that have excess capacity in their chemical recovery systems, such that black liquor produced from pretreatment could be regenerated by nearby pulp mills. However, a thorough energy and economic assessment of a given integrated biorefinery processes is still needed to determine its economical feasibility and to establish the most appropriate operating conditions.

## 6.4 CONCLUSIONS

The effect of pretreatment parameters on enzymatic hydrolysis of corn stover was investigated. It was concluded that the NaOH loading is the most dominant variable for enzymatic digestibility. Although alkali concentration (g NaOH/g pretreatment liquid) has been widely used as an indication of alkali strength in the literature, the experimental results suggest that alkali loading based on total solids (g NaOH/g dry biomass) governs the pretreatment efficiency. Supplementing cellulase with accessory enzymes such as α-arabinofuranosidases and β-xylosidase significantly improved the conversion of the hemicellulose by 6–17%. High chemical consumption can be one of the major hurdles for the commercialization of a biorefinery using alkaline pretreatment technology. However, repurposing or co-locating biorefinery with a paper mill can be a strategy to lower the operating cost as well as total capital investment.

## REFERENCES


1. Carrillo F, Lis M J, Colom X, López-Mesasa M, Valldeperas J: Effect of alkali pretreatment on cellulase hydrolyiss of wheat straw: Kinetic study. Process Biochem 2005, 40:3360-4.
2. Silverstein RA, Chen Y, Sharma-Shivappa RR, Boyette MD, Osborne J: A comparison of chemical pretreatment methods for improving saccharification of cotton stalks. Bioresource Technol 2007, 98:3000-11.
3. Chang VS, Nagwani M, Kim CH, Holtzapple MT: Oxidative lime pretreatment of high-lignin biomass - poplar wood and newspaper. Appl Biochem Biotechnol 2001, 94:1-28.
4. Kaar WE, Holtzapple MT: Using lime pretreatment to facilitate the enzymatic hydrolysis of corn stover. Biomass Bioenergy 2000, 18:189-99.
5. Chang VS, Holtzapple MT: Fundamental factors affecting biomass enzymatic reactivity. Appl Biochem Biotech 2000, 84–86:5-37.
6. Foster BL, Dale BE, Doran-Peterson JB: Enzymatic hydrolysis of ammonia-treated sugar beet pulp. Appl Biochem Biotechnol 2001, 91/93:269-82.
7. Kim TH, Kim JS, Sunwoo C, Lee YY: Petreatment of corn stover by aqueous ammonia. Bioresource Technol 2003, 90:39-47.
8. Prior BA, Day DF: Hydrolysis of ammonia-pretreated sugar cane bagasse with cellulase, beta-glucosidase, and hemicellulase preparations. Appl Biochem Biotechnol 2008, 146:151-64.

9. Saha BC, Cotta MA: Ethanol production from alkaline peroxide pretreated enzymatically saccharified wheat straw. Biotechnol Prog 2006, 22:449-53.
10. Saha BC, Cotta MA: Enzymatic saccharification and fermentation of alkaline peroxide retreated rice hulls to ethanol. Enzyme Microb Technol 2007, 41:528-32.
11. Hsu TA, In Handbook on bioethanol, production and utilization: Pretreatment of biomass. Washington DC: Taylor and Francis: Edited by Wyman, CE; 1996:179-212.
12. Zhang YHP, Lynd LR: Toward an aggregated understanding of enzymatic hydrolysis of cellulose: noncomplexed cellulase systems. Biotechnol Bioeng 2004, 88:797-824.
13. Cheng Y-S, Zheng Y, Yu C-W, Dooley TM, Jenkins BM, VanderGheynst JS: Evaluation of high solids alkaline pretreatment of rice straw. Appl Biochem Biotechnol 2010, 162:1768-84.
14. Chen BY, Chen SW, Wang HT: Use of different alkaline pretreatments and enzyme models to improve low-cost cellulosic biomass conversion. Biomass Bioenergy 2012, 39:182-91.
15. McIntosh S, Vancov T: Optimization of dilute alkaline pretreatment for enzymatic Saccharification of wheat straw. Biomass Bioenergy 2011, 35:3094-103.
16. Chen Y, Stevens MA, Zhu YM, Holmes J, Moxley G, Xu H: Reducing acid in dilute acid pretreatment and the impact on enzymatic Saccharification. J Ind Microbiol Biotechnol 2012, 39(5):691-700.
17. Sluiter A, Hyman D, Payne C, Wolfe J: Determination of total solids in biomass and total dissolved solids in liquid Process Samples. [http://www.nrel.gov/biomass/pdfs/42621.pdf]
18. Sluiter A, Hames B, Ruiz R, Scarlata C, Sluiter J, Templeton D: Determination of structural carbohydrates and lignin in biomass. [http://www.nrel.gov/biomass/pdfs/42618.pdf]
19. Zhu Y, Malten M, Torry-Smith M, McMillan JD, Stickel JJ: Calculating sugar yields in high solids hydrolysis of biomass. Bioresource Technol 2011, 102(3):2897-903.
20. Lai YZ, In Wood and Cellulose Chemistry 2nd edition: Chemical Degradation. New York: Marcel Dekker Inc: Edited by Hon DNS and Shiraishi N; 1991:455-73.
21. Gupta R, Lee YY: Investigation of biomass degradation mechanism in pretreatment of switchgrass by aqueous ammonia and sodium hydroxide. Bioresource Technol 2010, 101:8185-91.
22. Saha BC, Cotta MA: Lime pretreatment, enzymatic saccharification, and fermentation of rice hulls to ethanol. Biomass Bioenergy 2008, 32(10):971-7.
23. Pan XJ, Zhang X, Gregg DJ, Saddler JN: Enhanced enzymatic hydrolysis of steam-exploded Douglas fir wood by alkali-oxygen post-treatment. Appl Biochem Biotechnol 2004, 113–116:1103-14.
24. McDonough TJ, In Pulp Bleaching—Principles and Practice: Oxygen delignification. DW, Atlanta: TAPPI: Edited by Dence CW and Reeve; 1996:213-39.
25. Chakar FS, Ragauskas AJ: Review of current and future softwood kraft lignin process chemistry. Ind Crop Prod 2004, 20:131-41.
26. Sjöholma E, Gustafssona K, Bertholda F, Colmsjöb A: Influence of the carbohydrate composition on the molecular weight distribution of kraft pulps. Carbohyd Polym 2000, 41(1):1-7.

27. Jeoh T, Ishizawa CI, Davis MF, Himmel ME, Adney WS, Johnson DK: Cellulase digestibility of pretreated biomass is limited by cellulose accessibility. Biotechnol Bioeng 2007, 98:112-22.
28. Qing Q, Yang B, Wyman CE: Xylooligomers are strong inhibitors of cellulose hydrolysis by enzymes. Bioresource Technol 2010, 101:9624-30.
29. Buchmann SL, McCarthy AJ: Purification and cooperative acidity of enzymes constituting the xylan-degrading system of Thermomosospora fusca. Appl Environ Microbiol 1991, 57:2121-30.
30. Lee SF, Forsberg CW: Purification and characterization of an α-arabinofuranosidase from Clostridium acetobutylicum ATCC 824. Can J Microbiol 1987, 33:1011-16.
31. Rydholm SA: Pulping processes. New York: Wiley-Interscience; 1965.
32. Humbird D, Davis R, Tao L, Kinchin C, Hsu D, Aden A, Schoen P, Lukas J, Olthof B, Worley M, Sexton D, Dudgeon D: Process design and economics for biochemical conversion of lignocellulosic biomass to ethanol. Dilute-acid pretreatment and enzymatic hydrolysis of corn stover; [http://www.nrel.gov/biomass/pdfs/47764.pdf]
33. Cantrell J: Simulation of kraft black liquor gasification - A comparative look at performance and economics. TAPPI J 2001, 84(6):71-71.
34. Hamaguchi M, Vakkilainen EK: Influence of chlorine and potassium on operation and design of chemical recovery equipment. TAPPI J 2011, 10(1):33-9.
35. Fallavollita JA, Avedesiam MM, Mujumdar AS: Kraft black liquor recovery in a fluidized bed: part I - a review. Can J Chem Eng 1987, 65(5):812-17.
36. Katofsky R, Consonni S, Larson ED: A cost-benefit analysis of black liquor gasification combined cycle systems. Chicago. TAPPI Press: Engineering, Pulping & PCE&I; 2003:22-7.
37. Jin Y, Hassan J, Chang H-M: Green Liquor pretreatment of mixed hardwood for ethanol production in a repurposed kraft pulp mill. J Wood Chem Technol 2010, 30(1):86-104.
38. Fornell R, Berntsson T: Process integration study of a kraft pulp mill converted to an ethanol production plant. Part A: Potential for heat integration of thermal separation units. Appl Thermal Eng 2012, 35:81-90.

# CHAPTER 7

# SIMULTANEOUS SACCHARIFICATION AND CO-FERMENTATION FOR BIOETHANOL PRODUCTION USING CORNCOBS AT LAB, PDU AND DEMO SCALES

RAKESH KOPPRAM, FREDRIK NIELSEN, EVA ALBERS, ANNIKA LAMBERT, SUNE WÄNNSTRLIM, LARS WELIN, GUIDO ZACCHI, AND LISBETH OLSSON

## 7.1 BACKGROUND

The global $CO_2$ emissions in 2010 from fossil energy use grew at the fastest rate since 1969. The year 2010 also witnessed that the global oil production did not match the rapid growth in consumption [1]. These recent data further intensify worldwide concerns about greenhouse gas emissions and energy security for a sustained economic development. For a reduced dependence on oil from fossil reserves, use of biofuels such as bioethanol from abundantly available lignocellulosic biomass is of great interest nowadays because they will count towards meeting the mandate of 10% binding target for biofuels from renewable sources in the transport for all European member states by 2020 [2]. Along with this interest comes increased interest in commercializing ethanol production technology from inexpensive lignocellulosic feedstocks which includes wood biomass, agricultural and forestry residues, biodegradable fraction of industrial and municipal wastes. Irrespective of type, the basic structural composition of lignocellulosic biomass consists of cellulose, hemicellulose and lignin. The cellulose and hemicellulose that form the polysaccharide fraction are embedded in a recalcitrant and inaccessible arrangement [3] and therefore requires a pretreatment step to disrupt the structure and make it accessible for subsequent steps. Since lignocellulosic materials are very complex, not one pretreatment method can apply for all the materials. Several methods that are classified in to physical, physico-chemical, chemical and biological pretreatment have been investigated and an elaborate review on each of these methods has been presented by Taherzadeh and Karimi [4]. One of the most commonly used pretreatment methods is steam explosion, with the addition of $H_2SO_4$ or $SO_2$, which removes most of the hemicellulose, followed by enzymatic hydrolysis to convert cellulose to glucose [5,6].

The release of hexose and pentose sugars during pretreatment and enzymatic hydrolysis is often accompanied by liberation of compounds such as furans, weak organic acids and phenolics compounds [7] that inhibits growth, ethanol yield and productivity of fermenting microorganism, *Saccharomyces cerevisiae* [8-10]. Traditionally and industrially relevant microorganism for ethanol fermentation is *S. cerevisiae*, but its inability to consume pentose sugars like xylose and arabinose has led to intensive research on metabolic and evolutionary engineering to develop strains that can tolerate high concentration of inhibitors and ferment xylose and arabinose [11-15]. However, it has been shown that recombinant *S. cerevisiae* strain utilizing pentose sugar may lose its xylose consuming ability in a long term evolutionary engineering for inhibitor tolerance [15]. Consequently, to ensure that all properties are retained during evolutionary engineering requires careful design of the selection pressure.

The enzymatic hydrolysis can be performed simultaneously with the co-fermentation of glucose and xylose in a process referred to as simultaneous saccharification and co-fermentation (SSCF). Besides reduced capital cost [16], SSCF process offers several advantages which include continuous removal of end-products of enzymatic hydrolysis that inhibit cellulases or β-glucosidases [17] and higher ethanol productivity and yield than separate hydrolysis and fermentation [18,19]. It is required to operate a SSCF process at high content of water-insoluble-solids (WIS) to achieve high concentrations of ethanol. However, it has been shown that at high WIS content ethanol yield was decreased due to increased mass transfer resistance and inhibitors concentration [20]. Operating SSCF in a fed-batch mode at high WIS content not only assists ease of mixing and produces high ethanol concentrations [21] but also offers a possibility to maintain glucose at low levels allowing efficient co-fermentation of glucose and xylose [22]. Lowering of glucose concentration can be achieved by initially fermenting free hexoses before adding enzymes to a SSCF process in a concept referred as prefermentation enhanced xylose uptake irrespective of batch or fed-batch SSCF [23]. These flexibilities offered by a SSCF process makes it a promising process option for bioethanol production from lignocellulosic materials.

The heterogeneity of raw materials together with a variety of pretreatment methods, lack of detailed understanding of dynamic changes of substrate during enzymatic hydrolysis and unavailability of microorganisms that can ferment a wide range of carbohydrates and can tolerate high concentrations of inhibitors produced from pretreated biomass makes SSCF a highly researched area yet to reach the commercial status. There come additional technical challenges when operating at larger scales which include longer times to add material into the reactor, longer mixing times and therefore concentration gradients are inevitable. On-site propagation of yeast in large volumes is needed which also increases the probability of contamination since lignocellulosic ethanol plants will not employ aseptic operating conditions. Moving cellulosic ethanol technology from the laboratory to a commercial scale biorefinery is an expensive proposition and requires process data at sufficient scale to obtain engineering and process guarantees. Some prominent players that are working on this proposition include Chemtex, Inbicon, DuPont cellulosic ethanol, POET-DSM advanced biofuels, Iogen, Abengoa Bioenergy, Mascoma and SEKAB. A category of feedstock that is of considerable interest is corn derived residues due to that it is inexpensive and available in abundance. Corncob is an agricultural residue and a byproduct of corn production. Currently, 12.1 billion tons and 120 million tons of corn are being produced in the US and China, respectively. About 70 million metric tons of corncobs are available annually accounting only from the US and China markets [21,24]. Removal of corncobs from the agricultural grounds does not contribute to decreased soil organic matter since corncobs are low in nutrients.

In this work, a xylose fermenting *S. cerevisiae* strain was used in SSCF of pretreated corncobs with the objective of determining suitable conditions for co-consumption of glucose and xylose. Fed-batch mode of SSCF in combination with prefermentation was investigated at high WIS content. To validate the designed SSF process and verify the reproducibility at different scales, the process was scaled up from lab conditions to process development unit (PDU) (30 liters) and further to demo scale (10 $m^3$).

## 7.2 RESULTS AND DISCUSSION

The SSCF concept is one of the interesting process options and the potential of such process for biological conversion of lignocellulosic raw materials to bioethanol in large scales has to the best of our knowledge not been reported previously. A promising xylose consuming strain of *S. cerevisiae* was selected from screening seven different recombinant *S. cerevisiae* strains. The glucose influence on xylose consumption of the selected strain was investigated by model SSCF with glucose or hydrolysate feed. The potential of fed-batch SSCF process in combination with prefermentation was finally demonstrated in 10 $m^3$ demo scale bioreactors.

### *7.2.1 SCREENING AND SELECTION OF* S. CEREVISIAE *STRAIN*

#### *7.2.1.1 ANAEROBIC FERMENTATION OF CORNCOB HYDROLYSATE*

The seven different *S. cerevisiae* strains (Table 1) were evaluated on their fermentation performance in corncobs hydrolysate in shake flasks equipped with glycerol loops. Since, xylose constitutes a significant proportion of monosaccharides in corncobs hydrolysate xylose consumption and xylitol yield together with ethanol yield were determined (Figure 1) and used as parameters for strain selection. The strains, AD2-10, KE6-12 and RHC-15, RHD-15 displayed similar ethanol concentration, ethanol yield and performed better than their respective parental strains with regard to xylose consumption. The strain RHD-15 displayed the highest ethanol yield and xylose consumed. The strain KE6-12 stands alone among other strains in xylitol yield producing the lowest amount of xylitol from consumed xylose. Even though the screening revealed significant differences in fermentation of hydrolysate, it is important to evaluate the microbial performance in the whole slurry in a SSCF process. The strains RHD-15 and KE6-12, due to their high ethanol and low xylitol yields, were therefore, selected as the preferred strains for subsequent investigations in the SSCF process.

**TABLE 1:** *S cerevisiae* strains used in this study

| Parental strain | Evolved strain |
|---|---|
| KE4-22 | AD2-10 |
| KE6-12 | |
| AD1-13 | RHA-15 |
| | RHC-15 |
| | RHD-15 |

## *7.2.2 SSCF OF CORNCOBS WHOLE SLURRY*

To assess the fermentation performance, the strains RHD-15 and KE6-12 were evaluated in a base case batch SSCF of corncobs whole slurry at 7.5% WIS for ethanol production. During the SSCF process, the glucose concentration was quickly reduced to less than 1 g $l^{-1}$ within 10 h and thereafter, it was maintained at this level throughout the process (Figure 2a & 2b). Immediately after inoculation, both the strains started to consume xylose for a period of 72 h after which the xylose concentration in the reactor started to level off. After 96 h, the strain KE6-12 had consumed 37%

of the available xylose and 30% of the consumed xylose was converted to xylitol (2.8 g $l^{-1}$) whereas, the strain RHD-15 had consumed 42% of the available xylose and 66% of the consumed xylose was converted to xylitol (6.4 g $l^{-1}$). An ethanol concentration of 21.9 and 21.5 g $l^{-1}$ were achieved corresponding to a yield of 0.28 g $g^{-1}$ and 0.27 g $g^{-1}$ based on total available sugars for the strains KE6-12 and RHD-15, respectively. In comparison to RHD-15, strain KE6-12 consumed marginally lower amount of xylose but, produced 56% less xylitol. Since, bioconversion of xylose to ethanol is one of the predominant requirements for an economical lignocellulosic bioethanol production, further fermentation and SSCF experiments were carried out with the strain KE6-12, unless otherwise stated. In screening experiments using corncobs hydrolysate, the strain RHD-15 performed relatively better than KE6-12, however, in SSCF using corncobs whole slurry both the strains resulted in similar ethanol yields and RHD-15 was clearly outperformed by KE6-12 in lower xylitol yields. The differences in results from the two screening experimental systems could be attributed to the differences in experimental conditions. The effect of pH on xylose consumption by a recombinant xylose utilizing *S. cerevisiae* has been previously shown that increasing the pH from 5.0 to 5.5 resulted in 46% increase in xylose consumption rate [25]. Screening using corncobs hydrolysate in shake flasks were performed at an initial pH 6.0 and 30°C which clearly resulted in higher xylose consumption compared to screening in SSCF where the pH was controlled at 5.0 and sub-optimal temperature of 35°C. It should be noted that often strain engineering and development results in a numerous strains and a high throughput screening of these strains in SSCF process in shake flasks could be impractical due to difficulties in mixing at high WIS content. The difference in two screening systems illustrate the importance of choice of experimental setup and conditions for screening to be as close as possible to that used in the actual experiments.

### *7.2.3 MODEL SSCF AS A TOOL TO DESIGN THE SSCF PROCESS*

In order to understand the effect of glucose on xylose consumption and to optimally design the SSCF process with effective xylose consumption a model SSF [26] with prefermentation [23] was performed. A model SSCF is a SSCF process without the addition of enzymes but fed with pure glucose solution or hydrolysate to the reactor mimicking the release of glucose during enzymatic hydrolysis of cellulose. Prefermentation is a concept where initially available free glucose was fermented before starting the feed.

### 7.2.4 LAB SCALE

Model SSCF in lab scale was performed in corncobs hydrolysate with a feed of 100 g $l^{-1}$ glucose solution at a constant rate. A glucose feed corresponding to the amount of glucose from 7.5% WIS was started after 2 h of inoculation and terminated at 96 h. During the prefermentation period of 2 h, the glucose concentration was reduced to nearly 0 g $l^{-1}$ and maintained at this level until 72 h (Figure 3a). Immediately after prefermentation, xylose was rapidly consumed until 48 h and thereafter, the concentration started to level off. After 96 h, 79% of xylose was consumed and 37% of consumed xylose was converted to xylitol (6.4 g $l^{-1}$). An ethanol concentration of 31.2 g $l^{-1}$ was achieved corresponding to a yield of 0.38 g $g^{-1}$ based on total available sugars (75% of the theoretical yield). Higher ethanol concentration in model SSCF compared to batch SSCF may possibly be due to higher xylose consumption and also points to a direction that cellulose fibers were not completely hydrolyzed in batch SSCF to yield similar ethanol concentrations as that obtained in model SSCF.

### 7.2.5 PDU SCALE

A model SSCF in PDU scale similar to lab scale model SSCF was performed with a feed of hydrolysate from enzymatic hydrolysis. In order to completely hydrolyze cellulose fibers, enzymatic hydrolysis of solid fraction of corncobs slurry was carried out at 50°C with enzyme loading of 6 FPU $gWIS^{-1}$. The liquid fraction remaining after filtering the enzymatically hydrolyzed solid fraction was used as a feed. Prefermentation in corncobs hydrolysate was initiated by adding yeast and an enzyme solution corresponding to 3 FPU $gWIS^{-1}$. The glucose was rapidly consumed during the initial 10 h of prefermentation, reduced to near 0 g $l^{-1}$ and maintained at this level until 24 h (Figure 3b). A sharp increase in xylose concentration was observed immediately after the addition of enzyme solution indicating the hydrolysis of xylan. Thereafter, the xylose was consumed quickly for 10 h, however, when the glucose was completely consumed the xylose consumption dramatically slowed down. This indicates that the consumption of glucose with maintained low concentration of glucose is beneficial for efficient xylose consumption. Previous study on fed-batch SSCF using xylose rich wheat straw has highlighted that maintaining low levels of glucose consequently increased consumed xylose twice as compared to a batch SSCF [26]. It also has been discussed that presence of glucose at high concentrations may inhibit xylose uptake due to competition for transporters [27,28]. Feeding of the liquid fraction from enzymatic hydrolysis was started after 24 h of prefermentation and was maintained for 24 h corresponding to a final WIS content of 7.5%. The glucose concentration gradually increased during the 24 h feeding phase until 48 h and thereafter was completely consumed. The xylose started to accumulate when glucose concentration reached a peak of 10 g $l^{-1}$ and thereafter no xylose was consumed and no change in ethanol concentration was observed indicating the end of fermentation. The increase in xylose concentration after 50 h could be due to enzymatic hydrolysis of xylan. After 96 h, an ethanol concentration of 32 g $l^{-1}$ was produced corresponding to 77% of the theoretical yield based on available sugars. This ethanol yield is well in accordance with ethanol yields of model SSCF in lab scale. Evidences from model SSCF with prefermentation clearly suggest that fermentation of initial free glucose and thereafter, maintenance of glucose at low levels are crucial factors for efficient xylose consumption.

### 7.2.6 FED-BATCH SSCF

#### 7.2.6.1 PDU SCALE

It was also possible to achieve similar ethanol yields in a fed-batch SSCF as that in the model SSCF. Fed-batch SSCF in PDU was carried out using the whole slurry with a total WIS of 7.9%. Initially, prefermentation was carried out for 2 h by adding 6 g dry cell weight $l^{-1}$ of yeast from cell suspension. To maintain glucose concentrations at a minimum level in the reactor and thereby facilitate effective xylose consumption, a strategy to add enzyme solution at multiple time points to ensure controlled release of glucose was investigated. An enzyme solution corresponding to 3 FPU $gWIS^{-1}$ was added at 2 h, 24 h, and 48 h. The glucose concentration was maintained around 5 g $l^{-1}$ until 72 h before it was completely consumed at 96 h (Figurc 4). A steady co-consumption of glucose and xylose was observed throughout the SSCF. After 96 h, 50% of the available xylose was consumed producing xylitol with a concentration of only 1.5 g $l^{-1}$. An ethanol concentration of 32 g $l^{-1}$ was achieved corresponding to 76% of the theoretical ethanol yield based on available sugars.

In a commercial bioethanol production process it is desirable that the substrate load is higher than 7% WIS to achieve 4% (w/v) ethanol concentration to yield a subsequent economical distillation process [29]. It has been shown that working at high WIS content increases the concentration of inhibitors and results in inhibition of yeast and lower ethanol yields [26]. Therefore, along with split addition of enzymes, fed-batch SSCF at higher WIS was investigated with split addition of substrate resulting in lower amount of inhibitors for each addition. Fed-batch SSCF experiment was performed with a corncobs slurry addition at 0 h, 5 h, 27 h and 49 h to a total final WIS of 10%. Enzyme solution was added at multiple time points of 2 h, 24 h, 48 h, 72 h and 96 h to a total of 15 FPU $gWIS^{-1}$. During the first 2 h of prefermentation, the glucose concentration was reduced to nearly 0 g $l^{-1}$ and reached around 5 g $l^{-1}$ after the first addition of enzyme (Figure 5a). The glucose concentration was then maintained below 5 g $l^{-1}$ throughout the SSCF process. The xylose was co-consumed along with glucose for more than 100 h. At the end of fed-batch SSCF, 55% of the available xylose was consumed and 11% of the consumed xylose was converted to xylitol (3.4 g $l^{-1}$). An ethanol concentration of 47 g $l^{-1}$ was achieved corresponding to a yield of 0.35 g $g^{-1}$ based on total available sugars (69% of the theoretical yield). Higher xylose consumption and ethanol yield at 10% WIS clearly suggest that the combination of prefermentation and a feed of

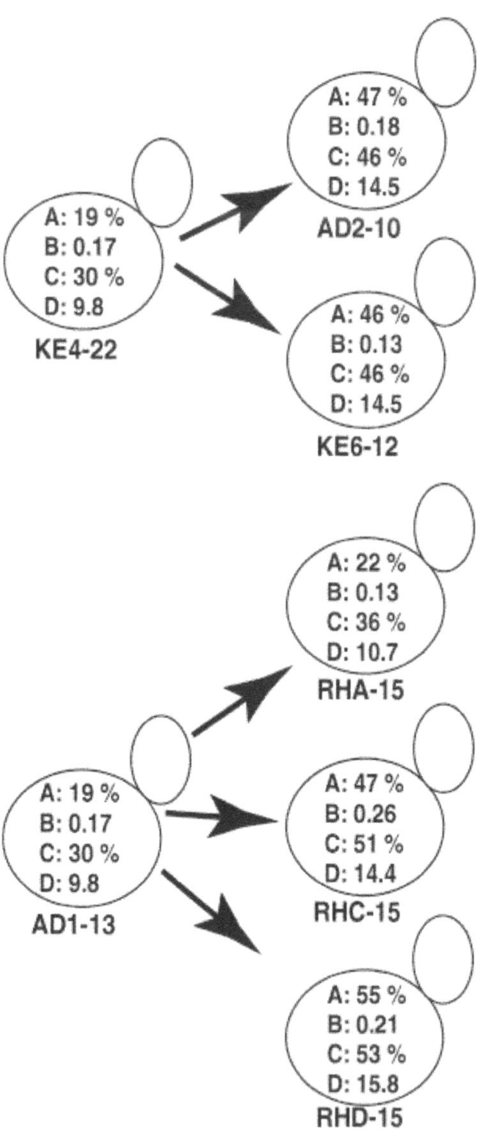


**FIGURE 1:** Screening of *S. cerevisiae* strains in corncobs hydrolysate. Xylose consumption, xylitol and ethanol yields, ethanol concentration in corncobs hydrolysate after 96 h of fermentation in anaerobic shake flasks. KE4-22 is the parental strain of AD2-10 and KE6-12. AD1-13 is the parental strain of RHA-15, RHC-15 and RHD-15. A: xylose consumed (%), B: xylitol yield on consumed xylose (g $g^{-1}$), C: ethanol yield (%, based on maximum theoretical ethanol yield on available glucose and xylose), D: ethanol concentration (g $l^{-1}$) at the end of 96 h.

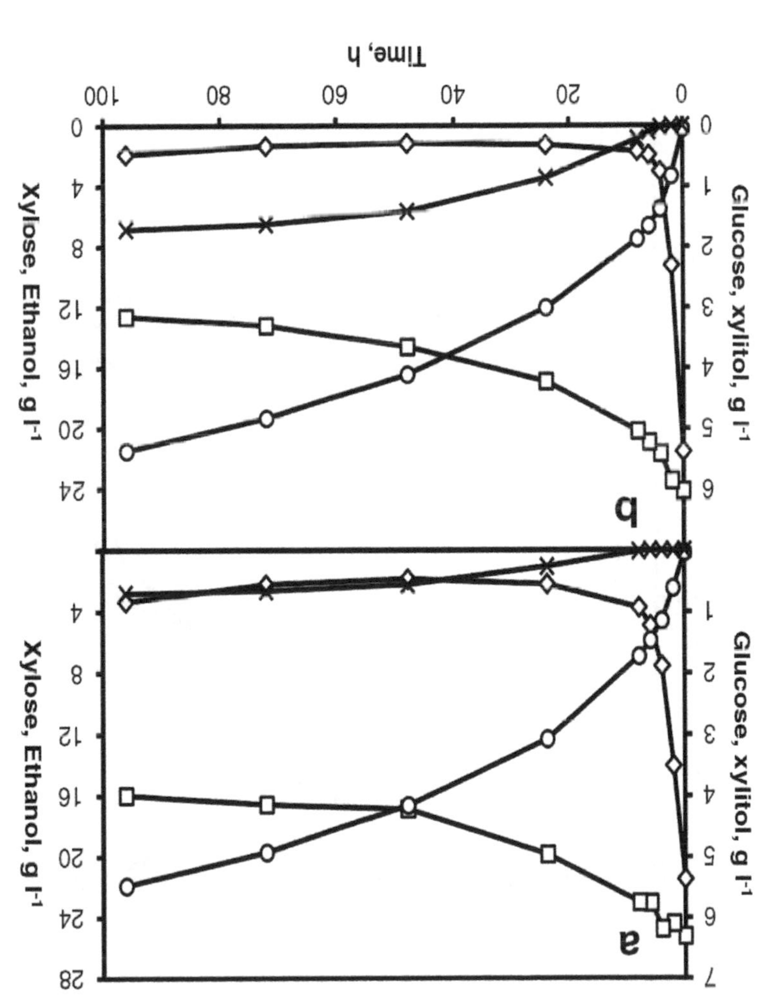


**FIGURE 2:** Screening of *S. cerevisiae* strains in corncobs whole slurry. Glucose (diamonds) and xylose (squares) consumption, ethanol (circles) and xylitol (crosses) production in SSCF at 7.5% WIS content, 5 g $l^{-1}$ of yeast loading and 5 FPU gWIS-1 of enzyme loading using KE6-12 (a) and RHD-15 (b).

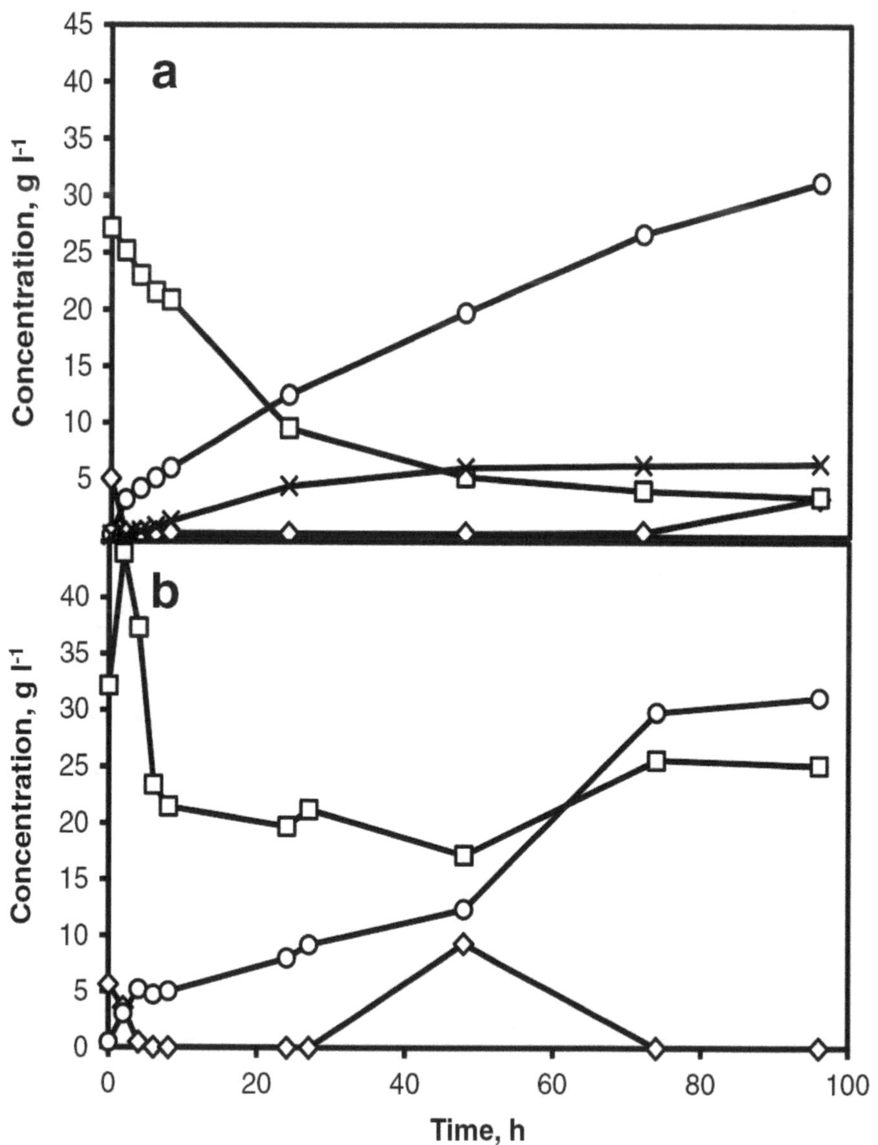


**FIGURE 3:** Model SSCF. Glucose (diamonds) and xylose (squares) consumption, ethanol (circles) and xylitol (crosses) production in a model SSCF in corncobs hydrolysate with 5 g $l^{-1}$ of KE6-12 at lab scale using a feed of glucose solution (a) and at PDU using a feed of liquid fraction after enzymatic hydrolysis (b). Amount of glucose fed is corresponding to 7.5% WIS content.

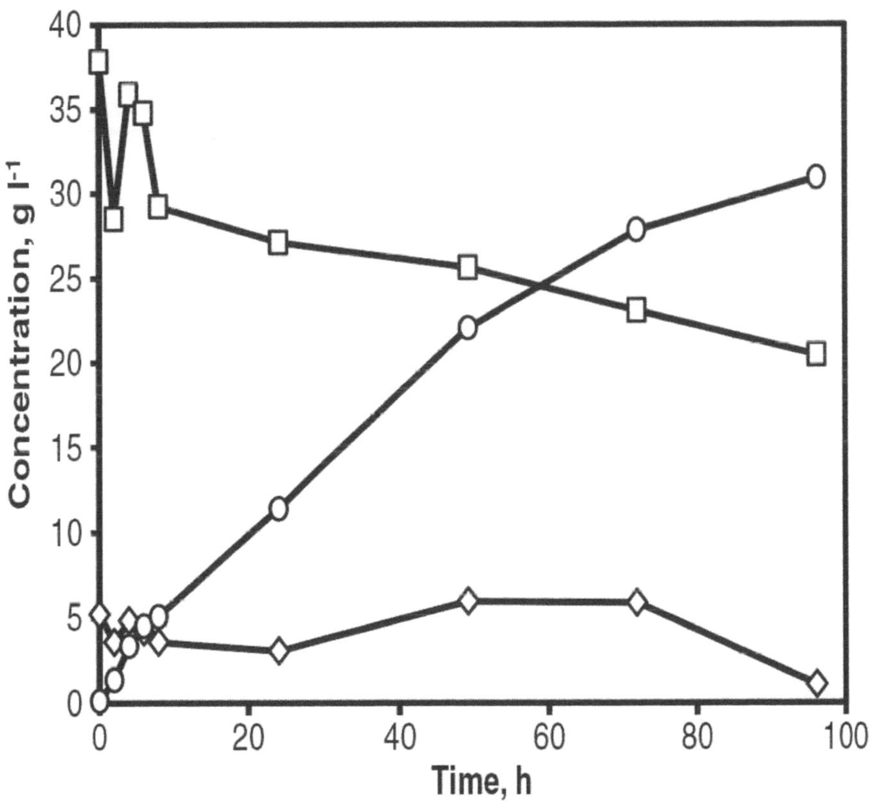


**FIGURE 4:** Fed-batch SSCF with prefermentation and split addition of enzyme at PDU. Glucose (diamonds) and xylose (squares) co-consumption, ethanol (circles) and xylitol (crosses) production using corncobs whole slurry at 7.9% WIS, 6 g $l^{-1}$ of KE6-12 with 3 FPU gWIS-1 of enzyme loading at each time points of 2 h, 24 h and 48 h.

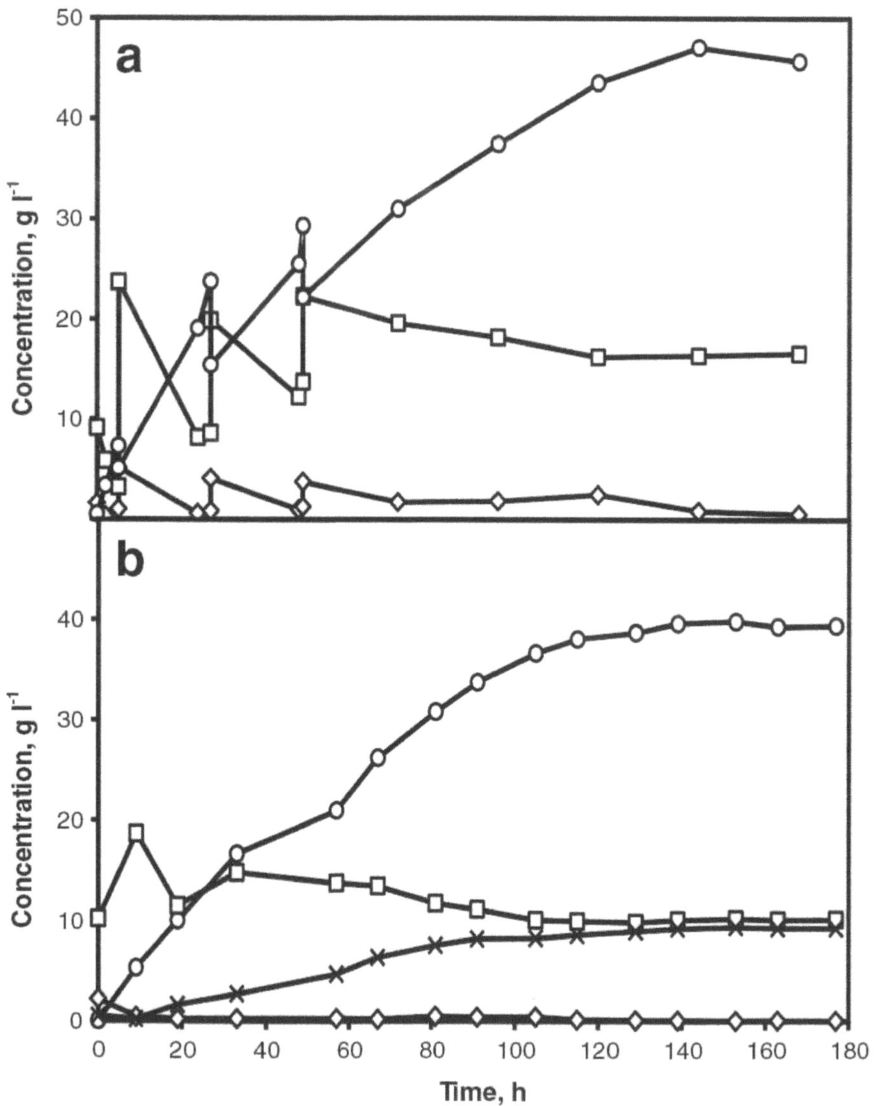


**FIGURE 5:** SSCF with prefermentation and fed-batch addition of substrate and enzyme. Glucose (diamonds) and xylose (squares) co-consumption, ethanol (circles) and xylitol (crosses) production using corncobs whole slurry at 10.5% WIS, 5 g l-1 of KE6-12 and 15 FPU gWIS-1 of enzyme loading. Split addition of substrate at 0 h, 5 h, 27 h, 49 h and enzyme solution at 2 h, 24 h, 48 h, 72 h, 96 h in PDU (a). Fed-batch addition of substrate for 48 h and split addition of enzyme solution at 2 h, 24 h, 48 h, 72 h, 96 h in demo scale (b).

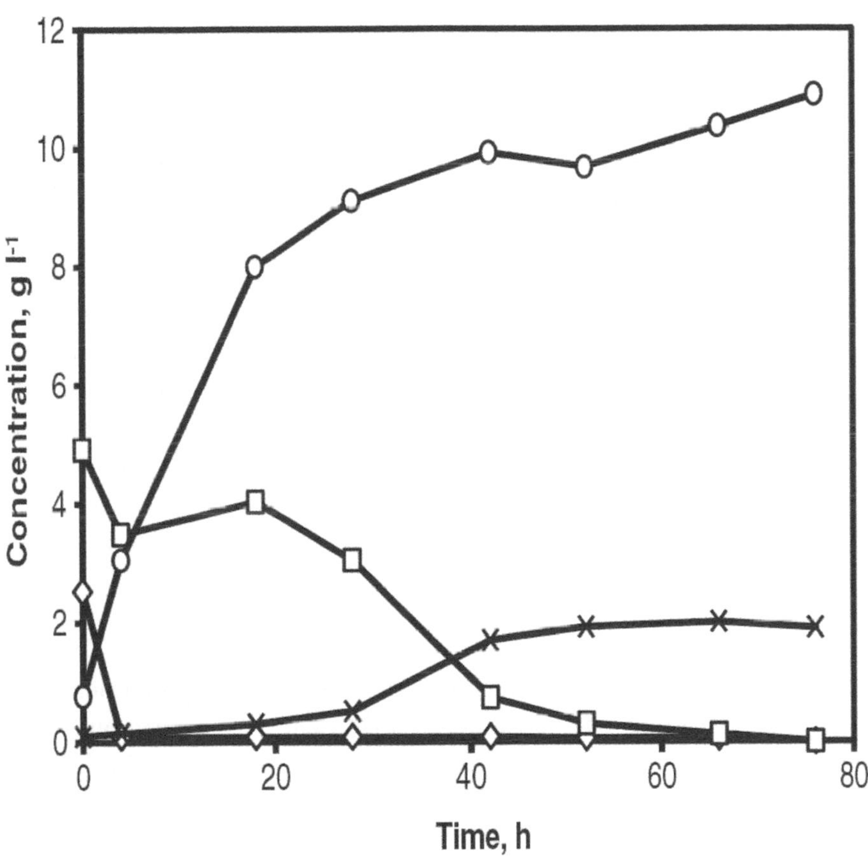


**FIGURE 6:** Fed-batch fermentation in corncobs hydrolysate at demo scale. Glucose (diamonds) and xylose (squares) consumption, ethanol (circles) and xylitol (crosses) production using 5 g l-1 of KE6-12. Corncobs hydrolysate corresponding to 6% WIS content was fed for 24 h.

enzymes and substrates as one of the possible SSCF strategies for demo scale execution.

### 7.2.7 DEMO SCALE

#### 7.2.7.1 XYLOSE FERMENTATION IN HYDROLYSATE

A time span of 24 to 48 h was used to pump a substrate in to the demo scale reactor of 10 $m^3$. In order to address the potential of the strain KE6-12 on xylose consumption, fed-batch fermentation of corncobs hydrolysate corresponding to a WIS content of 6% was evaluated. The corncobs hydrolysate was fed into the reactor for 24 h. The glucose concentration was reduced to nearly 0 g $l^{-1}$ within 5 h and all xylose was consumed in 76 h (Figure 6). Only 10.6% of the consumed xylose was converted to xylitol (2.0 g $l^{-1}$). An ethanol concentration of 10.9 g $l^{-1}$ was achieved corresponding to a yield of 0.46 g $g^{-1}$ on total available sugars (90% of the theoretical yield).

#### 7.2.7.2 FED-BATCH SSCF

A fed-batch SSCF with substrate and enzyme feed and prefermentation for 2 h similar to the one performed at PDU scale was carried out in the demo scale. The corncobs slurry was fed into the reactor for 48 h resulting in a total WIS of 10.5%. Enzyme solution was added at five different time points, 2 h, 24 h, 48 h, 72 and 96 h corresponding to a total of 15 FPU $gWIS^{-1}$. The glucose concentration was quickly reduced to nearly 0 g $l^{-1}$ within 10 h and thereafter, it was maintained at low concentration throughout the process (Figure 5b). Co-consumption of xylose and glucose was observed for more than 100 h similar to the fed-batch SSCF at PDU scale. After 150 h, 65% of the available xylose was consumed and 24.7% of the consumed xylose was converted to xylitol (9.3 g $l^{-1}$). Surprisingly, in comparison to fed-batch SSCF at PDU scale, higher amount of xylose was consumed in demo scale, however, also higher amount of xylitol was produced. An ethanol concentration of 39.8 g $l^{-1}$ was achieved corresponding to a yield of 0.29 g $g^{-1}$ based on total available sugars (58% of the theoretical yield). More controlled conditions of temperature, pH and homogenous mixing in PDU scale resulted in higher final ethanol concentration and yield compared to demo scale conditions with higher mass transfer limitations.

## 7.3 CONCLUSION

The performance of recombinant xylose utilizing *S. cerevisiae* strains varied in two different screening experiments, which highlights the importance of experimental setup and conditions for screening of strains to be highly similar to that of the actual experiments. The choice of the strain KE6-12 seems well justified when xylose was completely consumed at demo scale during the fermentation of hydrolysate with 90% of the theoretical ethanol yield. Different feeding profiles of glucose and its influence on xylose consumption was studied using model SSCF and it proved to be a valuable tool to optimally design a SSCF process. The potential of the fed-batch SSCF process is more vivid and we demonstrated that with prefermentation and substrate and enzyme feed it is possible to produce ethanol from corncobs as high as 40 g $l^{-1}$ and more, with relatively high WIS content at both 30 l (PDU scale) and 10 $m^3$ (demo scale). Using such a strategy it was possible to maintain low levels of glucose concentration, which facilitated co-consumption of glucose and xylose. We also confirmed that the results of fed-batch SSCF were similar at PDU and demo scales and the experimental system was reproducible at both the scales. However, at higher WIS content an optimal feeding strategy is required to ferment all xylose and avoid glucose accumulation.

## 7.4 MATERIALS AND METHODS

### *7.4.1* SACCHAROMYCES CEREVISIAE *STRAINS*

The seven *S. cerevisiae* strains used in this study (Table 1) were developed by a combination of different evolutionary engineering strategies and random mutagenesis (Albers et al., manuscript in preparation) on *S. cerevisiae* TMB 3400 [30] that harbours the xylose reductase gene and xylitol dehydrogenase from *Scheffersomyces stipitis* (formerly known as *Pichia stipitis*) and endogenous xylulokinase overexpressed. All the strains were stored at −80°C in culture aliquots containing 20% sterile glycerol. Volumes of 100 μl from the vials were used to inoculate precultures.

## 7.4.2 MEDIA

### 7.4.2.1 CORNCOBS SLURRY

Corncobs slurry with a water-insoluble-solids (WIS) content of 15% was received from SEKAB-E-Technology AB (Örnsköldsvik, Sweden) and was stored at −20°C. The corncobs were pretreated at 185°C for 5 min with 0.6% dilute sulfurous acid ($SO_2$ in water). Two batches were pretreated and the composition of which are presented in the Table 2. Batch 1 was used for screening and selection experiments. Batch 2 was used in the demo scale experiments. The corncobs hydrolysate (liquid fraction of corncobs slurry), pH adjusted to 5.0 with 10 M NaOH, was used in yeast cell cultivations when required.

**TABLE 2:** Composition of the pretreated corncobs

| Content in solid fraction (% of WIS) | | | Content in liquid fraction (g $l^{-1}$) | | |
|---|---|---|---|---|---|
| | Batch 1 | Batch 2 | | Batch 1 | Batch 2 |
| Glucan | 66.9 | 61.4 | Glucose* | 17.0 | 15.0 |
| Mannan | 0 | 0 | Mannose* | 0 | 0 |
| Galactan | 0 | 2.9 | Galactose* | 6.5 | 0 |
| Xylan | 5.8 | 8.2 | Xylose* | 79.4 | 74.4 |
| Arabinan | 1.0 | 1.4 | Arabinose* | 11.8 | 14.0 |
| Lignin | 27.6 | 28.9 | HMF | 2.0 | 1.9 |
| | | | Furfural | 3.8 | 4.0 |
| | | | Acetic acid | 10.4 | 8.3 |

**Both monomeric and oligomeric form is included.*

#### 7.4.2.2 MOLASSES

Molasses was obtained from SEKAB-E-Technology AB (Örnsköldsvik, Sweden) and was either used alone or mixed with liquid fraction of corncobs slurry for cultivating yeast cells that was then used for SSCF experiments.

#### 7.4.2.3 MINIMAL MEDIUM

The initial inoculum for screening yeast strains and SSCF experiments were cultivated in minimal medium containing 20 g $l^{-1}$ glucose and xylose, respectively and enriched with salts, two folds of vitamins and trace elements according to Verduyn et al.[31]. The pH of the medium was set to 6.0 with 1 M NaOH for all shake flask cultivations.

### 7.4.3 CULTIVATION OF YEAST

In order to improve inhibitor tolerance by adaptation, yeast cells were grown briefly in presence of corncobs hydrolysate during the cultivation for screening and SSCF experiments (as described below). It has been previously shown that the cultivation procedure of yeast significantly influences the performance in SSF and small-scale fermentations of hydrolysate liquor [32].

The precultures for screening *S. cerevisiae* strains for ethanol production were cultivated in 150 ml shake flasks with 50 ml of minimal medium. The cultures were inoculated to an initial OD650 of 0.005, incubated at 30°C on an orbital shaker at 180 rpm. After 18 h of incubation, corncobs hydrolysate supplemented with 23.5 g $l^{-1}$ $(NH_4)_2SO_4$, 3.0 g $l^{-1}$ $KH_2PO_4$ and 2.25 g $l^{-1}$ $MgSO_4 \cdot 7H_2O$ was added to the preculture cultivation flask to a final volume of 35% (v/v) and incubated for another 24 h.

The yeast cells for SSCF experiments in lab and PDU scales were cultivated in aerobic batch on molasses, followed by an aerobic fed batch on corncobs hydrolysate and molasses. In the demo scale molasses was used as the medium in aerobic batch and fed batch cultivation. The yeast strain was inoculated in to 50 ml (lab scale), 150 ml (PDU) of minimal medium contained in a 150 ml (lab scale) and 300 ml (PDU) shake flasks, respectively; incubated at 30°C on an orbital shaker at 180 rpm for 24 h. Aerobic batch cultivation was performed in 50 g $l^{-1}$ molasses supplemented with 23.5 g $l^{-1}$ $(NH_4)_2SO_4$, 3.0 g $l^{-1}$ $KH_2PO_4$, 2.25 g $l^{-1}$ $MgSO_4 \cdot 7H_2O$, 33 μg $l^{-1}$ biotin, 125 ppm vitahop (Betatech Gmbh, Schwabach, Germany) (to suppress bacterial growth) and 0.5 ml $l^{-1}$ antifoam. The yeast cultivation was carried out in 3.6 l Infors HT-Labfors bioreactor (lab scale), 30 l bioreactor (PDU) and 10 $m^3$ bioreactor (demo scale). The cultivation was initiated in the bioreactors by adding 50 ml or 150 ml of minimal medium culture to a working volume of 500 ml (lab scale) or 1.5 l (PDU) of molasses medium, respectively. The cultivation was carried out until all sugars are consumed which was indicated by $CO_2$ evolution in the gas-out and dissolved oxygen concentration in the culture. Upon depletion of sugars in batch phase, a feed solution containing corncobs hydrolysate and molasses was fed linearly for 20 h to a final volume of 1.5 l (lab scale) or 4.5 l (PDU). The concentration of corncobs hydrolysate in the feed solution was same as that of concentration of corncobs slurry in the SSCF experiments. Molasses concentration was 100 g $l^{-1}$ in the feed solution. The stirrer speed during the batch phase in lab scale was 700 rpm and increased linearly to 1000 rpm during the fed batch phase; whereas, the stirrer speed was maintained at 700 rpm throughout the cultivation in PDU scale; aeration rate was maintained at 1 vvm and the pH was maintained at 5.0 by automatic addition of 2 M NaOH.

After the cultivation, cells were harvested by centrifugation for 8 min at 4°C, 1800 g and the cell pellet was resuspended in 0.9 % sterile NaCl solution to yield a cell suspension with a dry weight of 80 g $l^{-1}$.

### 7.4.4 ANAEROBIC FERMENTATION IN SHAKE FLASKS

The pH of corncobs hydrolysate was set to 6.0, supplemented with 0.5 g $l^{-1}$ $(NH_4)_2HPO_4$, 125 ppm vitahop and filter sterilized using 0.45 μm cellulose acetate filter. This fermentation medium was inoculated using the cell suspension to reach a yeast concentration of 3 g dry cell weight $l^{-1}$. The fermentations were carried out in 50 ml working volume in 100 ml shake flasks fitted with glycerol loops providing anaerobic condition. The flasks were incubated at 30°C on an orbital shaker at 180 rpm for 96 h and samples were withdrawn for $OD_{650}$ measurement and extracellular metabolite analysis. Possible contamination during the shake flask fermentation was checked by ocular inspection in microscope.

**TABLE 3:** Brief list of SSCF experiments carried out in lab, PDU and demo scales

| Mode of operation_Scale | Initial Pre-fermentation time, h | Amount of solids, %WIS | Strain | Total cell amount, g $l^{-1}$ | Total enzyme amount, FPU $gWIS^{-1}$ |
|---|---|---|---|---|---|
| Batch SSCF_Lab | None | 7.5 | RHD-15 | 5 | 5 |
| Batch SSCF_Lab | None | 7.5 | KE6-12 | 5 | 5 |
| Fed-batch Model SSCF_Lab1 | 2 | 7.5 | KE6-12 | 5 | None |
| Fed-batch Model SSCF_PDU2 | 24 | 7.5 | KE6-12 | 5 | None* |
| Fed-batch SSCF_PDU | 2 | 7.9 | KE6-12 | 6 | 9 |
| Fed-batch SSCF_PDU | 2 | 10 | KE6-12 | 5 | 15 |
| Fed-batch SSCF_Demo | 2 | 10.5 | KE6-12 | 5 | 15 |

*1Model SSCF in lab scale with a feed of glucose solution with glucose amounts corresponding to 7.5%WIS.*
*2Model SSCF in PDU scale with a feed of filtered hydrolysate from enzymatic hydrolysis of whole slurry at 7.5% WIS.*
**No enzyme added during the model SSCF. However, filtered hydrolysate from enzymatic hydrolysis of whole slurry using an enzyme solution of 6 FPU gWIS-1 was used as a feed solution.*

### *7.4.5 SSCF*

The SSCF experiments were carried out in lab scale (3.6 l Infors HT-Labfors), PDU scale (30 l), and demo scale (10 m$^3$) bioreactors with a total working weight of 1.5 kg, 20 kg and 4000 kg, respectively. In the lab and PDU scale experiments the corncobs slurry was pH adjusted to 5.0 with

10 M NaOH and supplemented with 0.5 g $l_{-1}$ $(NH_4)_2HPO_4$. In the demo scale the pH was adjusted using $NH_3$ solution and supplemented with 0.25 g $l^{-1}$ $H_3PO_4$. To avoid possible contamination and foam formation 125 ppm of Vitahop solution and 0.5 ml $l^{-1}$ antifoam, respectively were added to the medium. In order to obtain the desired WIS content the supplemented medium was diluted with water and used for SSCF experiments. Unless otherwise stated, all the experiments were initiated by adding 5 g dry cell weight $l^{-1}$ of yeast from cell suspension. An enzyme preparation, Cellic Ctec-2 from Novozymes A/S, Denmark with filter paper activity of 95 FPU $g^{-1}$ enzyme, β-glucosidase activity of 590 IU $g^{-1}$ enzyme was added to SSCF experiments corresponding to the desired cellulase activity. All SSCF experiments were carried out at 35°C; pH was maintained at 5.0 by automatic addition of 3 M NaOH and the stirrer speed was maintained at 400 rpm in lab and PDU scales, respectively. A brief summary of all SSCF experiments carried out is listed in the Table 3. All SSCF experiments performed in duplicates in lab scale and one of them is represented in the results and discussion section.

### *7.4.6 ANALYSIS OF METABOLITES*

Samples for extracellular metabolites were analyzed by high performance liquid chromatography using Aminex HPX-87H column with 30 × 4.6 mm Cation-H Biorad micro-guard column maintained at 45°C. 5 mM H2SO4 was used as an eluent at a flow rate of 0.6 ml $min^{-1}$. Ethanol, xylitol, and acetic acid were detected using RI detector maintained at 35°C and HMF, furfural and lactic acid were detected using UV detector at 210 nm. The sugars in corncobs hydrolysate and samples from shake flasks and SSCF experiments were analyzed by high performance anion exchange chromatography using 4 × 250 mm Dionex CarboPac PA1 column with 4 × 50 mm guard column maintained at 30°C. Eluent A: 300 mM NaOH, eluent B: 100 mM NaOH+85 mM sodium acetate were used for elution at a flow rate of 1 ml $min^{-1}$. Monosaccharides including arabinose, galactose, glucose, xylose and mannose were detected using pulsed amperometric detector. Optical density (OD) was used as an estimate of cell concentration in shake flask experiments. OD was measured at 650 nm using the cell free medium at the point of sampling as background.

### 7.4.7 YIELD CALCULATIONS

#### 7.4.7.1 ETHANOL YIELD (% OF MAXIMUM THEORETICAL YIELD)

The sum of available fermentable sugars including glucose, mannose, galactose, and xylose in liquid fraction and glucan and xylan fibers in the WIS was calculated. Due to the addition of water during hydrolysis, the theoretical weight of glucose and xylose released are 1.11 and 1.13 times the weight of glucan and xylan, respectively. By using the maximum theoretical ethanol yield of 0.51 g $g^{-1}$ sugar, the maximum ethanol that can be produced from total available sugars was calculated. The percentage of the theoretical ethanol yield is defined as $Y_{SE}$ = 100*produced amount of ethanol (g)/maximum theoretical amount of ethanol (g).

#### 7.4.7.2 XYLOSE CONSUMED (%)

The percentage xylose consumed = 100*amount of xylose consumed (g)/ total amount of available xylose in liquid and WIS fraction (g).

#### 7.4.7.3 XYLITOL YIELD (%)

The percentage xylitol yield = 100*amount of xylitol produced (g)/amount of xylose consumed (g).

## REFERENCES

1. GlobalBP: [http://www.bp.com/statisticalreview] BP Statistical review of world energy June 2011. 2011.
2. Cockerill S, Martin C: Are biofuels sustainable? The EU perspective. Biotechnology for Biofuels 2008, 1:9.

3. Bidlack J, Malone M, Benson R: Molecular structure and component integration of secondary cell walls in plants. Proceedings of the Oklahoma Academy of Science 1992, 72:51-56.
4. Taherzadeh MJ, Karimi K: Pretreatment of lignocellulosic wastes to improve ethanol and biogas production: A review. Int J Mol Sci 2008, 9:1621-1651.
5. Sun Y, Cheng JY: Hydrolysis of lignocellulosic materials for ethanol production: a review. Bioresour Technol 2002, 83:1-11.
6. Öhgren K, Galbe M, Zacchi G: Optimization of steam pretreatment of SO2-impregnated corn stover for fuel ethanol production. Appl Biochem Biotechnol 2005, 121:1055-1067.
7. Palmqvist E, Hahn-Hägerdal B: Fermentation of lignocellulosic hydrolysates. II: inhibitors and mechanisms of inhibition. Bioresour Technol 2000, 74:25-33.
8. Larsson S, Palmqvist E, Hahn-Hägerdal B, Tengborg C, Stenberg K, Zacchi G, Nilvebrant NO: The generation of fermentation inhibitors during dilute acid hydrolysis of softwood. Enzyme Microb Technol 1999, 24:151-159.
9. Palmqvist E, Grage H, Meinander NQ, Hahn-Hägerdal B: Main and interaction effects of acetic acid, furfural, and p-hydroxybenzoic acid on growth and ethanol productivity of yeasts. Biotechnol Bioeng 1999, 63:46-55.
10. Larsson S, Quintana-Sainz A, Reimann A, Nilvebrant NO, Jönsson LJ: Influence of lignocellulose-derived aromatic compounds on oxygen-limited growth and ethanolic fermentation by Saccharomyces cerevisiae. Appl Biochem Biotechnol 2000, 84–6:617-632.
11. Eliasson A, Christensson C, Wahlbom CF, Hahn-Hägerdal B: Anaerobic xylose fermentation by recombinant Saccharomyces cerevisiae carrying XYL1, XYL2, and XKS1 in mineral medium chemostat cultures. Appl Environ Microbiol 2000, 66:3381-3386.
12. Kuyper M, Harhangi HR, Stave AK, Winkler AA, Jetten MSM, de Laat WTAM, den Ridder JJJ, Op den Camp HJM, van Dijken JP, Pronk JT: High-level functional expression of a fungal xylose isomerase: the key to efficient ethanolic fermentation of xylose by Saccharomyces cerevisiae? FEMS Yeast Res 2003, 4:69-78.
13. Wisselink HW, Toirkens MJ, Wu Q, Pronk JT, van Maris AJA: Novel evolutionary engineering approach for accelerated utilization of glucose, xylose, and arabinose mixtures by engineered Saccharomyces cerevisiae strains. Appl Environ Microbiol 2009, 75:907-914.
14. Bettiga M, Bengtsson O, Hahn-Hagerdal B, Gorwa-Grauslund MF: Arabinose and xylose fermentation by recombinant Saccharomyces cerevisiae expressing a fungal pentose utilization pathway. Microb Cell Fact 2009, 8:40.
15. Koppram R, Albers E, Olsson L: Evolutionary engineering strategies to enhance tolerance of xylose utilizing recombinant yeast to inhibitors derived from spruce biomass. Biotechnology for Biofuels 2012, 5:32.
16. Wingren A, Galbe M, Zacchi G: Techno-economic evaluation of producing ethanol from softwood: Comparison of SSF and SHF and identification of bottlenecks. Biotechnol Prog 2003, 19:1109-1117.
17. Olofsson K, Bertilsson M, Liden G: A short review on SSF - an interesting process option for ethanol production from lignocellulosic feedstocks. Biotechnology for Biofuels 2008, 1:7.

18. Tomás-Pejó E, Oliva JM, Ballesteros M, Olsson L: Comparison of SHF and SSF processes from steam-exploded wheat straw for ethanol production by xylose-fermenting and robust glucose-fermenting Saccharomyces cerevisiae strains. Biotechnol Bioeng 2008, 100:1122-1131.
19. Alfani F, Gallifuoco A, Saporosi A, Spera A, Cantarella M: Comparison of SHF and SSF processes for the bioconversion of steam-exploded wheat straw. J Ind Microbiol Biotechnol 2000, 25:184-192.
20. Hoyer K, Galbe M, Zacchi G: Production of fuel ethanol from softwood by simultaneous saccharification and fermentation at high dry matter content. J Chem Technol Biotechnol 2009, 84:570-577.
21. Zhang MJ, Wang F, Su RX, Qi W, He ZM: Ethanol production from high dry matter corncob using fed-batch simultaneous saccharification and fermentation after combined pretreatment. Bioresour Technol 2010, 101:4959-4964.
22. Öhgren K, Bengtsson O, Gorwa-Grauslund MF, Galbe M, Hahn-Hägerdal B, Zacchi G: Simultaneous saccharification and co-fermentation of glucose and xylose in steam-pretreated corn stover at high fiber content with Saccharomyces cerevisiae TMB3400. J Biotechnol 2006, 126:488-498.
23. Bertilsson M, Olofsson K, Liden G: Prefermentation improves xylose utilization in simultaneous saccharification and co-fermentation of pretreated spruce. Biotechnology for Biofuels 2009, 2:8.
24. Lee JW, Houtman CJ, Kim HY, Choi IG, Jeffries TW: Scale-up study of oxalic acid pretreatment of agricultural lignocellulosic biomass for the production of bioethanol. Bioresour Technol 2011, 102:7451-7456.
25. Ask M, Olofsson K, Di Felice T, Ruohonen L, Penttilä M, Lidén G: Challenges in enzymatic hydrolysis and fermentation of pretreated Arundo donax revealed by a comparison between SHF and SSF. Process Biochem 2012, 47:1452-1459.
26. Olofsson K, Rudolf A, Liden G: Designing simultaneous saccharification and fermentation for improved xylose conversion by a recombinant strain of Saccharomyces cerevisiae. J Biotechnol 2008, 134:112-120.
27. Kötter P, Ciriacy M: Xylose fermentation by Saccharomyces cerevisiae. Appl Microbiol Biotechnol 1993, 38:776-783.
28. Meinander N, Hahn-Hägerdal B: Influence of cosubstrate concentration on xylose conversion by recombinant, XYL1-expressing Saccharomyces cerevisiae: a comparison of different sugars and ethanol as cosubstrates. Appl Environ Microbiol 1997, 63:1959-2023.
29. Zacchi G, Axelsson A: Economic-evaluation of preconcentration in production of ethanol from dilute sugar solutions. Biotechnol Bioeng 1989, 34:223-233.
30. Wahlbom CF, van Zyl WH, Jönsson LJ, Hahn-Hägerdal B, Otero RRC: Generation of the improved recombinant xylose-utilizing Saccharomyces cerevisiae TMB 3400 by random mutagenesis and physiological comparison with Pichia stipitis CBS 6054. FEMS Yeast Res 2003, 3:319-326.
31. Verduyn C, Postma E, Scheffers WA, Vandijken JP: Effect of benzoic-acid on metabolic fluxes in yeasts - a continuous-culture study on the regulation of respiration and alcoholic fermentation. Yeast 1992, 8:501-517.
32. Alkasrawi M, Rudolf A, Liden G, Zacchi G: Influence of strain and cultivation procedure on the performance of simultaneous saccharification and fermentation of steam pretreated spruce. Enzyme Microb Technol 2006, 38:279-286.

CHAPTER 8

# COMPARISON OF ENZYMATIC REACTIVITY OF CORN STOVER SOLIDS PREPARED BY DILUTE ACID, AFEX™, AND IONIC LIQUID PRETREATMENTS

XIADI GAO, RAJEEV KUMAR, SEEMA SINGH, BLAKE A. SIMMONS, VENKATESH BALAN, BRUCE E. DALE, AND CHARLES E. WYMAN

## 8.1 BACKGROUND

Lignocellulosic biomass, including agricultural and forestry residues and herbaceous and woody crops [1], provides the only sustainable resource with potential for large-scale and low-cost production of liquid fuels and organic chemicals that are currently produced from dwindling and non-renewable fossil resources that are major contributors to greenhouse gas emissions [1,2]. Enzymatic hydrolysis is a key step in the biological conversion of lignocellulosic biomass into fuels and chemicals, with the high product yields important to commercial success [1-5]. Endoglucanases, exoglucanases, and β-glucosidase as well as supplementary enzymes such as xylanases and β-xylosidase are generally required to complete enzymatic hydrolysis effectively and efficiently [6-10]. However, to realize the high yields vital to commercial success of enzymatic conversion [11], most cellulosic biomass must be pretreated prior to enzymatic hydrolysis, and the choice of pretreatment not only affects enzymatic digestion performance but impacts upstream and downstream processing as well [1,12]. To overcome the natural recalcitrance of cellulosic biomass, several biological, chemical, thermochemical, and physical pretreatment methods have been applied, but thermochemical pretreatments are often preferred due to a more favorable combination of capital costs, operating costs, and performance [12].

Among thermochemical pretreatments, hemicellulose or lignin removal and/or alternation by dilute acids, with just hot water, or base promise reasonable costs [11,13,14]. In particular, dilute sulfuric acid (DA) and ammonia fiber expansion (AFEX™) pretreatments are currently among the most promising from a combined cost and performance perspective [1]. DA and hydrothermal pretreatments effectively remove and recover as sugars a large portion of hemicellulose as well as disrupting and dislocating lignin, while increasing cellulose digestibility [15-17]. The AFEX process pretreats biomass with anhydrous liquid ammonia at high pressure and moderate to high temperatures. Following pretreatment for a given time, the pressure is rapidly released resulting in biomass structure disruption and partial cellulose decrystallization that presumably enhance cellulose digestibility [18-20]. Lately, certain ionic liquids (ILs) such as the IL 1-ethyl-3-methylimidazolium acetate have been employed for pretreatment followed by addition of an anti-solvent to precipitate biomass [21]. Such ILs remove most of the lignin from biomass and disrupt the native cellulose crystalline structure and hydrogen networks to form cellulose II, thus reducing biomass recalcitrance [22-24].

Various biomass physicochemical changes resulting from the action of different leading pretreatments have been reported to enhance cellulose enzymatic digestibility, such as surface area, pore volume, hemicellulose removal, lignin removal and/or dislocation, crystallinity reduction, and reduced cellulose degree of polymerization (DP) [25-31]. Of these many possible factors, hemicellulose removal and lignin dislocation are mainly credited with enhancing digestion by DA pretreatment [16,17]. Creation of pore structures in biomass and disruption of lignin-carbohydrate complex (LCC) linkages enhance digestion for AFEX pretreatment [32,33]. Increased surface roughness, reduced cellulose crystallinity, and expansion or even a transformation of the cellulose lattice have been reported to account for the beneficial effects of IL pretreatments on biomass [22,34].

With several pretreatments being effective in achieving high sugar yields, it is vital to develop comparative data on sugar release from enzymatic hydrolysis of solids produced by each to facilitate identification of promising technologies. It is also important to measure key features influenced by each pretreatment to determine if enzymatic digestion can be related to any common characteristics, especially at the low enzyme loadings needed to be commercially viable [35]. In line with these objectives, researchers from the University of California, Riverside (UCR; Riverside, CA, USA), supported by the BioEnergy Science Center (BESC; Oak Ridge, TN, USA), Michigan State University (East Lansing, MI, USA) supported by the Great Lakes Bioenergy Research Center (GLBRC; Michigan State University), and the Joint BioEnergy Institute (JBEI; Emeryville, CA, USA) collaborated to better understand how biomass pretreatments by AFEX, DA, and IL with much different deconstruction patterns impact glucose and xylose release from enzymatic digestion over a wide range of enzyme loadings. To provide comparative information, a single source of corn stover was used as the feedstock for all three pretreatments, and one source of commercial cellulases and accessory enzymes was applied to the pretreated solids. In addition to sugar release data, changes in composition of solids and enzyme adsorption capacity of pretreated biomass solids were also measured to determine if they could help explain performance differences. Other papers by this team are in progress to examine changes in other features of pretreated biomass and more fully follow key features that could account for the performance differences reported here.

**TABLE 1:** Pretreatment conditions, corresponding solids compositions, and component removals for pretreatment of corn stover by dilute acid (DA), ammonia fiber expansion (AFEX™), and ionic liquid (IL)

| | Pretreatment | | | |
|---|---|---|---|---|
| | None | DA | AFEX | IL |
| Pretreatment conditions | | | | |
| Chemicals | NA | Dilute sulfuric acid | Anhydrous ammonia | 1-ethyl–3–methyl–imidazoliumacetate |
| Loadings | NA | 0.5% wt | 1:1 (Biomass: NH3) | 1:9 (Biomass: IL) |
| L/S ratio | NA | 9:1 | 1:1 | 9:1 |
| Temperature (°C) | NA | 160 | 140 | 140 |
| Time (min) | NA | 20 | 15 | 180 |
| Component (%) | | | | |
| Glucan | 33.4 | 59.1 | 33.5 | 46.9 |
| Xylan | 24.9 | 6.5 | 24.8 | 29.8 |
| Arabinan | 3.7 | 3.6 | 3.3 | 0.3 |
| Lignin (AIL) | 17.2 | 32.2 | 12.2 | 2.7 |
| Component removal (%)* | | | | |
| Solid | | 51.0 | 0 | 36.0 |
| Xylan | - | 87.0 | 0.4 | 23.4 |
| Lignin | - | 8.2 | 2.8 | 89.9 |

**Solid removal (%) = 100% × (1- Total solid after pretreatment (g)/Total solid before pretreatment (g)). Xylan removal (%) = 1- (%Xylan in pretreated solids / % Xylan in raw biomass) × Solid yield. Lignin removal (%) = 1- (%Lignin in pretreated solids / % Lignin in raw biomass) × Solid yield. L/S - liquid to solid.*

## 8.2 RESULTS AND DISCUSSION

### *8.2.1 COMPOSITIONAL ANALYSIS OF DA, AFEX, AND IL PRETREATED CORN STOVER*

Table 1 summarizes the pretreatment conditions applied, compositions of raw and pretreated corn stover solids, and the amount of xylan and lignin removed by each pretreatment. It can be seen that DA removed about 87% of the xylan, resulting in solids with a very low xylan content of 6.5% and increased glucan content of 59.1%. AFEX pretreatment of corn stover, as a dry to dry process, did not solubilize much of the biomass during pretreatment, resulting in a negligible compositional change. However, IL pretreatment removed >90% of the lignin and >20% of the xylan originally in corn stover, resulting in solids with a lignin content of only 2.7% and glucan and xylan contents of 46.9% and 29.8%, respectively. These changes in composition are consistent with results from previously reported studies [31,36,37].

### 8.2.2 ENZYMATIC HYDROLYSIS OF PRETREATED CORN STOVER SOLIDS AND MODEL COMPOUNDS

#### 8.2.2.1 GLUCOSE YIELDS FROM ENZYMATIC DIGESTION OF PRETREATED SOLIDS

Enzymatic digestion was performed at solids loading of 1% (w/w) glucan to allow focus on how pretreatments impacted substrate digestion and avoid confusion by end-product inhibition of the enzymes; this approach is consistent with the National Renewable Energy Laboratory (NREL) Laboratory Analytical Procedure (LAP) [38]. Enzyme loadings of 3, 6, 12, and 30 mg of total protein/g glucan in the raw biomass were applied at ratios of cellulase (Cellic® CTec2, Novozymes North America, Inc, Franklinton, NC, USA), hemicellulase (Cellic® HTec2, Novozymes North America, Inc), and pectinase (Multifect® Pectinase, DuPont™ Genencor® Science, DuPont Industrial Biosciences, Palo Alto, CA, USA), as shown in Table 2, that GLBRC had previously shown to give the highest sugar release from pretreated solids using their high-throughput microplate-based hydrolysis method [39]. Although different enzyme mixtures were required to achieve the highest sugar yields in enzymatic hydrolysis of solids from each pretreatment, the total mass loadings of the enzyme combinations applied were kept the same. In addition, optimal ratios of enzyme components in the mixtures were not simply related to the composition of the pretreated solids. For example, DA pretreated corn stover had negligible amounts of xylan, but substitution of hemicellulases for some of the cellulase still improved yields, possibly due to synergies or other effects among the enzymes [40].

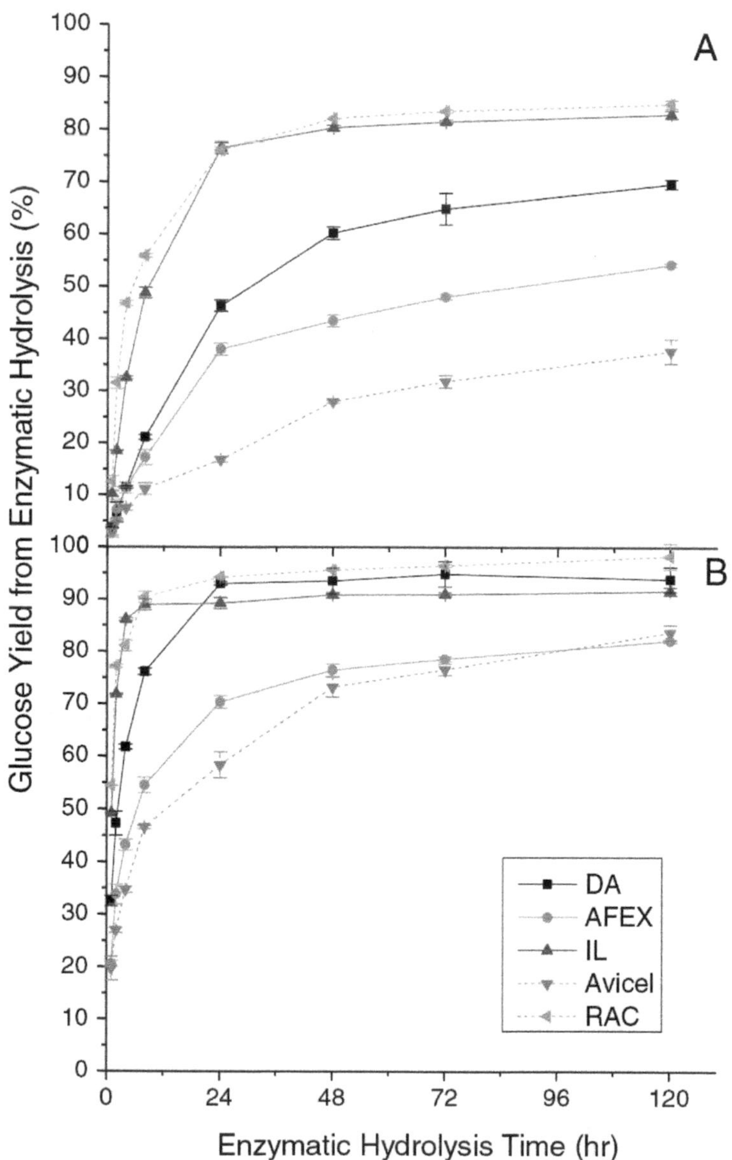


**FIGURE 1:** Glucose yields in solution following enzymatic hydrolysis of DA, AFEX, and IL pretreated corn stover solids, Avicel cellulose, and RAC. Enzyme loadings of (A) 3 mg and (B) 30 mg protein/g glucan in the raw corn stover. AFEX, ammonia fiber expansion; DA, dilute sulfuric acid; IL, ionic liquid; RAC, regenerated amorphous cellulose.

**TABLE 2:** Optimized enzyme mixtures on protein mass percents for DA, AFEX, and IL pretreated corn stover as determined by GLBRC with their high-throughput system

| Pretreatment | Cellic® CTec2 | Cellic® HTec2 | Multifect® Pectinase |
|---|---|---|---|
| Dilute acid | 67% | 33% | 0 |
| AFEX | 67% | 16.5% | 16.5% |
| Ionic liquid | 39% | 33% | 28% |

All enzymatic hydrolysis reactions were conducted at 50°C for time periods up to 120 hours. Sugar release is reported in terms of the glucose and xylose yield calculations described in the Materials and Methods section, in which the amount of either sugar actually released is divided by the maximum amount of the respective sugar that could be released from a particular pretreated solid including the appropriate factors to account for mass changes in hydrolysis. The glucose yield calculation includes monomeric glucose plus cellobiose, while the xylose yield only takes into account the monomeric xylose released.

As shown in Figure 1, glucose yields from hydrolysis of DA and AFEX pretreated corn stover were rapid in the first 8 hours and then slowed down considerably to almost level off after 72 hours at both 3 and 30 mg protein/g glucan loadings. As expected, glucose yields were always higher at the higher enzyme loading, particularly for AFEX and DA pretreated solids. Corn stover solids from IL pretreatment had greater initial glucose yields compared to either DA or AFEX pretreated corn stover and nearly reached its maximum glucose yield in only 8 hours. Adding more enzyme increased glucose yields from IL pretreated solids less than from the other two pretreatments because the yields were already high for IL solids.

Figures 2 and 3 compare the 1-hour and 72-hour glucose yields from enzymatic hydrolysis at four enzyme loadings applied to solids produced from corn stover by the three pretreatments. Figure 2 shows that the 1-hour glucose yields of all substrates increased almost linearly with enzyme loading. Furthermore, IL pretreated corn stover solids displayed the highest 1-hour hydrolysis yield at all enzyme loadings followed by DA and then AFEX. Increasing the enzyme loading from 3 to 30 mg protein/g glucan increased glucose yields from IL pretreated solids by 40% in the first hour of hydrolysis. The 1-hour glucose yields from DA corn stover at the highest enzyme loading of 30 mg/g glucan were about six times higher than at the lowest enzyme loading of 3 mg/g glucan. The 1-hour glucose yield from solids from AFEX pretreatment was slightly lower compared to DA at the lowest enzyme loading but 10% lower at the high enzyme loading. Thus, the 1-hour glucose yield from AFEX pretreated corn stover did not benefit from higher enzyme loadings as much as solids from the other two pretreatments.

The 72-hour glucose yields from enzymatic hydrolysis of the pretreated solids were also compared for the three pretreatments at four enzyme loadings, as shown in Figure 3. Overall, higher enzyme loadings resulted in higher glucose yields, as expected, but the benefit of adding more enzymes varied with pretreatment. For example, as shown in Figure 3, when the enzyme loading was increased from 3 to 30 mg/g glucan in raw biomass, the 72-hour glucose yields for DA and AFEX corn stover increased by about 30% and 30.5%, respectively, while solids from IL pretreatment of corn stover had only about a 10% enhancement in glucose yield as it was already close to the maximum at the low enzyme loading. As seen in Figure 3, at 30 mg total protein for the optimized enzyme mixtures/g glucan in raw corn stover, 72-hour glucose yields from enzymatic hydrolysis were 93.8% for DA pretreated corn stover, 91.6% for IL, and 82.1% for AFEX. Avicel® and regenerated amorphous cellulose (RAC) model compounds realized about 42% and 16% increases in glucose yields, respectively, over the range of enzyme loadings applied.

### *8.2.3 XYLOSE MONOMER AND OLIGOMER YIELDS*

The liquid samples from 4, 24, and 72 hours of enzymatic hydrolysis of solids produced by DA, AFEX, and IL pretreatments of corn stover were analyzed for longer chain length glucooligomers (> cellobiose) and xylooligomers by post-hydrolysis of liquid samples from pretreatment with 4% w/w DA at 121°C for 1 hour [41]. Figure 4 reports both xylose monomer and xylooligomer yields following 72 hours of enzymatic hydrolysis of solids produced by DA, AFEX, and IL pretreatments of corn stover at total enzyme protein loadings of 3 and 30 mg/g glucan in raw biomass for the same optimized enzyme mixtures as used to obtain glucose yields. Results from enzymatic hydrolysis of the model compound beechwood xylan are also included over a range of 3 to 30 mg of just HTec2 xylanase protein/g xylan. Figure 4 shows that about a quarter of the total xylose in solution was released as xylooligomers during hydrolysis for all pretreated solids and remained at about the same fraction of the total for both enzyme loadings for these three solids. However, a much higher fraction of the total xylose in solution was as oligomers for beechwood xylan, particularly at the lower enzyme loading, most likely due to low β-xylosidase and other accessory activities [37]. When comparing solubilized xylan yields from enzymatic hydrolysis of pretreated solids at the 10 g glucan/L employed for evaluating substrate breakdown by enzymes, it should be kept in mind that compositional variations in the substrates from each pretreatment led to quite different ratios of enzyme protein to xylan content. As shown in Figure 4, at an enzyme loading of 30 mg total protein/g glucan in raw corn stover, the total xylose monomer plus oligomer yields for solids produced from corn stover by DA, AFEX, and IL were 83%, 68%, and 77%, respectively, with oligomers contributing about 9%, 14%, and 12%, respectively, to these amounts.

### *8.2.4 PERCENTAGE OF OLIGOMERS RELEASED DURING HYDROLYSIS*

As shown in Figure 5A for glucooligomers and Figure 5B for xylooligomers, longer hydrolysis times and increased enzyme loadings both reduced the proportion of gluco- and xylooligomers. At the early stage of hydrolysis or for lower enzyme loadings, the maximum glucooligomer percentage was less than 13% for AFEX followed by 12% for IL and 6% for DA. Figure 5A suggests that application of mixtures of CTec2, HTec2, and Multifect Pectinase enzymes optimized individually for each pretreatment achieved essentially complete digestion of the glucan to monomers for all three pretreated materials at high enzyme loadings.

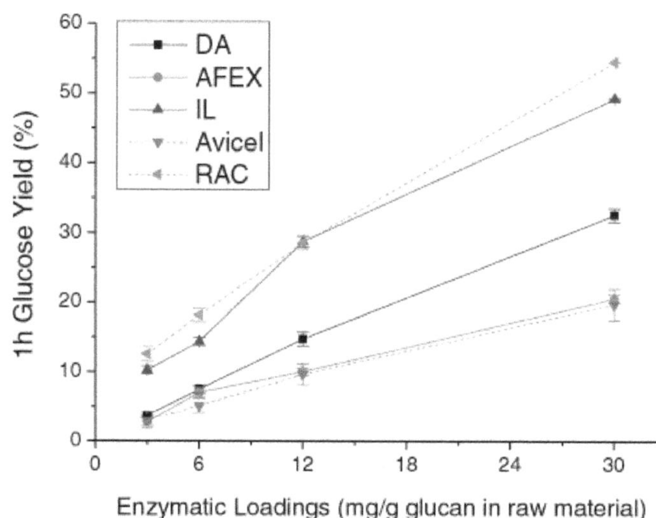


**FIGURE 2:** Effect of enzyme loadings on the 1-hour glucose yields from enzymatic hydrolysis of DA, AFEX, and IL pretreated corn stover solids, Avicel cellulose, and RAC. AFEX, ammonia fiber expansion; DA, dilute sulfuric acid; IL, ionic liquid; RAC, regenerated amorphous cellulose.

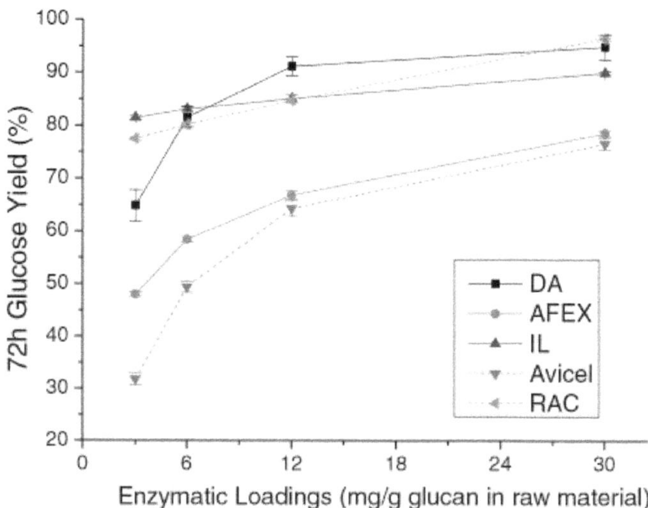


**FIGURE 3:** Effect of enzyme loadings on the 72-hour glucose yields from enzymatic hydrolysis of DA, AFEX, and IL pretreated corn stover solids, Avicel cellulose, and RAC. AFEX, ammonia fiber expansion; DA, dilute sulfuric acid; IL, ionic liquid; RAC, regenerated amorphous cellulose.

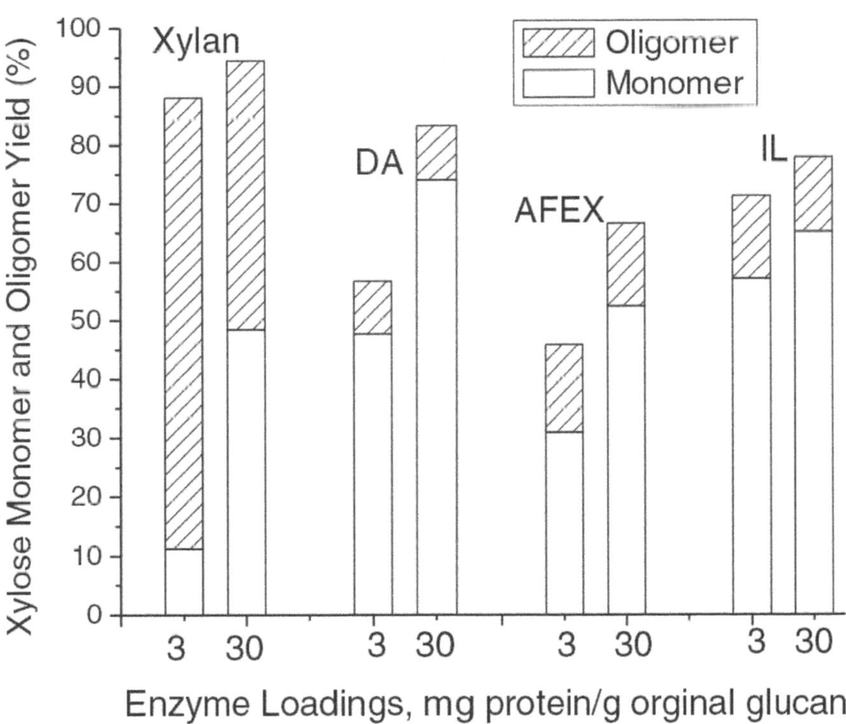


**FIGURE 4:** Xylose monomer and oligomers yields from 72-hour enzymatic hydrolysis of DA, AFEX, and IL pretreated corn stover, and beechwood xylan. Note only HTec2 was added to beechwood xylan. AFEX, ammonia fiber expansion; DA, dilute sulfuric acid; IL, ionic liquid.

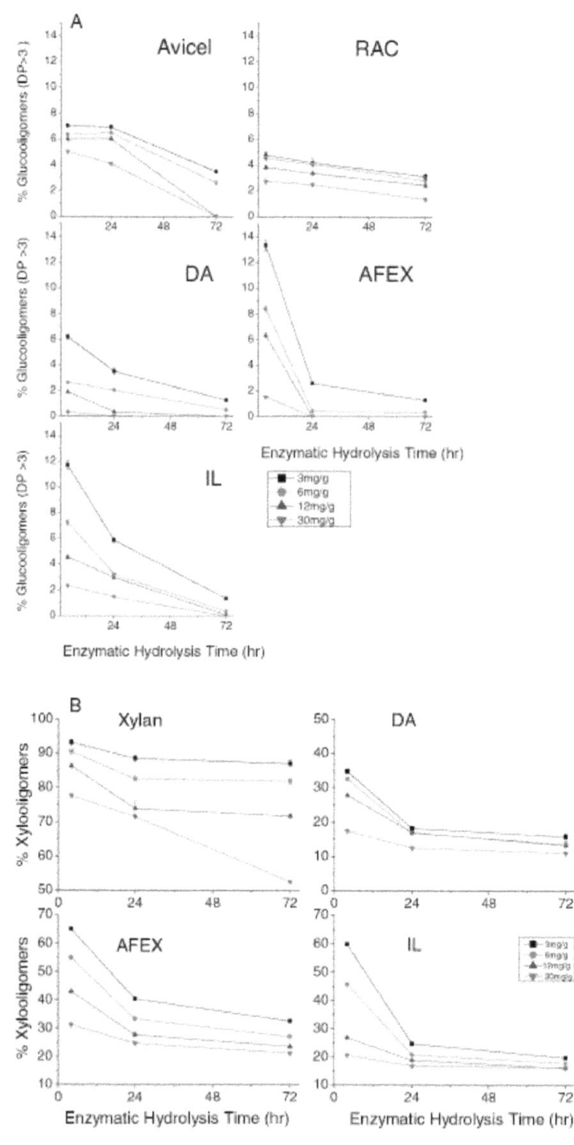


**FIGURE 5:** Soluble glucooligomers (DP >3) and xylooligomers as a percentage of the total glucose and xylose in solution following 72 hours of hydrolysis at enzyme loadings of 3, 6, 12, and 30 mg/g glucan in raw biomass for Avicel, RAC, and beechwood xylan and DA, AFEX, and IL pretreated corn stover solids. (A) Glucooligomers and (B) xylooligomers. AFEX, ammonia fiber expansion; DA, dilute sulfuric acid; DP, degree of polymerization; IL, ionic liquid; RAC, regenerated amorphous cellulose.

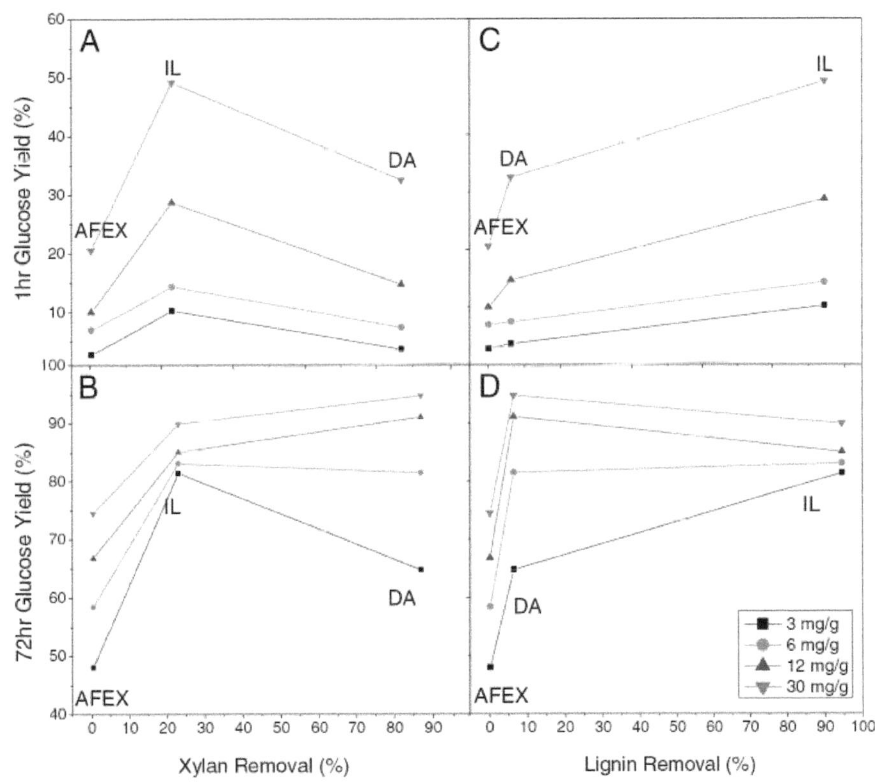


**FIGURE 6:** The 1-hour and 72-hour glucose yields from enzymatic hydrolysis of DA, AFEX, and IL pretreated corn stover solids plotted against xylan removal and lignin removal at total enzyme loadings of 3, 6, 12, and 30 mg/g glucan in raw corn stover. (A, B) Xylan removal and (C, D) lignin removal. AFEX, ammonia fiber expansion; DA, dilute sulfuric acid; IL, ionic liquid.

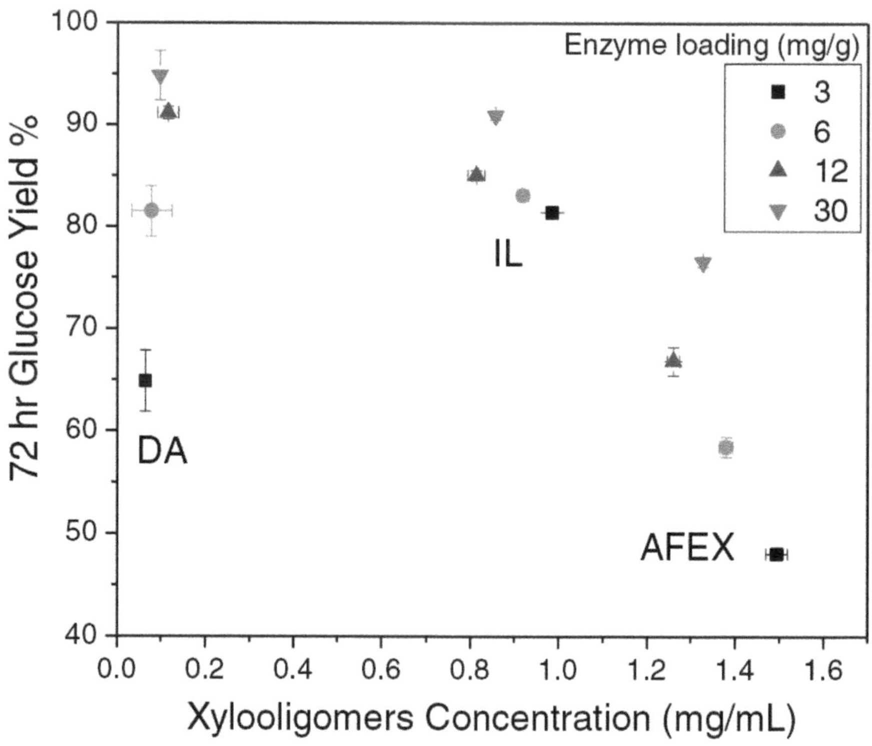


**FIGURE 7:** The 72-hour glucose yields at enzyme loadings of 3, 6, 12, and 30 mg enzyme protein/g glucan in the raw corn stover plotted versus xylooligomer concentration measured during enzymatic hydrolysis of DA, AFEX, and IL pretreated corn stover. AFEX, ammonia fiber expansion; DA, dilute sulfuric acid; IL, ionic liquid.

Figure 5B shows that hydrolysis of xylooligomers was less complete than for glucooligomers. For hydrolysis of AFEX and IL pretreated corn stover solids, approximately 20% or more of the total xylose in solution persisted as oligomers even after hydrolysis for 72 hours at an enzyme loading of 30 mg/g glucan in the raw material. Hydrolyzate from enzymatic hydrolysis of AFEX pretreated solids contained the highest amount of xylooligomers followed by IL and DA. Overall, enzymatic hydrolysis of DA pretreated corn stover released the lowest percentage of both glucooligomers and xylooligomers.

### *8.2.5 EFFECT OF XYLAN AND LIGNIN REMOVAL ON ENZYMATIC DIGESTION*

Figure 6 plots 1-hour and 72-hour glucose yields from enzymatic hydrolysis against lignin and xylan removal. Xylan removal has been reported to enhance glucan digestibility by improving cellulose accessibility [42] and/or reducing cellulase inhibition by xylooligomers produced from partial hydrolysis of xylan [43-45]. However, these results show that the 1-hour glucose yields did not follow a clear trend with xylan removal, and 72-hour yields only correlated with xylan removal at the two highest enzyme loadings.

From Figure 6, it is apparent that the 1-hour glucose yields correlated well with lignin removal at all enzyme loadings, while such a relationship was only apparent for 72-hour results at the lowest enzyme loading. This result is consistent with reports that lignin is one of the key biomass components impacting enzymatic digestion of cellulosic biomass [2,34]. Lignin is believed to not only hinder cellulose accessibility as a result of LCC linkages but also impact cellulase effectiveness by unproductive binding of enzymes [31,46,47].

It has been shown that xylobiose and xylooligomers with higher DP strongly inhibit enzymatic hydrolysis of pure cellulose, pure xylan, and pretreated corn stover [43], and that xylooligomers were more inhibitory to cellulase than xylose or xylan for an equivalent amount of xylose or than equal molar amounts of glucose or cellobiose [44]. Figure 7 reports glucose yields against the concentration of xylooligomers in solution following enzymatic hydrolysis for 72 hours over the full range of enzyme loadings employed in this study, with the concentration of xylooligomers calculated on an equivalent xylose mass basis. First, these results show that DA solids released much less xylooligomers during enzymatic hydrolysis than solids from IL or AFEX. In addition, the xylooligomer concentrations remained virtually constant after 72 hours of enzymatic hydrolysis for all three pretreated solids. These results also show that xylooligomer concentrations had no clear effect on the trends in 72-hour glucose yields from enzymatic hydrolysis of the three pretreated solids. In fact, 72-hour glucose yields from enzymatic hydrolysis of DA and AFEX pretreated solids increased significantly with enzyme loading even though the xylooligomer concentration remained virtually the same for each. On the other hand, 72-hour glucose yields from IL pretreated solids were high and changed little despite having a xylooligomer concentration that was about two-thirds of that for enzymatic hydrolysis of AFEX pretreated solids. Therefore, other factors such as lignin removal, unproductive binding of enzyme, or cellulose structure appeared to have a greater effect on glucose yields.

### 8.2.6 COMPARISON TO RESULTS WITH MODEL COMPOUNDS

Two model compounds, microcrystalline Avicel cellulose and RAC, were enzymatically hydrolyzed as well. Figure 1 shows that glucose release was greatest from enzymatic hydrolysis of RAC and lowest from Avicel cellulose at the highest and lowest enzyme loadings applied in this study over almost the entire hydrolysis time. These trends are generally reinforced for the 1-hour and 72-hour glucose yields in Figures 2 and 3, respectively. In all cases, glucose yields from enzymatic hydrolysis of IL pretreated solids closely paralleled those with RAC. RAC was a highly homogeneous substrate with disrupted hydrogen bonds [48] and consequently had much higher glucose yields. Other than xylan and lignin removal, cellulose crystallinity is believed to be one of the major factors limiting cellulose enzymatic hydrolysis [48,49]. The high glucose yields from enzymatic hydrolysis of IL corn stover and RAC were probably due to reduction in crystalline cellulose content and/or altered cellulose structure previously shown for corn stover [36,50]. The rapid hydrolysis of amorphous substrate could be explained as a homogeneous reaction that enabled cleavage of all β-glycosidic bonds randomly, resulting in a rapid reduction of DP [48]. This result is consistent with an earlier hypothesis that lower crystallinity has a particularly significant influence on initial glucose yields from enzymatic hydrolysis [46].

### *8.2.7 POSSIBLE CAUSES OF DIFFERENCES IN SUGAR RELEASE*

Based on the hydrolysis model reported by Zhang and Lynd [8], three processes occur simultaneously when enzymes act on insoluble cellulosic substrates: 1) chemical and physical changes in the solid residue; 2) primary hydrolysis in which the solid phase is hydrolyzed into soluble cellodextrins; and 3) secondary hydrolysis in which the soluble oligomers are further hydrolyzed into monomers. Given that the rate of primary hydrolysis is much slower than the rate of secondary hydrolysis, a substrate with greater primary hydrolysis sugar release should result in more rapid initial glucose release. The experimental observation of similar glucose yield patterns for IL pretreated corn stover and RAC solids suggest structural similarities. However, cellulose in corn stover from DA and AFEX pretreatment are expected to be more crystalline, while Avicel cellulose is known to be highly crystalline. This difference in crystallinity could at least partially explain why glucose yields from IL corn stover solids were higher than from the AFEX and DA pretreated solids and much higher than from Avicel, particularly at shorter times.

### *8.2.8 ADSORPTION OF CTEC2 AND HTEC2 ON PRETREATED CORN STOVER*

Enzyme adsorption onto the substrate is the primary step in enzymatic degradation of cellulose [9,51]. Cellulose accessibility to enzyme has long been recognized as an essential factor controlling enzymatic hydrolysis of cellulosic biomass [52-54], and enzymatic hydrolysis rates and yields are often claimed to be related to enzyme adsorption [42,52,53,55,56].

Adsorption parameters calculated for the Langmuir model, the maximum adsorption capacity σ and equilibrium constant Kd, are summarized in Table 3. These results show substantial variances in enzyme adsorption with pretreatment type. AFEX corn stover had the lowest maximum enzyme adsorption capacity for both CTec2 and HTec2, while IL corn stover had the highest values for both. In line with this reasoning, the initial hydrolysis yields of the three pretreated substrates reported in Figures 1 and 2 followed the same trend as their maximum enzyme adsorption capacities for both CTec2 and HTec2. Thus, enzyme adsorption onto solids and their effectiveness are affected by pretreatment type and biomass composition, consistent with other published information [47,53,57].

**TABLE 3:** Maximum CTec2 and HTec2 adsorption capacities, equilibrium constants, and correlation coefficients for solids resulting from pretreatments of corn stover by DA, AFEX, and IL pretreatments

| | Pretreatment | | |
|---|---|---|---|
| | DA | AFEX | IL |
| CTec2 | | | |
| Maximum adsorption capacity, σ (mg/g substrate) | 139 | 111 | 190 |
| Equilibrium constants, Kd (mg/mL) | 1.88 | 0.19 | 1.43 |
| $R^2$ | 0.92 | 0.96 | 0.92 |
| HTec2 | | | |
| Maximum adsorption capacity, σ (mg/g substrate) | 142 | 127 | 239 |
| Equilibrium constants, Kd (mg/mL) | 0.67 | 0.27 | 2.3 |
| $R^2$ | 0.97 | 0.95 | 0.98 |

## 8.3 CONCLUSIONS

No single factor absolutely dominated early and longer-term glucose yields from enzymatic hydrolysis of solids from AFEX, DA, and IL pretreatments. The high initial hydrolysis yields from IL corn stover correlated with high lignin removal, high change in crystallinity, and high enzyme adsorption, and was very similar in pattern to results from enzymatic digestion of RAC solids. The final glucose yields did not follow a consistent trend with concentration of xylooligomers released from xylan during hydrolysis. IL pretreated corn stover showed the highest initial glucose yields at low enzyme loadings, while DA pretreated corn stover, which removed the most xylan, achieved the highest glucose yields at high enzyme loadings.

## 8.4 MATERIALS AND METHODS

### 8.4.1 PRETREATED CORN STOVER AND MODEL COMPOUNDS

Corn stover was harvested in September 2008 at the Michigan State University Farms (East Lansing, MI, USA) from the corn hybrid NK 49-E3 (Syngenta, Basel, Switzerland), typical of that grown in the Great Lakes region. Solids resulting from pretreatment of the same source of corn stover were prepared by the collaboration partners of the Bioenergy Research Centers (BRCs), as follows: DA pretreatment by BESC at UCR, AFEX by GLBRC at Michigan State University, and IL by JBEI. Upon receipt, the AFEX and IL pretreated corn stover solids were immediately refrigerated at 4°C until further analysis. The pretreatment conditions summarized in Table 1 for all three pretreatments were selected based on highest total glucan plus xylan yields from both pretreatment and enzymatic digestion, but only the solids were employed in the enzymatic hydrolysis study reported here.

Pure cellulose (Avicel PH101, catalogue number 11365, lot number 1094627) was purchased from FMC Corporation (Philadelphia, PA, USA). RAC was prepared from Avicel PH101 according to the method reported by Zhang and coworkers [48]. Beechwood xylan (lot number BCBS8393V) was purchased from Sigma Chemicals (St Louis, MO, USA). Moisture content and compositional analysis of the corn stover solids and model compounds were determined according to the NREL LAP [58].

### 8.4.2 ENZYMES

Cellic CTec2 (batch number VCNI0001) and Cellic HTec2 (batch number VHN0001) enzymes were generously provided by Novozymes North

America, Inc., and Multifect Pectinase (batch number 4861295753) was from DuPont Industrial Biosciences. Table 4 shows the enzyme protein concentrations determined by the Kjeldahl method [59], with the nitrogen factor (NF) calculated by Equation 1:

$$NF = \% \text{ protein} / \% \text{ nitrogen} \quad (1)$$

in which the percentage of protein was calculated as:

$$\% \text{ protein} = \text{protein content (mg/mL)} / \text{solid concentration (mg/mL)} \quad (2)$$

**TABLE 4:** Enzyme nomenclatures, descriptions, protein concentrations, and nitrogen factors

| Enzyme | Description | Protein concentration (mg/mL) | Nitrogen factor |
|---|---|---|---|
| Cellic® CTec2 | Blend of cellulase, high level of β-glucosidase, and hemicellulases | 138 | 6.09 |
| Cellic® HTec2 | Blend of hemicellulases and cellulase background | 157 | 6.58 |
| Multifect® Pectinase | Pectinase, cellulase, and hemicellulases | 72 | - |

The nitrogen content was determined by following a previously described method [60]. The solids content of the enzyme solution was determined according to the NREL LAP [61].

### *8.4.3 ENZYMATIC HYDROLYSIS*

In accordance with the NREL LAP [38], enzymatic hydrolysis was conducted in triplicate at a solids loading corresponding to 1% (w/w) glucan in 0.05 M citrate buffer (pH=4.9) containing 10 mg/mL sodium azide in 50 mL Erlenmeyer flasks. The slurries were incubated at 50°C for 120 hours in a shaker incubator (Multitron Infors-HT, ATR Biotech, Laurel, MD, USA) at 150 rpm. Enzyme loadings were 3, 6, 12, and 30 mg of total protein/g glucan in the raw biomass. Enzyme combinations of Cellic CTec2, Cellic HTec2, and Multifect Pectinase to achieve maximum sugar release for solids from DA, AFEX, and IL pretreatments were determined by GLBRC using their novel high-throughput microplate hydrolysis method [39], and are shown in Table 4.

Hydrolysis samples were collected at 1, 2, 4, 8, 24, 48, 72, and 120 hours. To determine the amount of sugar generated from enzymatic hydrolysis, 400 μL samples were drawn, filtered through 0.2 μm nylon filter vials (Alltech Associates Inc., Deerfield, IL, USA), pipetted into 500 μL polyethylene HPLC vials, and then stored at 4°C until analysis. Glucan to glucose and xylan to xylose hydrolysis yields were calculated according to the following two equations, respectively:

$$\%\ \text{Glucuse yield} = 100 \times ((\text{GH(g)} + \text{CB(g)} \times 1.503)) / (1.111*\text{GP(g)}) \quad (3)$$

$$\%\ \text{Xylose yield} = 100 \times \text{XH(g)} / (1.136*\text{XP(g)}) \quad (4)$$

in which GH, CB, and XH are the measured masses of glucose, cellobiose, and xylose released from enzymatic hydrolysis; GP and XP are the masses of glucan and xylan available in the pretreated biomass; and the factors 1.111, 1.136, and 1.053 account for the mass gained during hydrolysis of glucan to glucose, xylan to xylose, and cellobiose to glucose, respectively.

### 8.4.4 BASIS FOR ENZYME PROTEIN LOADING PER GRAM OF GLUCAN IN RAW BIOMASS

Consistent with the approach used by our team in prior research at the recommendation of the Consortium for Applied Fundamentals and Innovation (CAFI) Advisory Board, enzyme loadings for all enzymatic digestion experiments in this study were based on glucan content in the original raw material [1,35]. Loading enzyme based on glucan content in raw biomass better represents the cost of enzyme per amount of potential ethanol, and also benefits removing more glucan in pretreatment with more enzyme per unit of glucan left in the solids to enzymatic hydrolysis. This comparison is particularly important for enzymatic hydrolysis at commercially viable low enzyme loadings. Because of the differences in glucan removal by the three pretreatments of interest here, the enzyme loadings per gram of glucan in the pretreated biomass solids varied as shown in Table 5. The result is that solids from DA and IL pretreatments had higher enzyme loadings per gram of glucan in the pretreated solids compared to solids from AFEX.

**TABLE 5:** Glucan recovery in solids following pretreatment and enzyme loadings for hydrolysis of pretreated solids based on glucan content in raw corn stover

| Pretreatment | Glucan yield (%) | Enzyme loading (mg protein/g glucan in raw) | Enzyme loading (mg/g glucan in pretreated) |
|---|---|---|---|
| Dilute acid | 87 | 30 | 34 |
| AFEX | 100 | 30 | |
| Ionic liquid | 90 | | 33 |

*Glucan yield in solids = Glucan in pretreated biomass (g)/ Glucan in the starting material (g). Enzyme loading per g glucan in pretreated biomass = Enzyme loading per g glucan in raw/ glucan yield in solids.*

### 8.4.5 ESTIMATION OF THE AMOUNTS OF OLIGOMERS

To determine the total amount of glucose and xylose oligomers generated by enzymatic hydrolysis, liquid samples following enzymatic hydrolysis for 4, 24, and 72 hours were subjected to post-hydrolysis according to the NREL LAP [41]. In particular, slurries after enzymatic hydrolysis were centrifuged to separate solids from liquid. Then, the liquid was incubated for 1 hour with 4% sulfuric acid at 121°C in an autoclave (model HA300MII; Hirayama Manufacturing Corporation, Saitama, Japan) along with sugar recovery standards. It is important to note that post-hydrolysis was carried out in 1.5 mL high recovery glass HPLC vials (Agilent, Santa

Clara, CA, USA) and scaled down to 1 mL reaction volume instead of applying the conventional method in 125 mL pressure bottles with 5 to 20 mL liquid [62]. Following post-hydrolysis, about 400 μL samples were withdrawn, pipetted into 500 μL polyethylene HPLC vials, and kept at 4°C or frozen at −20°C until further analysis. From this information, the percentage of the total glucose in solution that was glucooligomers with a DP>cellobiose, $G_{3+}$, was calculated by:

$$\%G_{3+} = 100 \times [GH'(g) - GH(g) - 1.053 \times CB(g)] / GH'(g) \quad (5)$$

$X_{2+}$, the percentage yield of xylooligomers containing two or more xylose units, was calculated as:

$$\%X_{2+} = 100 \times [XH'(g) - XH(g)] / [XP(g)*1.136] \quad (6)$$

In addition, the percentage of the total xylose in solution that was oligomers containing two or more xylose units was calculated as:

$$\%XS_{2+} = 100 \times [XH'(g) - XH(g)] / XH'(g) \quad (7)$$

GH’ and XH’ represent the masses of glucose and xylose measured after post-hydrolysis and adjusting for losses by the sugar recovery standard [41].

### *8.4.6 SUGAR ANALYSIS*

Samples along with appropriate calibration standards were run on a Waters Alliance HPLC system (Model e-2695; Waters Corporation, Milford, MA, USA) employing an Aminex® HPX-87H column (Bio-Rad Laboratories, Life Science Research, Hercules, CA, USA). Samples were processed at an eluent (5 mM sulfuric acid) flow rate of 0.60 mL/min using a refractive index

(RI) detector (Model 2414; Waters Corporation). The chromatograms were recorded and processed with Empower® 2 software (Waters Corporation).

### 8.4.7 ENZYME ADSORPTION

Adsorption experiments were performed at 4°C in 0.05 M citrate buffer ($pH = 4.8 \pm 0.2$) in 15 mL test tubes (catalogue number 430055; Thermo Fisher Scientific Inc., Waltham, MA, USA) with a biomass loading to achieve 1% w/w glucan with enzyme loadings of 0 to 2,000 mg protein/g glucan (0 to 20 mg protein/mL). The tubes containing biomass slurry and enzyme proteins were mounted on a variable speed rugged rotator (Glass-Col, LLC, Terre Haute, IN, USA) and equilibrated for 6 hours at 40 rpm. Following equilibration, the tubes were centrifuged (model Allegra X-15R; Beckman Coulter, Fullerton, CA, USA) for 15 minutes at 3,500 rpm for solid–liquid separation, the liquid was decanted, and the tubes dried overnight at 105°C. The adsorbed protein amount was directly determined by the nitrogen factor method described above [60]. The nitrogen content of the dried and homogenized biomass solids was measured using a Flash EATM 112 N/Protein plus CHNS/O Analyzer (CE Elantech, Lakewood, NJ, USA) with atropine as a standard (catalogue number 33835210; CE Elantech). Adsorption data was non-linearly fit to a Langmuir model according to Equation 8 [7,63]:

$$[CE] = (\sigma[S_t][E_f]) / (K_d + [E_f]) \tag{8}$$

in which [CE] is the amount of adsorbed enzyme in mg/mL, $[E_f]$ is the free enzyme concentration in mg/mL, $\sigma$ is the maximum adsorption capacity in mg/mg substrate, $[S_t]$ is the substrate concentration mg/mL, and $K_d$ is the equilibrium constant equal to [C] [E]/[CE].

## REFERENCES

1. Wyman CE, Dale BE, Elander RT, Holtzapple M, Ladisch MR, Lee YY: Coordinated development of leading biomass pretreatment technologies. Bioresour Technol 2005, 96:1959-1966.

2. Himmel ME, Ding SY, Johnson DK, Adney WS, Nimlos MR, Brady JW, Foust TD: Biomass recalcitrance: engineering plants and enzymes for biofuels production. Science 2007, 315:804-807.
3. Kamm B, Kamm M: Principles of biorefineries. Appl Microbiol Biotechnol 2004, 64:137-145.
4. Jorgensen H, Kristensen JB, Felby C: Enzymatic conversion of lignocellulose into fermentable sugars: challenges and opportunities. Biofuel Bioprod Bioresour 2007, 1:119-134.
5. Ragauskas AJ, Williams CK, Davison BH, Britovsek G, Cairney J, Eckert CA, Frederick WJ Jr, Hallett JP, Leak DJ, Liotta CL, Mielenz JR, Murphy R, Templer R, Tschaplinski T: The path forward for biofuels and biomaterials. Science 2006, 311:484-489.
6. Henrissat B: Celluases and their interaction with cellulose. Cellulose 1994, 1:169-196.
7. Lynd LR, Weimer PJ, van Zyl WH, Pretorius IS: Microbial cellulose utilization: fundamentals and biotechnology. Microbiol Mol Biol Rev 2002, 66:506-739.
8. Zhang YHP, Lynd LR: Toward an aggregated understanding of enzymatic hydrolysis of cellulose: noncomplexed cellulase systems. Biotechnol Bioeng 2004, 88:797-824.
9. Bansal P, Hall M, Realff MJ, Lee JH, Bommarius AS: Modeling cellulase kinetics on lignocellulosic substrates. Biotechnol Adv 2009, 27:833-848.
10. Kumar R, Wyman CE: Effects of cellulase and xylanase enzymes on the deconstruction of solids from pretreatment of poplar by leading technologies. Biotechnol Prog 2009, 25:302-314.
11. Mosier N, Wyman C, Dale B, Elander R, Lee YY, Holtzapple M, Ladisch M: Features of promising technologies for pretreatment of lignocellulosic biomass. Bioresour Technol 2005, 96:673-686.
12. Yang B, Wyman CE: Pretreatment: the key to unlocking low-cost cellulosic ethanol. Biofuel Bioprod Bioresour 2008, 2:26-40.
13. Hsu TA: Pretreatment of biomass. In Handbook on Bioethonal Production and Utilization. Edited by Wyman CE. Washington, DC: Taylor & Francis; 1996.
14. Hu F, Ragauskas A: Pretreatment and lignocellulosic chemistry. Bioenergy Res 2012, 5:1043-1066.
15. Knappert D, Grethlein H, Converse A: Partial acid hydrolysis of poplar wood as a pretreatment for enzymatic-hydrolysis. Biotechnol Bioeng 1981, 11:67-77.
16. Yang B, Wyman CE: Effect of xylan and lignin removal by batch and flowthrough pretreatment on the enzymatic digestibility of corn stover cellulose. Biotechnol Bioeng 2004, 86:88-95.
17. Ohgren K, Bura R, Saddler J, Zacchi G: Effect of hemicellulose and lignin removal on enzymatic hydrolysis of steam pretreated corn stover. Bioresour Technol 2007, 98:2503-2510.
18. Dale BE, Moreira MJ: A freeze-explosion technique for increasing cellulose hydrolysis. Biotechnol Bioeng 1982, 12:31-43.
19. Holtzapple MT, Jun JH, Ashok G, Patibandla SL, Dale BE: The ammonia freeze explosion (AFEX) process. Appl Biochem Biotechnol 1991, 28–9:59-74.
20. Dale BE, Henk LE, Shiang M: Fermentation of lignocellulosic materials treated by ammonia freeze-explosion. Dev Ind Microbiol 1985, 26:223-234.

21. Swatloski RP, Spear SK, Holbrey JD, Rogers RD: Dissolution of cellose with ionic liquids. J Am Chem Soc 2002, 124:4974-4975.
22. Cheng G, Varanasi P, Li C, Liu H, Menichenko YB, Simmons BA, Kent MS, Singh S: Transition of cellulose crystalline structure and surface morphology of biomass as a function of ionic liquid pretreatment and its relation to enzymatic hydrolysis. Biomacromolecules 2011, 12:933-941.
23. Dadi AP, Schall CA, Varanasi S: Mitigation of cellulose recalcitrance to enzymatic hydrolysis by ionic liquid pretreatment. Appl Biochem Biotechnol 2007, 137–140:407-421.
24. Zhao H, Jones CL, Baker GA, Xia S, Olubajo O, Person VN: Regenerating cellulose from ionic liquids for an accelerated enzymatic hydrolysis. J Biotechnol 2009, 139:47-54.
25. Michalowicz G, Toussaint B, Vignon MR: Ultrastructural changes in poplar cellwall during steam explosion treatment. Holzforschung 1991, 45:175-179.
26. Grethlein HE: Pretreatment for enhanced hydrolysis of cellulosic biomass. Biotechnol Adv 1984, 2:43-62.
27. Clark TA, Mackie KL, Dare PH, McDonald AG: Steam explosion of the softwood Pinus radiata with sulfur dioxide addition. 2. process characterization. J Wood Chem Tech 1989, 9:135-166.
28. Wong KKY, Deverell KF, Mackie KL, Clark TA, Donaldson LA: The relationship between fiber porosity and cellulose digestibility in steam-exploded Pinus radiata. Biotechnol Bioeng 1988, 31:447-456.
29. Selig MJ, Viamajala S, Decker SR, Tucker MP, Himmel ME, Vinzant TB: Deposition of lignin droplets produced during dilute acid pretreatment of maize stems retards enzymatic hydrolysis of cellulose. Biotechnol Prog 2007, 23:1333-1339.
30. Grous WR, Converse AO, Grethlein HE: Effect of steam explosion pretreatment of pore size and enzymatic-hydrolysis of poplar. Enzyme Microb Technol 1986, 8:274-280.
31. Kumar R, Mago G, Balan V, Wyman CE: Physical and chemical characterizations of corn stover and poplar solids resulting from leading pretreatment technologies. Bioresour Technol 2009, 100:3948-3962.
32. Chundawat SPS, Bellesia G, Uppugundla N, Sousa LC, Gao D, Cheh AM, Agarwal UP, Bianchetti CM, Phillips GN Jr, Langan P, Balan V, Gnanakaran S, Dale BE: Restructuring the crystalline cellulose hydrogen bond network enhances its depolymerization rate. J Am Chem Soc 2011, 133:11163-11174.
33. Chundawat SPS, Donohoe BS, Sousa LC, Elder T, Agarwal UP, Lu F, Ralph J, Himmel ME, Balan V, Dale BE: Multi-scale visualization and characterization of lignocellulosic plant cell wall deconstruction during thermochemical pretreatment. Energy Environ Sci 2011, 4:973-984.
34. Arora R, Manisseri C, Li C, Ong M, Scheller H, Vogel K, Simmons B, Singh S: Monitoring and analyzing process streams towards understanding ionic liquid pretreatment of switchgrass (Panicum virgatum L.). Bioenergy Res 2010, 3:134-145.
35. Wyman CE, Dale BE, Elander RT, Holtzapple M, Ladisch MR, Lee YY: Comparative sugar recovery data from laboratory scale application of leading pretreatment technologies to corn stover. Bioresour Technol 2005, 96:2026-2032.

36. Li C, Knierim B, Manisseri C, Arora R, Scheller HV, Auer M, Vogel KP, Simmons BA, Singh S: Comparison of dilute acid and ionic liquid pretreatment of switchgrass: biomass recalcitrance, delignification and enzymatic saccharification. Bioresour Technol 2010, 101:4900-4906.
37. Shi J, Ebrik MA, Yang B, Garlock RJ, Balan V, Dale BE, Pallapolu VR, Lee YY, Kim Y, Mosier NS, Ladisch MR, Holtzapple MT, Falls M, Sierra-Ramirez R, Donohoe BS, Vinzant TB, Elander RT, Hames B, Thomas S, Warner RE, Wyman CE: Application of cellulase and hemicellulase to pure xylan, pure cellulose, and switchgrass solids from leading pretreatments. Bioresour Technol 2011, 102:11080-11088.
38. Selig M, Weiss N, Ji Y: Enzymatic Saccharification of Lignocellulosic Biomass. Golden, CO: National Renewable Energy Laboratory; 2008. [Technical Report NREL/TP-510-42629]
39. Chundawat SPS, Balan V, Dale BE: High throughput microplate technique for enzymatic hydrolysis of lignocellulosic Biomass. Biotechnol Bioeng 2008, 99:1281-1294.
40. Hu J, Arantes V, Saddler J: The enhancement of enzymatic hydrolysis of lignocellulosic substrates by the addition of accessory enzymes such as xylanase: is it an additive or synergistic effect? Biotechnol Biofuels 2011, 4:36.
41. Sluiter A, Hames B, Ruiz R, Scarlata C, Sluiter J, Templeton D: Determination of Sugars, Byproducts, and Degradation Products in Liquid Fraction Process Samples. Golden, CO: National Renewable Energy Laboratory; 2008. [Technical Report NREL/TP-510-42623]
42. Jeoh T, Ishizawa CI, Davis MF, Himmel ME, Adney WS, Johnson DK: Cellulase digestibility of pretreated biomass is limited by cellulose accessibility. Biotechnol Bioeng 2007, 98:112-122.
43. Kumar R, Wyman CE: Effect of enzyme supplementation at moderate cellulase loadings on initial glucose and xylose release from corn stover solids pretreated by leading technologies. Biotechnol Bioeng 2009, 102:457-467.
44. Qing Q, Yang B, Wyman CE: Xylooligomers are strong inhibitors of cellulose hydrolysis by enzymes. Bioresour Technol 2010, 101:9624-9630.
45. Qing Q, Wyman CE: Supplementation with xylanase and beta-xylosidase to reduce xylo-oligomer and xylan inhibition of enzymatic hydrolysis of cellulose and pretreated corn stover. Biotechnol Biofuels 2011, 4:18.
46. Chang V, Holtzapple M: Fundamental factors affecting biomass enzymatic reactivity. Appl Biochem Biotechnol 2000, 84–86:5-37.
47. Kumar R, Wyman CE: Cellulase adsorption and relationship to features of corn stover solids produced by leading pretreatments. Biotechnol Bioeng 2009, 103:252-267.
48. Zhang YHP, Cui JB, Lynd LR, Kuang LR: A transition from cellulose swelling to cellulose dissolution by o-phosphoric acid: evidence from enzymatic hydrolysis and supramolecular structure. Biomacromolecules 2006, 7:644-648.
49. Zhang S, Wolfgang DE, Wilson DB: Substrate heterogeneity causes the nonlinear kinetics of insoluble cellulose hydrolysis. Biotechnol Bioeng 1999, 66:35-41.
50. Singh S, Simmons BA, Vogel KP: Visualization of biomass solubilization and cellulose regeneration during ionic liquid pretreatment of switchgrass. Biotechnol Bioeng 2009, 104:68-75.

51. Stahlberg J, Johansson G, Pettersson G: A new model for enzymatic-hydrolysis of cellulose based on the two-domain structure of cellobiohydrolyase I. Nat Biotechnol 1991, 9:286-290.
52. Rollin JA, Zhu Z, Sathitsuksanoh N, Zhang YHP: Increasing cellulose accessibility is more important than removing lignin: a comparison of cellulose solvent-based lignocellulose fractionation and soaking in aqueous ammonia. Biotechnol Bioeng 2011, 108:22-30.
53. Kumar R, Wyman CE: Physical and chemical features of pretreated biomass that influence macro-/micro-accessibility and biological processing. In Aqueous Pretreatment of Plant Biomass for Biological and Chemical Conversion to Fuels and Chemicals. Edited by Wyman CE. Chichester: John Wiley & Sons Ltd; 2013:281-310.
54. Arantes V, Saddler J: Cellulose accessibility limits the effectiveness of minimum cellulase loading on the efficient hydrolysis of pretreated lignocellulosic substrates. Biotechnol Biofuels 2011, 4:3.
55. Kotiranta P, Karlsson J, Siika-aho M, Medve J, Viikari L, Tjerneld F, Tenkanen M: Adsorption and activity of Trichoderma reesei cellobiohydrolase I, endoglucanase II, and the corresponding core proteins on steam pretreated willow. Appl Biochem Biotechnol 1999, 81:81-90.
56. Lou H, Zhu JY, Lan TQ, Lai H, Qiu X: pH-Induced lignin surface modification to reduce nonspecific cellulase binding and enhance enzymatic saccharification of lignocelluloses. ChemSusChem 2013, 6:919-927.
57. Kumar R, Wyman CE: Access of cellulase to cellulose and lignin for poplar solids produced by leading pretreatment technologies. Biotechnol Prog 2009, 25:807-819.
58. Sluiter A, Hames B, Ruiz R, Scarlata C, Sluiter J, Templeton D, Crocker D: Determination of Structural Carbohydrates and Lignin in Biomass. Golden, CO: National Renewable Energy Laboratory; 2011. [Technical Report NREL/TP-510-42618]
59. Association of Official Analytical Chemists (AOAC): Protein (crude) determination in animal feed: copper catalyst Kjeldahl method (984.13). In Official Methods of Analysis. 15th edition. Gaithersburg, MD: AOAC International; 1990.
60. Kumar R, Wyman CE: An improved method to directly estimate cellulase adsorption on biomass solids. Enzyme Microb Technol 2008, 42:426-433.
61. Sluiter A, Hames B, Hyman D, Payne C, Ruiz R, Scarlata C, Sluiter J, Templeton D, Wolfe J: Determination of Total Solids in Biomass and Total Dissolved Solids in Liquid Process Samples. Golden, CO: National Renewable Energy Laboratory; 2008. [Technical Report NREL/TP-510-42621]
62. DeMartini JD, Studer MH, Wyman CE: Small-scale and automatable high-throughput compositional analysis of biomass. Biotechnol Bioeng 2011, 108:306-312.
63. Beldman G, Voragen AGJ, Rombouts FM, Searlevanleeuwen MF, Pilnik W: Adsorption and kinetic-behavior of purified endoglucanases and exoglucanases from Trichoderma viride. Biotechnol Bioeng 1987, 30:251-257.

# PART III

# CONSIDERATIONS FOR POLICY FORMATION

CHAPTER 9

# TECHNOECONOMIC AND POLICY ANALYSIS FOR CORN STOVER BIOFUELS

RYAN PETTER AND WALLACE E. TYNER

## 9.1 INTRODUCTION

This study is a technoeconomic analysis of the fast pyrolysis process. The main objective is to evaluate the effect of fuel price and technical uncertainty on the economic feasibility of biofuels created by catalytic fast pyrolysis using a circulating fluidized bed reactor. Previous studies have focused on creating reliable estimates of the initial capital and operating costs of a biorefinery. This study analyzes the effect of uncertainty in capital cost, hydrogen price, bio-oil yield, feedstock cost, and final product selling price on the appeal of project investment to a private investor. It directly evaluates the effect of two policy instruments—a reverse auction, which effectively fixes the price of the biofuel over the life of the project, and a capital subsidy. This research provides policy makers with information on how these policies could affect biofuel investment decisions.

Supply of advanced biofuels is expected to be a growing part of future liquid supply. US government investment in renewable power and fuels was $36 billion in 2012 [1]. Total US investment in the clean energy sector was $268 billion in 2012, a 500% increase since 2004 [2]. Imported petroleum products as a percentage of US petroleum consumption have decreased from 60% to 40% between 2005 and 2012 [3]. Supply of advanced biofuels is expected to be a growing part of future liquid supply [4, 5]. Research and development of economically attractive alternative fuel sources are ongoing in many nations around the world.

Catalytic fast pyrolysis is a process used to convert biomass to a bio-oil. This process is chosen for this study because of the relatively low cost per gallon [6–9]. Pyrolysis oil can be refined to diesel, gasoline, or jet fuel. Detailed descriptions of the fast pyrolysis process are available in the previous literature [9–15]. In this study, we use corn stover (corn residue), but the process can handle most cellulosic feedstocks. The remaining char and off-gas are burned in the production of electricity. Excess electricity can be sold [6].

A review of literature found no strong naming convention for pyrolysis derived biofuels. Cellulosic biofuels, biomass derived liquids, hydrocarbon fuels, and "drop-in" fuels all refer to renewable fuel derived from biomass. We use the term "biofuels" to describe biofuels produced via fast pyrolysis, although it can have many different meanings.

Technoeconomic analysis of fast pyrolysis is a method of forecasting the potential returns to an investment in a fast-pyrolysis production facility. It has been used extensively for studying feasibility of biofuel production. The National Renewable Energy Laboratory and the Pacific Northwest National Laboratory have conducted a series of technoeconomic analyses on a range of advanced biofuels and production methods [6, 8, 14]. This research produced detailed projections of the total cost of large capital investments, although there is some variability in the results [15]. The result of technoeconomic analysis is a cost break-even selling fuel price at which the future sales of transportation liquids and byproducts are equal to the present value of capital and operating expenditures, that is, a minimum fuel selling price. Research has shown that pyrolysis derived biofuels could be produced competitively, with estimated fuel selling prices ranging from $2.00 to $2.71 per gallon ($0.53 to $0.72 per liter) [8, 9, 16]. These studies suggest that commercial production of biofuel via fast pyrolysis is a competitive method of increasing domestic supply of transportation fuels.

According to research by Brown and Brown (2013), 215 million gallons per year (814 million liters per year) of cellulosic biofuel facilities are expected to be produced in the US by 2014 [15]. KiOR, ClearFuels, and Sundrop Fuels are three of the nine companies with commercial-scale projects expected to begin production by 2014 [15]. Meier et al. (2013) summarized current research, interest, and production of biofuels from pyrolysis in the six member countries of the IEA Bioenergy Task 34 [17]. They highlight Envergent and KiOR as commercial interests in the United States.

Although this study focuses on the economics of biofuel production, there are environmental effects as well [18–22]. The impact of biofuel expansion on greenhouse gas emissions is debated. Snowden-Swan and Male found that GHG reduction estimates from the petroleum baseline range from 62 to 68% [23]. However, if dedicated energy crops are used as feedstock, increased demand for land might cause an overall increase in greenhouse gas emissions [24]. It is generally agreed that the corn stover to biofuels pathway has no land use change impacts. There are other societal and economic effects from constructing and managing large production facilities and from harvesting, storing, processing, and transporting large amounts of biomass.

This research focuses on the effect of reverse auctions on the variance of net present value and internal rate of return. The government has many options for affecting biofuel prices, but these are not discussed here because they are covered in the previous literature [25, 26]. In a reverse auction, biofuel suppliers bid against each other for the opportunity to supply a certain volume of biofuel [27]. The result is a long-term forward contract with a competitive known price. The winning (lowest) bidder would have the biofuel price fixed for the duration of the contract, which is here assumed to be 20 years. The government would decrease long-term risk and obtain a competitive long-term price. In addition to examining the risk reduction impacts of the reverse auction, we also compare it to a capital subsidy with the same expected cost to government as a reverse auction.

## 9.2 METHODOLOGY

The method of production of biofuel used in this study is based on a technoeconomic analysis done at Iowa State University [6]. That study analyzed the production of biofuel via catalytic fast pyrolysis and hydroprocessing, a similar process to the commercial implementation by KiOR [28]. This study uses the same assumptions to validate the base case but then uses financial and economic assumptions that more closely represent reality in the marketplace. Additionally, this report adds uncertainty distributions to some parameters.

In this section, we first recreate the analysis done by Tristan Brown et al. [6]. We then detail the data used in our analysis and show how it differs from Tristan Brown et al. Next, we describe how uncertainty is modeled. Finally, we describe the eight experimental cases examined in this study.

Much of the research on fast pyrolysis has been done by chemists and engineers. In order to recreate the analysis done, we initially used the same spreadsheet model as some chemical engineers: a discounted cash flow rate of return analysis [14, 29]. Many economists employ a different treatment of inflation, debt repayment, taxes, and other factors. Of course, not all engineers use only the discounted case flow rate of return method, nor have all economists used the more complete economic and financial analysis. However, for simplicity in presentation, we will differentiate the two types of spreadsheets by labeling them as engineering analysis and economic analysis, respectively.

### *9.2.1 ENGINEERING ANALYSIS*

To be certain that our spreadsheet was an accurate representation of the original Brown data, we first recreated the analysis by Tristan Brown et al. [6] using an engineering analysis. Some of the data needed to recreate the analysis was not available in that paper. For those parameters we assumed that they match those in a previous Iowa State University study [14].

The economic assumptions used by Wright et al. [14] or Tristan Brown et al. [6] are documented in Table 1.

**TABLE 1:** Economic assumptions.

| Parameter | Input value | Units | Source |
|---|---|---|---|
| Real discount rate | 10% | % | [6] |
| Nominal interest rate | 7.5% | % | [6] |
| 10 year depreciation | 200% | % | [14] |
| Plant depreciable life | 7 | yrs | [14] |
| Equity | 50% | % | [6] |
| Financing | 50% | % | [6] |
| Loan term | 10 | yrs | [6] |
| Construction time | 3 | yrs | [14] |
| % spent in year 1 | 8% | % | [14] |
| % spent in year 2 | 60% | % | [14] |
| % spent in year 3 | 32% | % | [14] |
| Income taxes | 35% | % | [14] |

Some of the key economic assumptions are a total project timeline of 23 years, 50% debt financing, a 10% discount rate, a loan payback period of 10 years, a tax rate of 35% of net income, and a construction time of 3 years. No inflation rate was specified in either analysis.

In addition to the economic variables, we used many of the technical variable values as Tristan Brown et al. [6]. The technical assumptions we used in our recreation are summarized in Table 2. More details, including the number and type of employees and other assumptions are available from the authors on request.

Besides technical parameters, financing assumptions also affect the analysis. Tristan Brown et al. [6] do not provide details, but we assume that their engineering spreadsheet is similar to work done by Humbird et al. [29]. Humbird et al. [29] assume that land is paid immediately with equity, interest is paid off during the first three construction years, and that tax benefits from losses on operations are claimed in the current year. Finally, Humbird et al. [29] do not include loan or interest payments when computing tax payments.

**TABLE 2:** Technical assumptions.

| Parameter | Input value | Units | Source |
|---|---|---|---|
| Working capital | 15% | % | [6] |
| Project contingency | $61,490,157 | $ | Calculated from data in [14] |
| Working capital expenditure | $55,341,142 | $ | Calculated from data in [14] |
| Total capital investment | $429,000,000 | $ | [6] |
| Land | $4,717,915 | $ | Calculated from data in [14] |
| Nominal fixed capital Investment | $394,799,061 | $ | Calculated from data in [14] |
| Plant operation/online time | 329.5 | days/yr | Calculated from data in [6] |
| Startup production rate | 75% | % | [14] |
| Startup variable expense | 87.5% | % | [14] |
| Input capacity | 2,000 | MT/day | [6] |
| Annual feedstock use | 659,000 | MT/yr | [6] |
| Bio-oil yield | 63.00% | Mg/Mg | [6] |
| Gas conversion rate | 21.0% | Mg/Mg bio-oil | [14] |
| Diesel conversion rate | 21.0% | Mg/Mg bio-oil | [14] |
| Feedstock cost | 83 | $/MT | [6] |
| Catalyst replacement costs | 1,767,000 | $/yr | [14] |
| Electricity produced | 223,000,000 | kwh/yr | [6] |
| Electricity use | 11,490 | kw/hour | [14] |
| Electricity price | $0.0540 | $/kwh | [14] |
| Hydrogen use | 2,041 | kg/hour | [14] |
| Hydrogen price | $1.33 | $/kg | [6] |

Using the engineering analysis and the above data, we found a facility fuel output of 58.6 millions of gallons per year (222 millions of liters). Wright et al. [14] reported a fuel yield of 58.2 MGY, and 57.4 MGY was reported by Tristan Brown et al. [6]. We found a minimum fuel selling price of $2.55 per gallon ($0.67 per liter) compared to $2.57 ($0.68 per liter) in Tristan Brown et al. [6]. Thus, we can be confident that we have accurately reconstructed the Tristan Brown et al. [6] assumptions and results.

The fuel price was determined using the goal seek add-in from Excel to set net present value equal to zero. Minimum fuel selling price is the price that makes the net present value with a 10% real discount rate equal to zero. The 10% discount rate is sometimes called a hurdle rate. We do not know what discount rate a private investor would require. However, choosing a different discount rate would not change the conclusions from this study. The conclusions from this study are based on variability not break-even fuel price. We have chosen to use the 10% real discount rate used in Tristan Brown et al. [6]. Thus, the term "minimum fuel selling price" is analogous to "break-even price."

**TABLE 3:** Differences in financing assumptions.

| Engineering analysis | Economic analysis |
|---|---|
| Land paid with equity | Land included in capital investment and cost is divided between financing and equity |
| Tax benefits or losses carry over | Tax benefits or losses occur in the year they occur |
| Interest is paid with equity during construction | Interest is compounded annually during construction (interest is capitalized) |
| Loan and interest are not included in computing taxes | Loan interest is deductible for purposes of tax calculations |
| Inflation is assumed to be zero | Inflation is assumed to be 2.5% |

### *9.2.2 ECONOMIC SPREADSHEET ANALYSIS*

We did our economic analysis in an Excel spreadsheet using most of the same technical data values from Tristan Brown et al. [6]. This section provides details on changes made to the technical and economic parameters. First, we describe financing differences. Then, we describe the difference in assumed tax rate. Finally, we examine changes in parameter values and their effect on expected return.

In the previous section we described financing assumptions present in the engineering analysis. These were obtained from a report by Humbird et al. [29] and seem to be used in Table D-3 in Wright et al. [14]. We assume that they are also used in Tristan Brown et al. [6]. We chose different financing assumptions in our economic analysis which have material effects on the return on investment. For example, the engineering spreadsheets assumed the inflation rate to be zero. However, average inflation as shown in the US consumer price index has been 2.9% over the past 30 years (Bureau of Labor Statistics 2013a). We assumed a 2.5% inflation rate. The differences in financing assumptions are summarized in Table 3. Depreciation for tax purposes was the same in both the engineering and economic analyses.

The combined effect on investment return of the assumption differences was calculated to see the importance of the approach. As described previously, our recreation of the previous studies' assumptions found a 10.0% internal rate of return at a fuel price of $2.55 per gallon ($0.67 per liter). With our financing assumptions but not our price or tax assumptions, a fuel price of $2.55 per gallon ($0.67 per liter) results in a 12.0% return. Using the economic financing assumptions, we find a break-even fuel price of $2.40 per gallon ($0.63 per liter). In other words, there is a 6.3% difference in the break-even price between the economic and engineering analyses and a 2.0% difference in the internal rate of return.

In addition to these financing assumption differences, three technical parameter values were changed: hydrogen price, feedstock price, and fuel yield. These variables have min, mode, and max values, which make up the Pert distribution input parameters for the stochastic analysis. Details on the changes to hydrogen price, feedstock price and fuel yield are explained in Section 2.3. However, for reference these changes are summarized in Table 4.

**TABLE 4:** Differences in stochastic parameter estimates.

| Parameter | Old value Tristan Brown et al. [6] | New value | | | Mean value | Unit | Source |
|---|---|---|---|---|---|---|---|
| | | Min | Mode | Max | | | |
| Hydrogen price | 1.33 | 1.33 | 2.02 | 2.94 | 2.06 | $/kg | [30, 31] |
| Feedstock price | 83 | 55 | 83 | 110 | 82.83 | $/MT | [6] |
| Fuel yield | 63 | 49 | 63 | 70 | 61.8 | % | [6] |

Hydrogen price, feedstock price, and fuel yield each have new mean values. Hydrogen has a much higher price, and fuel yield is slightly lower. The combined effect shifts the break-even fuel price from $2.40 to $2.65 per gallon ($0.63 to $0.70 per liter). Remember that our recreation of Tristan Brown et al. [6] had a break-even price of $2.55 per gallon ($0.67 per liter).

Besides the adjustments made to variables in Table 4, income tax was adjusted lower. The tax rate in Tristan Brown et al. [6] is 35%, which is the statutory corporate federal tax rate. Effective tax rates measure the proportion of taxes paid to economic income. For profitable firms filing a Schedule M-3, the effective federal tax rate was about 13% in 2010 [32]. Including foreign, state, and local taxes for all firms filing a Schedule M-3 raises the effective tax rate to about 23% [32]. Therefore, for this study, we assume a tax rate of 24% because it is closer to the tax rate paid by corporations.

The lower tax rate has a less severe effect than the previous changes. Lowering the tax rate from 35% to 24% in our recreation of the Tristan Brown et al. [6] study lowers the break-even price from $2.65 to $2.62 per gallon ($0.70 to $0.69 per liter). Table 5 summarizes the changes made and their effect on break-even price.

**TABLE 5:** Alternative measures of break-even price.

| Description of conditions | $ per gallon | $ per liter |
|---|---|---|
| As reported in Tristan Brown et al. [6] | $2.57 | $0.68 |
| Tristan Brown et al. [6] as recreated by authors with engineering financing assumptions, higher tax rate | $2.55 | $0.67 |
| Old variable values, economic financing assumptions, higher tax rate | $2.40 | $0.63 |
| New variable values, economic financing assumptions, higher tax rate | $2.65 | $0.70 |
| New variable values, economic financing assumptions, lower tax rate | $2.62 | $0.69 |

### *9.2.3 TECHNICAL UNCERTAINTY*

The uncertainty inherent in a biofuel production facility stems from a multitude of sources, but four are chosen because of their importance to project viability. The four parameters modeled as uncertain are grouped into technical uncertainty and fuel price uncertainty. The parameters labeled with technical uncertainty are the price of the feedstock, the bio-oil yield, and hydrogen purchase price. Table 6 shows the share in total cost of capital investment, feedstock, hydrogen, and other operating cost. Bio-oil yield clearly could have an important impact on economic viability, as the higher the yield, the lower the unit cost. Later, in one of the simulation cases, we will test the importance of the capital cost assumption.

**TABLE 6:** Cost shares for key cost components.

| Item | NPV cost ($) | Cost share (%) |
|---|---|---|
| Capital cost (with working capital) | $303,129,655 | 30.3% |
| Feedstock | $344,499,049 | 34.4% |
| Hydrogen | $209,663,508 | 21.0% |
| Other operating cost | $142,824,530 | 14.3% |
| Total | $1,000,116,742 | 100.0% |

*Source: author's calculations.*

In related studies, researchers conduct uncertainty analysis by showing the impact of changing certain variables one at a time and determining the maximum and minimum rate of return or fuel selling price [6, 14]. These analyses do not provide probability of the high or low values or the assessment of combined uncertainty. These studies use sensitivity analysis to identify the variables that are most critical to the cash flow. However, sensitivity analysis does not consider variable interaction, so it provides a less quantifiable measure of risk. These studies determine that two of the most important variables in terms of their expected effect on the break-even price are bio-oil and fuel yield and biomass cost. Other important parameters reported are electricity selling price, interest rate, and project investment cost.

This study uses Monte Carlo simulation, which uses input distributions rather than fixed values. The Palisades risk and decision analysis software,

@Risk, is used to account for variability in the technical parameters: biomass cost, bio-oil yield, and hydrogen price. Monte Carlo simulation creates input distributions based on data from the literature, econometric estimates, or other sources. Common distributions that are used for input costs are the Triangular and Pert. Both of these distributions have as parameters the max, mode, and min values. Experts can more easily answer expected, maximum, and minimum parameter values than selecting a mean and standard deviation as for a normal distribution. The main difference between the two is that the Pert has more of the probability density closer to the mean, and the triangular has more towards the max and min values. Since Pert has more probability near the mean, it was chosen for this study.

The high, mode, and low values for bio-oil yield are the same as used by previous studies [6, 14]. Bio-oil yield varies depending on factors such as the type of catalyst and the rate of temperature change during the pyrolysis reaction. Yields are expressed as weight of oil as a percentage of the weight of starting biomass. Bio-oil yields of 70%, 63%, and 49% set out the max, mode, and min for a Pert distribution. The mean value for this Pert distribution, as reported earlier, is 61.8%. This changes the annual biofuel production from 58.6 MGY to 57.5 MGY, nearly the same as the 57.4 MGY reported in Tristan Brown et al. [6].

The feedstock price varies depending on material and the cost of preparing the material for pyrolysis. Prices of $110, $83, and $55/MT are reported in the same studies [6, 14]. We use these prices as the maximum, mode, and minimum for a Pert distribution. The mean value, as shown in Table 4, is $82.83/MT.

The other technical parameter modeled with uncertainty is hydrogen price. The National Research Council studied conventional and advanced hydrogen production related to their study of hydrogen fuel cell vehicles [30]. Using the H2A model, they found costs of $1.60 and $1.90 in 2005 dollars, respectively, for natural gas reforming and coal gasification. Adjusted to 2011 dollars using the Producer Price Index for Industrial Commodities, the natural gas reforming cost increases to $2.02 [33]. A 2010 study conducted by Iowa State and Texas A&M summarized the cost estimates of various studies in the production of hydrogen [31]. They found coal gasification with current technology and carbon sequestration costing between \$1.25 and \$1.83 and natural gas steam methane reforming costing between \$2.33 and \$3.17, in 2007 dollars. The low estimate for steam methane reforming using natural gas was \$2.33. Adjusted to 2011 dollars using the Producer Price Index for Industrial Commodities this cost increases to \$2.94 [33]. We use a low price of \$1.33, mode price of \$2.02, and high price of \$2.94 per kilogram. These three technical parameters were summarized in Table 4.

### *9.2.4 FUEL PRICE UNCERTAINTY*

Diesel and gasoline compete with biofuel, so the prices of those fuels directly impact the economic feasibility of biofuel production. The selling price of fuel in this study is assumed to be the same as wholesale gasoline and diesel prices [6]. The DOE projects real increases in the price of crude oil but with wide uncertainty as shown by the differences among the high, reference, and low price scenarios in Figure 1 [4]. Figure 1 displays projections from the 2013 Annual Energy Outlook for Brent crude spot prices from 1990 to 2040 in constant 2011 dollars per barrel.

To capture the uncertainty regarding fuel prices and the impact of selling price uncertainty on the expected return on investment, we employ three fuel price projections: (1) a fixed price, (2) a price that fluctuates but remains at about the same level on average over the time period, and (3) a price that fluctuates but increases at a rate similar to the DOE reference scenario.

In the first price projection, the fuel price is fixed in real terms. In nominal terms the price increases because of the 2.5% inflation rate. This fixed fuel price represents a reverse auction outcome. If the government created a long-term forward contract based on a reverse auction, from a private perspective the fuel selling price would be fixed.

The nondeterministic steady price is useful for understanding the importance of variability in fuel selling price on the economic profitability of investment in a biofuel refinery. It has about the same average price as the fixed price, but the uncertainty in fuel price is expected to increase the risk of investment.

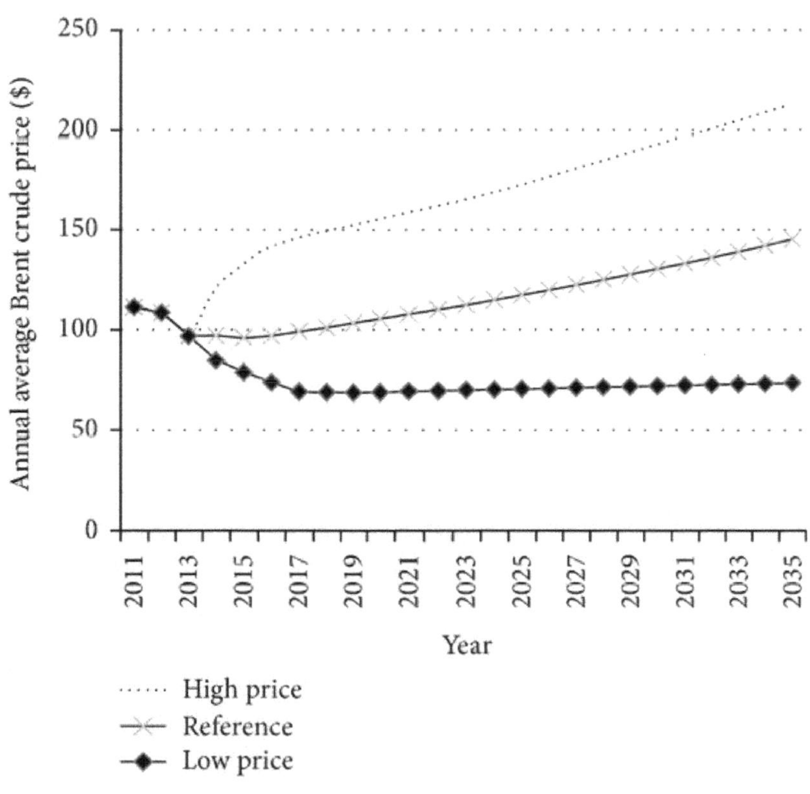


**FIGURE 1:** Projected Brent crude prices from Annual Energy Outlook 2013 [4].

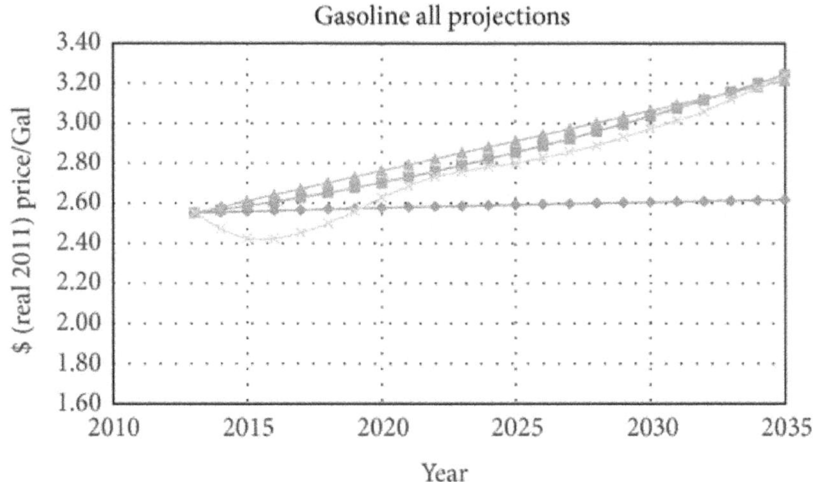


**FIGURE 2:** Comparison of all gasoline price projections.

The nondeterministic price with upward trend was created to be similar to DOE projections from 2013 to 2035. The starting gasoline price in our analysis is the DOE projection for 2013 (2011 dollars) but subtracting $0.70 per gallon ($0.18 per liter) to convert retail prices to wholesale prices [34]. This is the same starting price used for the fixed price. Our forecast for the uncertain price with upward trend also matches the DOE forecast less $0.70 per gallon ($0.18 per liter) for expected fuel price in 2035, which is the final project year.

We tested both mean reversion and Brownian motion with trend models. The Brownian motion with trend model had a lower standard deviation and did not follow the DOE projections as closely. Figure 2 compares the projections: static DOE projections, mean reversion, mean reversion with upward drift, and Brownian motion with upward drift. The three gasoline prices used in the experiments are static price, mean reversion with no upward trend, and mean reversion with upward trend. Tristan Brown et al. [6] assume that half of the fuel output from the plant is gasoline and half is diesel. These fuels have different selling price, although their prices are highly correlated with each other. Regressing the historic wholesale prices of gasoline and diesel for 2004–2012, we found an R-square of 97%. We used the intercept and slope of this regression, which were −0.322 and 1.201, to find the diesel projected prices based on gas prices. This is shown in the following equation:

$$\text{Diesel price}_t = -0.322 + 1.201 * \text{Gasoline price}_t \quad (1)$$

The fuel produced in this project is 50% each of diesel and gasoline. Therefore, we forecast a combined price by using the average of the diesel and gasoline price for each year. The cumulative results are fuel price forecasts for the next 23 years, from 2013 to 2035.

### 9.2.5 EXPERIMENTAL CASES

There is a high expected variance in gasoline and diesel prices over the time period. Thus, the experiments are chosen to help identify the effect of government intervention on the investment decision given that price uncertainty. Together, these cases provide an understanding of the impact of fuel price uncertainty on investment in a biofuel production facility.

1. A base case with a fixed fuel price of $2.68 per gallon ($0.71 per liter) and no uncertain variables.
2. A stochastic case (for all three technical uncertain variables) plus an uncertain fuel price that has no upward trend.
3. A stochastic case (for all three technical uncertain variables) plus an uncertain fuel price that has an upward trend keyed to the DOE reference price forecast.
4. A stochastic case (for all three technical uncertain variables) but with a break-even fuel price. The fuel price is fixed in real terms to provide the same expected net present value as in case two ($2.68 per gallon or $0.71 per liter). This price is a possible outcome of a reverse auction.
5. A stochastic case (for all three technical uncertain variables) but with a fuel price fixed to provide the same expected economic net present value as in case three ($2.92 per gallon $0.77 per liter). This price is also a possible outcome of a reverse auction.
6. A stochastic case that matches case three but with 20% higher capital investment cost.
7. A stochastic case that matches case three but with a fixed fuel selling price ($2.92 per gallon $0.77 per liter) for 45 million gallons (170 million liters) of fuel per year (for project years 5–23).
8. A stochastic case that matches case three but with a $5 million capital subsidy. This is designed to compare public intervention in the form of a capital subsidy with a reverse auction. Cases (7) and (8) have the same expected government cost.

## 9.3 RESULTS AND DISCUSSION

This section presents and summarizes the results of the analysis. The next section draws conclusions from these results.

### 9.3.1 BASE CASE

Case 1, the base case, is a completely deterministic model with the hydrogen price, bio-oil yield, feedstock price, and fuel selling price values fixed at their expected mean levels. It is the only case where the parameters are all deterministic. The values of the technical parameters are fixed at their mean levels which, for the Pert distribution, are calculated as shown in the following equation:

$$\text{Pert Mean} = (\min + 4 * \text{mode} + \max) / 6 \tag{2}$$

The mean values are shown in Table 7.

**TABLE 7:** Case 1 technical variable values.

| Variable name | Input value |
|---|---|
| Hydrogen price | \$2.06/gal or \$0.71/liter |
| Feedstock price | \$82.83/MT |
| Fuel yield | 61.8% |

Results are shown in net present value (NPV), internal rate of return (IRR), and benefit cost ratio (B/C). Net present value provides a discounted measure of project worth. The project is accepted when NPV is greater than zero. One problem with NPV is the large numbers that can come from large projects. Another problem is choosing an appropriate discount rate. IRR measures the rate of interest that makes the NPV equal zero. IRR assumes that profits can be reinvested at the IRR, but this may not be possible. Table 8 summarizes the results from the base case.

Table 8 is separated into a financial analysis and an economic analysis. The economic analysis is before financing and taxes while the financial is after accounting for those. The economic analysis is also known in the financial literature as asset based analysis. There are two competing causes for differences in the financial and economic NPV and IRR. The first is that the nominal loan interest rate (7.5%) is lower than the real discount rate (10%). When the interest rate is lower than the discount rate, as the equity ratio decreases, the return increases. The rational private investor under these conditions would seek as much financing as possible for capital investments when the loan interest rate is below the discount rate. We assumed a 50% equity structure. Thus, including financing improves the project return.

**TABLE 8:** Case 1 results.

| | NPV | IRR (real) | B/C |
|---|---|---|---|
| Economic | $0.00 | 10.0% | 1.00 |
| Financial | $16,641,078 | 11.0% | 1.09 |

Second, tax payments are included in financing, which lowers project return. All of the cases with positive expected IRR pay taxes during the project time period. When returns are higher, tax payments are greater. In some cases that follow, when returns are higher, the tax payment cost exceeds the benefit from financing. In those cases, the economic NPV will be higher than financial NPV. In Case 1, however, the benefit from financing exceeds the cost from tax payments.

In the base case, the project has an economic NPV of zero but a positive financial NPV. We chose to use an economic break-even fuel price for this case. At a price of $2.68 ($0.71 per liter), the financial IRR is 11.0% and the economic IRR is 10.0%. The financial break-even price is $2.62 ($0.69 per liter). The financial IRR at this price is 10%, and the economic is 9.2%. The economic break-even price is $2.68 ($0.71 per liter), higher than the $2.57 ($0.68 per liter) break-even price in the Tristan Brown et al. [6] study.

Investors are risk averse, so without knowing the riskiness of an investment one cannot discern its attractiveness. In the next scenario, we examine the impact of uncertainty.

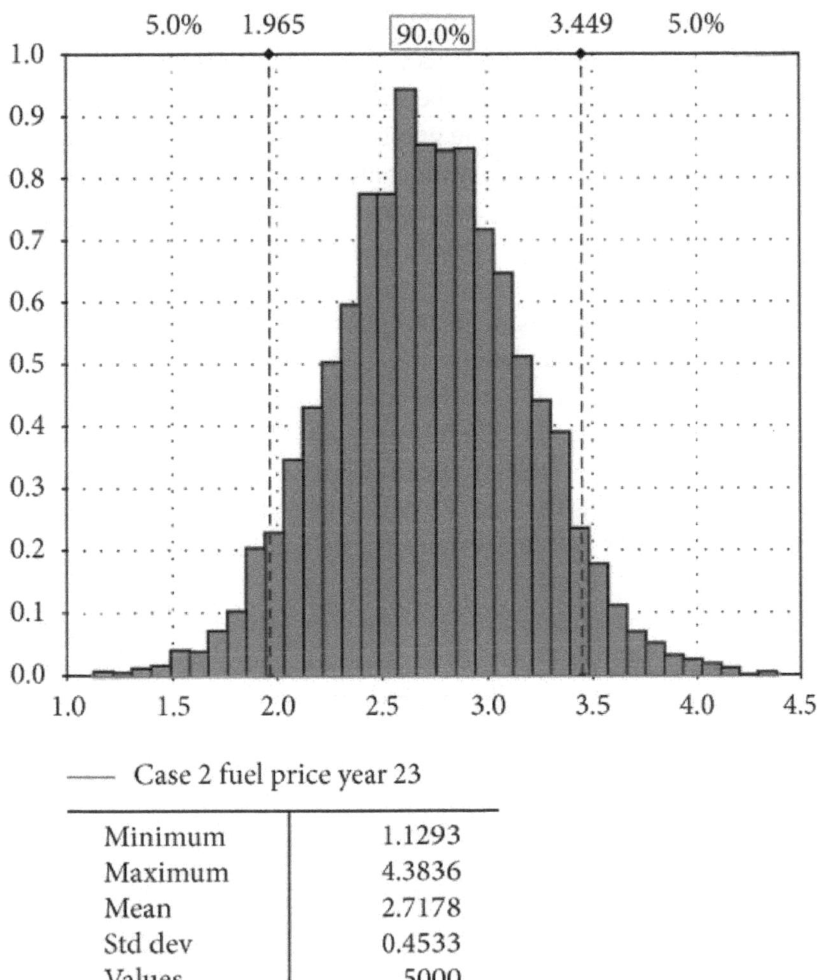


**FIGURE 3:** steady fuel price distribution.

### 9.3.2 STOCHASTIC STEADY PRICE CASE

In Case 2, we employ Monte Carlo analysis to predict uncertainty in NPV given uncertainty in technical and market parameters. The break-even fuel selling price in the base case was $2.68 per gallon ($0.71 per liter). This case has the same economic NPV as in Case 1 but adds uncertainty.

While the previous case could be shown in a small table, the results of Case 2 are better understood using graphs of input and output distributions. As noted earlier, the technical variables are assumed to have Pert probability distributions. The technical parameters for each year are considered to be independent of each other, so the distribution curves are nearly identical for each project year.

The fuel selling price is nondeterministic in a different way because of the mean reversion calculations. The fuel selling price distribution will have variance from year to year because of the mean reverting process. Figure 3 shows the outcome of a fuel price distribution in the final project year. There is an 80% probability that the ending price was between $2.34 and $3.10 per gallon ($0.62 and $0.82 per liter). The maximum fuel price in year 23 was $4.38 ($1.16 per liter) and the minimum was $1.13 ($0.30 per liter).

On average, the economic NPV was near zero. However, there was a total range of $837 million for the economic case. Table 9 summarizes some of the key results from this experiment.

**TABLE 9:** Case 2.

| | Economic | | | Financial | | |
|---|---|---|---|---|---|---|
| | NPV | IRR (nominal) | B/C | NPV | IRR (nominal) | B/C |
| Expected value | $0 | 10.0% | 1.00 | $16,641,078 | 11.0% | 1.09 |
| Standard deviation | $100,993,377 | 3.9% | 0.29 | $76,669,403 | 4.8% | 0.40 |
| Max | $427,930,098 | 22.8% | 2.24 | $339,121,179 | 26.9% | 2.76 |
| Min | ($408,886,573) | −15.6% | −0.19 | ($295,088,669) | −13.6% | −0.53 |
| Probability of loss | 50.3% | | | 40.8% | | |

The financial case had a higher standard deviation for IRR but lower standard deviation for NPV. In calculating the economic IRR, there were 105 errors out of the 5,000 iterations during the simulation. The financial case IRR had 121 errors. Errors occur in the IRR because, under some combination of random draws from the input distributions, the flows are so skewed in the negative or positive direction (usually negative) that no IRR can be found. Under that condition, the IRR function returns an error, and the result is reported as such in the Monte Carlo outputs. For example, for the financial case, the mean IRR is based on 4879 iterations instead of 5000. Errors are not found in NPV calculations. For this reason NPV is generally a better measure of risk distributions.

The economic case had a wider NPV distribution than the financial case, and this will remain true for all of the experiments. This is because of the tax payments which reduce gains in highly profitable iterations and reduce losses (due to negative taxes) in low profitability or loss iterations. One standard deviation in the financial case is \$77 million, while it is \$101 million in the economic case. These standard deviations are based on the input distributions from the uncertain variables. Since the underlying input distributions will not change in the stochastic cases, the standard deviations should be about the same from case to case except when uncertain variables are held constant. The financial NPV was slightly better than the economic NPV.

The economic NPV ranged from −\$409 million to +\$428 million. There was a 50% chance of a loss for the economic case, as the mean of the distribution was near zero. The financial NPV had an 80% probability of being between −\$81 million and \$113 million. The maximum was \$339 million and the minimum was −\$295 million. There was a 41% probability that the financial NPV was less than zero.

This case enforces the wide variability and high amount of risk inherent in an investment of this magnitude. Even if the expected IRR was 20% instead of 11%, the wide spread of possible outcomes would make investment risky. Remember, the initial capital investment is \$429 million. In this result there is a spread in the financial NPV of \$634 million. This experiment only had three technically uncertain variables. Including others would make the spread of possibilities even greater. The next case creates an increasing fuel selling price that more closely matches DOE projected fuel prices.

### 9.3.3 STOCHASTIC INCREASING CASE

Case 3 has an uncertain fuel price that increases to, on average, meet DOE projections in 2035. It is the same as in the second case, except that it has a rising fuel selling price. The ending price is higher in Case 3, increasing to $3.41 per gallon ($0.90 per liter) of fuel on average in 2011 real dollars. One standard deviation is $0.40 ($0.11 per liter). Table 10 summarizes the results from this scenario.

**TABLE 10:** Case 3 results.

| | Economic | | | Financial | | |
|---|---|---|---|---|---|---|
| | NPV | IRR (nominal) | B/C | NPV | IRR (nominal) | B/C |
| Expected value | $84,252,474 | 12.6% | 1.24 | $80,672,958 | 14.2% | 1.42 |
| Standard deviation | $100,675,911 | 3.1% | 0.29 | $76,305,318 | 3.8% | 0.40 |
| Max | $447,048,535 | 22.7% | 2.30 | $349,991,319 | 26.4% | 2.81 |
| Min | ($268,437,824) | −5.0% | 0.22 | ($190,303,825) | −8.1% | 0.01 |
| Probability of loss | 20.7% | | | 14.5% | | |

The increasing fuel price had a positive effect on the expected NPV of the investment, but Case 3 still has a risk of loss. The economic NPV had a 21% probability of loss, while it was 50% in the second case. The financial NPV likelihood of loss decreased from 41% to 15%. Increasing the product selling price increased probability of having a positive NPV and increased the expected return.

While in Case 2 the economic NPV was greater than the financial NPV, in Case 3 the expected financial NPV of $81 million was lower than the expected economic NPV of $84 million. The higher fuel selling price results in higher profits from project investment. The negative cost of taxes had a larger impact than the benefit gained from the spread between debt and discount rate in the financing of capital debt. The NPV of taxes for case two was $18 million, but in Case 3 the NPV of taxes was $39 million.

At the same time, the financial IRR of 14% was slightly better than the economic IRR of 13%. This is because of the initial capital investment being financed through debt. The financial case had a lower initial equity investment. The IRR had a minimum return of −2% in this case, but the middle 80% of results were between 9% and 19%. The expected IRR increased by 3% from the second to the third case.

Cases 1 to 3 have shown a range of possibilities. The following cases will isolate the effect of changes in either the product selling price or the initial capital investment.

### 9.3.4 FORWARD CONTRACT: FIXED SELLING PRICE

Case 4 is the same as Case 2 except for the selling price of fuel. While Case 2 had a variable selling price, this case has a fixed fuel price. The fuel price in this scenario is set to $2.68 ($0.71 per liter) in real terms. This scenario isolates the uncertainty in return on investment caused by the technical variables. Table 11 shows the results.

**TABLE 11:** Case 4 results.

| | Economic | | | Financial | | |
|---|---|---|---|---|---|---|
| | NPV | IRR (nominal) | B/C | NPV | IRR (nominal) | B/C |
| Expected value | $0 | 10.0% | 1.00 | $16,641,078 | 11.0% | 1.09 |
| Standard deviation | $20,684,124 | 0.7% | 0.06 | $15,051,080 | 0.9% | 0.08 |
| Max | $79,292,123 | 12.9% | 1.23 | $73,091,401 | 14.5% | 1.38 |
| Min | ($86,537,333) | 6.8% | 0.75 | ($43,554,644) | 7.3% | 0.77 |
| Probability of loss | 50.2% | | | 13.2% | | |

The economic NPV is made to be zero by finding the break-even fuel price, while expected economic NPV did not change from Case 2. The expected economic NPV was equal to zero in both the second and the fourth scenarios, so although standard deviation decreased, half of the outcomes were positive and half were negative. Standard deviation decreased dramatically, from $101 million to $21 million. The probability of loss did not decrease.

The total range between the maximum and minimum financial NPV decreased from $634 million to $117 million from Case 2 to Case 4. One standard deviation decreased from $76 million to $15 million. The probability of a loss in NPV decreased from 41% to 13%. When risk is reduced to technical parameters, there is much less risk to the private investor.

If the government were to guarantee biofuel producers a minimum or a fixed price for their product, private investors would become more interested in this industry. Under these circumstances, it is apparent that reducing fuel price uncertainty would encourage investment.

**TABLE 12:** Case 5 results.

| | Economic | | | Financial | | |
|---|---|---|---|---|---|---|
| | NPV | IRR (nominal) | B/C | NPV | IRR (nominal) | B/C |
| Expected value | $84,252,474 | 12.9% | 1.24 | $80,672,958 | 14.7% | 1.42 |
| Standard deviation | $21,955,577 | 0.8% | 0.06 | $15,980,391 | 1.0% | 0.08 |
| Max | $157,793,602 | 15.5% | 1.46 | $133,789,946 | 17.8% | 1.69 |
| Min | $3,099,769 | 10.1% | 1.01 | $19,716,164 | 11.1% | 1.10 |
| Probability of loss | 0.0% | | | 0.0% | | |

### *9.3.5 FORWARD CONTRACT: INCREASING SELLING PRICE*

Case 5 is similar to Case 4 in that the fuel price is fixed while the technical parameters are still forecast with uncertainty. The fuel price in Case 5 is fixed to meet the financial NPV in Case 3. Fuel price is fixed to $2.92 ($0.77 per liter) in this case. It makes intuitive sense that the fixed fuel price is much higher in this scenario than in Cases 2 and 4 because of the higher product selling price. The results from this simulation are in Table 12.

Eliminating the fuel price uncertainty reduced the standard deviation in the financial NPV from $76 million to $16 million. The total range of financial NPVs decreased from $540 million in case three to $114 million in this case.

As in Case 3, the higher tax payments related to the higher fuel price result in a lower financial NPV than economic NPV. The financial IRR was better than the economic IRR because the financial IRR had a smaller initial investment cost.

Similar to our conclusion from Case 4, we can say that fuel price uncertainty comprised a large part of the uncertainty in these experiments. The government could employ reverse auctions to reduce private risk and encourage investment in this infant industry.

### *9.3.6 STOCHASTIC WITH HIGHER CAPITAL COST*

For Case 6, we test the impact of higher-than-expected capital costs. Capital costs with new technologies and unknown production costs are difficult to predict. There has been a history of cost overruns for new technologies. Also, much of the literature casts the estimated capital cost as being for the th plant. We attempted to measure the impact on project investment with a 20% cost overrun.

Case 6 uses the scenario in Case 3 as its base. Total capital investment is modeled as deterministic in all cases. Total capital investment increased by 20% in this case, from $429 million to $514.8 million. Table 13 summarizes the results.

With an $86 million increase in upfront capital costs, the economic NPV changed by nearly the same amount, as would be expected. It was not exactly the same due to inflation. The financial NPV changed by about $60 million. This occurs because the increased capital investment was decreased by adding taxes and financing. Probability of a loss was more than twice as high as in Case 3. Financial NPV probability of loss increased from 15% to 34% with only the change in capital costs.

**TABLE 13:** Case 6.

| | Economic | | | Financial | | |
|---|---|---|---|---|---|---|
| | NPV | IRR (nominal) | B/C | NPV | IRR (nominal) | B/C |
| Expected value | $16,619,218 | 10.5% | 1.04 | $32,599,899 | 11.5% | 1.14 |
| Standard deviation | $101,765,400 | 2.9% | 0.25 | $77,252,890 | 3.5% | 0.33 |
| Max | $377,296,296 | 18.7% | 1.91 | $311,273,718 | 22.2% | 2.34 |
| Min | ($341,382,691) | −8.9% | 0.17 | ($238,976,396) | −11.8% | -0.03 |
| Probability of loss | 43.7% | | | 34.0% | | |

The standard deviation, middle 80% range, and total range remained about the same, because the shift occurred to a deterministic parameter. The economic NPV distribution is remarkably similar to Case 2, because the NPV loss from increased capital cost was about equal to the NPV of the benefit accrued from the higher selling price.

### 9.3.7 STOCHASTIC WITH FORWARD CONTRACT

Case 7 is a model of the impact of a realistic reverse auction outcome. We start with the assumptions used in Case 3. Then, we add a forward contract for 45 MGY for project years 5–23 at the fixed fuel price used in Case 5.

Project years 5–23 were chosen because this is the time period in which the facility is producing at full capacity. In project year four, the facility is starting up, so it has a 75% start-up production rate.

To determine the forward contract volume, we measured the minimum fuel output based on the input uncertainty during the project timeline. We assume that the biofuel producer is willing to contract for the minimum expected facility fuel output. The minimum facility fuel output is above 45 million gallons (170 million liters) per year. The expected volume is 57.5 MGY but changes each year. The remaining volume beyond 45 MGY is sold at the fuel price. The forecasted fuel price is nondeterministic and is the same as in Case 3.

The forward contract price used is \$2.92 (\$0.77 per liter). This is the price determined in Case 5 to provide the same expected return as in Case 3. \$2.92 per gallon (\$0.77 per liter) could be the estimated average price over the 20-year production period. The government benefits when the forward contract fuel price is below market fuel price. In other words, when the market fuel price is high, the government gains from a lower purchase price than what it could get on the open market. The opposite is also true. When the market fuel price falls below the contracted price, the government cannot choose to break the contract. It must purchase 45 MGY regardless of the market fuel price.

The total cost to government over the project timeline is expressed in NPV terms. Government cost is just the difference between the reverse auction contract price and the market price. The mean expected NPV is near zero because of the value of the forward contract fuel price. However, the variance of this cost is high. Case 7 shifts risk from the private sector to the public sector. The mean government cost is −\$4.8 million and one standard deviation is \$74 million. The government cost of the reverse auction could effectively range from −\$260 million to \$250 million.

**TABLE 14:** Case 7.

| | Economic | | | Financial | | |
|---|---|---|---|---|---|---|
| | NPV | IRR (nominal) | B/C | NPV | IRR (nominal) | B/C |
| Expected value | \$79,408,633 | 12.7% | 1.23 | \$76,991,639 | 14.3% | 1.40 |
| Standard deviation | \$33,119,073 | 1.1% | 0.10 | \$24,732,466 | 1.4% | 0.13 |
| Max | \$203,027,071 | 16.7% | 1.59 | \$171,025,408 | 19.3% | 1.89 |
| Min | (\$28,653,844) | 9.0% | 0.92 | (\$4,140,055) | 9.8% | 0.98 |
| Probability of loss | 0.73% | | | 0.04% | | |

Table 14 presents the results of this case. The expected NPV of this case is slightly lower than in Case 3, yet the variance is much lower. This is a result of the forward contract. In the first year—the startup year—of production, Case 3 had a fuel selling price of \$2.92 (\$0.77 per liter). The contract in Case 7 does not begin until the second year of production. The market price in the first year is on average much below this price so the benefit gained in the first production year of Case 3 is greater than in Case 7. The financial NPV probability of loss was reduced from 15% to 0%. The economic NPV probability of loss decreased from 21% to 1%.

This case quantifies how risk is transferred from the private to the public sectors through a reverse auction. Private investment in a biofuel production facility becomes much more appealing by having a forward contract. However, reverse auction is not commonly used in the United States. Much more common are tax breaks, low interest loans, or other capital subsidies.

### 9.3.8 STOCHASTIC WITH CAPITAL SUBSIDY

The government more often subsidizes the initial capital investment than it provides a forward contract. Case 8 uses the same expected public cost as in Case 7. However, while Case 7 had an uncertain public cost, the public cost in Case 8 is entirely deterministic: total capital investment is reduced in this case by $5 million. Therefore, Cases 7 and 8 have the same expected government cost. Table 15 summarizes the results of Case 8.

**TABLE 15:** Case 8.

| | Economic | | | Financial | | |
|---|---|---|---|---|---|---|
| | NPV | IRR (nominal) | B/C | NPV | IRR (nominal) | B/C |
| Expected value | $88,193,806 | 12.8% | 1.26 | $83,474,418 | 14.3% | 1.44 |
| Standard deviation | $101,965,463 | 3.2% | 0.30 | $77,349,466 | 3.9% | 0.41 |
| Max | $439,114,508 | 22.1% | 2.29 | $351,062,589 | 26.0% | 2.84 |
| Min | ($269,840,864) | −5.9% | 0.21 | ($191,571,723) | −11.1% | 0.00 |
| Probability of loss | 19.3% | | | 14.3% | | |

Although both cases have the same expected government cost, Case 8 has a specified cost while the government cost in Case 7 is uncertain. Case 8 has a slightly higher expected value than in Case 3 because of the lower capital cost. Still, the probability of loss is much higher than in Case 7: the probability of loss is 19% and 14% for the economic and financial cases, respectively.

## 9.4 CONCLUSIONS

The results reveal that product selling price uncertainty contributes greatly to the inherent risk of private investment in a biofuel production facility.

### 9.4.1 CASES 2 AND 4

Cases 2 and 4 have the same expected economic NPV. Case 2 includes fuel price uncertainty, while Case 4 does not. Based on our estimates, it appears that private investment in biofuel production facilities could provide an expected 11% annual rate of return at a fuel price of $2.68 ($0.71 per liter), but including market price uncertainty makes this return rate highly uncertain. Case 2 had a 41% probability of loss even with an 11% expected IRR. The financial NPV ranged by $634 million. Table 16 compares these two scenarios.

**TABLE 16:** Cases 2 and 4 comparison.

| Financial measure | Case 2 | Case 4 |
|---|---|---|
| Expected NPV | $16,641,078 | $16,641,078 |
| Standard deviation of NPV | $76,669,403 | $15,051,080 |
| Probability of loss | 40.8% | 13.2% |
| Total range | $634,209,848 | $116,646,045 |

### 9.4.2 CASES 3 AND 5

Case 3 had both technical uncertainty and a fuel price that followed DOE projections. Case 5 had technical uncertainty but not fuel price uncertainty. Fuel selling price in Case 5 was fixed to \$2.92 (\$0.77 per liter), which provides the same expected NPV as in Case 3. The results of these two scenarios are compared in Table 17. Probability of a loss is greatly reduced from Case 3 to Case 5. Comparing the scenarios reveals the impact of fuel price uncertainty.

**TABLE 17:** Cases 3 and 5 comparison.

| Financial measure | Case 3 | Case 5 |
|---|---|---|
| Expected NPV | $80,672,958 | $80,672,958 |
| Standard deviation of NPV | $76,305,318 | $15,980,391 |
| Probability of loss | 14.5% | 0.0% |
| Total range | $540,295,144 | $114,073,782 |

### *9.4.3 CASES 7 AND 8*

Case 7 forecasted the impact of a government originated reverse auction on profitability in investment. It had a forward contract for 45 MGY of fuel at a price of \$2.92 (\$0.77 per liter). The remaining facility fuel output was sold at wholesale gasoline and diesel prices. Under these conditions, the expected government cost was \$4.8 million, with a standard deviation of \$74 million.

Case 8 had a capital subsidy of \$5 million, slightly higher than the expected government cost in Case 7. Cases 7 and 8 are compared in Table 18.

**TABLE 18:** Cases 7 and 8 comparison.

| Financial measure | Case 7 | Case 8 |
|---|---|---|
| Expected NPV | $76,991,639 | $83,474,418 |
| Standard deviation of NPV | $24,732,466 | $77,349,466 |
| Probability of loss | 0.04% | 14.33% |
| Total range | $175,165,462 | $542,634,312 |

### 9.4.4 *POLICY IMPLICATIONS*

Investment in a fast pyrolysis commercial biofuel facility carries a high amount of risk. Case 3 shows that, with a total capital investment of $429 million, the NPV could range from –$190 million to $350 million. Uncertainty arises directly from new technology and from market prices. This investment uncertainty arises from technical uncertainty and fuel price selling price uncertainty.

In our study, fuel price uncertainty was shown to have a large effect on investment risk. The government has many options for stimulating growth in biofuel production. However, to encourage private investment in biofuel production at the lowest public cost, the government should first look for ways to reduce private risk. Conventional fuels dominate the marketplace, and their prices are a direct competitor for drop-in biofuels. To reduce risk, the government may consider ways by which it can control fuel price fluctuations for biofuel producers. A reverse auction would provide a competitive long-term price and would shift risk from the private sector to the public sector. In our study, a reverse auction was shown to be more effective at encouraging private investment than capital subsidies for the same expected cost.

## REFERENCES


1. A. McCrone, E. Usher, O. V. Sonntag-'Brien, U. Moslener, and C. Grüning, "Global Trends in Renewable Energy Investment," United Nations Environment Programme, 2013, http://www.unep.org/pdf/GTR-UNEP-FS-BNEF2.pdf.
2. Energy Efficiency and Renewable Energy, FY 2014 Budget Rollout. U.S. Department of Energy, 2013, http://www1.eere.energy.gov/office_eere/pdfs/budget/fy2014_eere_congressional_budget_request.pdf.
3. Bioenergy Technologies Office, "Replacing the Whole Barrel To Reduce U.S. Dependence on Oil," U.S. Department of Energy, 2013, https://www1.eere.energy.gov/bioenergy/pdfs/replacing_barrel_overview.pdf.
4. U.S. Energy Information Administration, Annual Energy Outlook, 2013, http://www.eia.gov/aeo.
5. Bioenergy Technologies Office, "US Billion-Ton Update: Biomass Supply for a Bioenergy and Bioproducts Industry," U.S. Department of Energy, http://www1.eere.energy.gov/bioenergy/pdfs/billion_ton_update.pdf.

6. R. Tristan Brown, R. Thilakaratne, R. C. Brown, and G. Hu, "Techno-economic analysis of biomass to transportation fuels and electricity via fast pyrolysis and hydroprocessing," Fuel, vol. 106, pp. 463–469, 2013. S. B. Jones, C. Valkenburg, C. W. Walton, et al., "Production of gasoline and diesel from biomass via fast pyrolysis, hydrotreating and hydrocracking: a design case," Pacific Northwest National Laboratory, http://www.pnl.gov/main/publications/external/technical_reports/PNNL-18284rev1.pdf.
7. R. P. Anex, A. Aden, F. K. Kazi et al., "Techno-economic comparison of biomass-to-transportation fuels via pyrolysis, gasification, and biochemical pathways," Fuel, vol. 89, no. 1, supplement, pp. S29–S35, 2010.
8. S. B. Jones and J. L. Male, "Production of Gasoline and Diesel from Biomass via Fast Pyrolysis, Hydrotreating and Hydrocracking: 2011 State of Technology and Projections to 2017," Pacific Northwest National Laboratory, http://www.pnnl.gov/main/publications/external/technical_reports/PNNL-22133.pdf.
9. A. V. Bridgwater, "Review of fast pyrolysis of biomass and product upgrading," Biomass and Bioenergy, vol. 38, pp. 68–94, 2012.
10. T. Bridgwater, "Biomass for energy," Journal of the Science of Food and Agriculture, vol. 86, no. 12, pp. 1755–1768, 2006.
11. S. Czernik and A. V. Bridgwater, "Overview of applications of biomass fast pyrolysis oil," Energy and Fuels, vol. 18, no. 2, pp. 590–598, 2004.
12. D. Mohan, C. U. Pittman Jr., and P. H. Steele, "Pyrolysis of wood/biomass for bio-oil: a critical review," Energy and Fuels, vol. 20, no. 3, pp. 848–889, 2006.
13. M. M. Wright, D. E. Daugaard, J. A. Satrio, and R. C. Brown, "Techno-economic analysis of biomass fast pyrolysis to transportation fuels," Fuel, vol. 89, no. 1, supplement, pp. S2–S10, 2010.
14. R. Tristan Brown and R. C. Brown, "Techno-economics of advanced biofuels pathways," RSC Advances, vol. 3, no. 17, pp. 5758–5764, 2013.
15. T. Brown and G. Hu, "Technoeconomic sensitivity of biobased hydrocarbon production via fast pyrolysis to government incentive programs," Journal of Energy Engineering, vol. 138, no. 2, pp. 54–62, 2011. D. Meier, B. van de Beld, A. V. Bridgwaterorg, et al., "State-of-the-art of fast pyrolysis in IEA bioenergy member countries," Renewable and Sustainable Energy Reviews, vol. 20, pp. 619–641, 2013. C. E. Ridley, C. M. Clark, S. D. Leduc et al., "Biofuels: network analysis of the literature reveals key environmental and economic unknowns," Environmental Science and Technology, vol. 46, no. 3, pp. 1309–1315, 2012.
16. W. E. Tyner, "Biofuels and agriculture: a past perspective and uncertain future," International Journal of Sustainable Development & World Ecology, vol. 19, no. 5, pp. 389–394, 2012.
17. D. Rajagopal and D. Zilberman, "Review of Environmental, Economic and Policy Aspects of Biofuels," Working Paper 4341, World Bank Policy Research, 2007.
18. National Research Council, Renewable Fuel Standard: Potential Economic and Environmental Effects of US Biofuel Policy, National Academies Press, 2011.
19. T. L. Richard, "Challenges in scaling up biofuels infrastructure," Science, vol. 329, no. 5993, pp. 793–796, 2010.

20. L. J. Snowden-Swan and J. L. Male, "Summary of Fast Pyrolysis and Upgrading GHG Analyses," Pacific Northwest National Laboratory, 2012, http://www.pnnl.gov/main/publications/external/technical_reports/PNNL-22175.pdf.
21. T. Searchinger, R. Heimlich, R. A. Houghton et al., "Use of U.S. croplands for biofuels increases greenhouse gases through emissions from land-use change," Science, vol. 319, no. 5867, pp. 1238–1240, 2008.
22. W. E. Tyner, F. Taheripour, and D. Perkis, "Comparison of fixed versus variable biofuels incentives," Energy Policy, vol. 38, no. 10, pp. 5530–5540, 2010.
23. W. Tyner, "Description of 2011 Biofuels Policy Alternatives," Policy Briefs 1, Global Policy Research Institute (GPRI), 2011.
24. W. E. Tyner, "The US ethanol and biofuels boom: its origins, current status, and future prospects," BioScience, vol. 58, no. 7, pp. 646–653, 2008.
25. KiOR, "Technology," 2013, http://www.kior.com/content/?s=11&t=Technology.
26. D. Humbird, R. Davis, L. Tao, et al., "Process Design and Economics for Biochemical Conversion of Lignocellulosic Biomass to Ethanol: Dilute-Acid Pretreatment and Enzymatic Hydrolysis of Corn Stover," National Renewable Energy Laboratory, 2011, http://www.afdc.energy.gov/pdfs/47764.pdf.
27. National Research Council, Transitions to Alternative Transportation Technologies: A Focus on Hydrogen, National Academies Press, 2008.
28. J. R. Bartels, M. B. Pate, and N. K. Olson, "An economic survey of hydrogen production from conventional and alternative energy sources," International Journal of Hydrogen Energy, vol. 35, no. 16, pp. 8371–8384, 2010.
29. U.S. Government Accountability Office, "Corporate Income Tax: Effective Tax Rates Can Differ Significantly from the Statutory Rate," 2013, http://www.gao.gov/products/gao-13-520.
30. Bureau of Labor Statistics, "Producer Price Index-Commodities," U.S. Department of Labor, 2013, http://www.bls.gov/ppi.
31. S. Irwin and D. Good, "Farmdocdaily: What Combination of Corn and RINs Prices Makes E85 Competitive?" University of Illinois, http://farmdocdaily.illinois.edu/2013/06/combination-corn-rins-prices-e85.html.

## CHAPTER 10

# LAND USAGE ATTRIBUTED TO CORN ETHANOL PRODUCTION IN THE UNITED STATES: SENSITIVITY TO TECHNOLOGICAL ADVANCES IN CORN GRAIN YIELD, ETHANOL CONVERSION, AND CO-PRODUCT UTILIZATION

RITA H. MUMM, PETER D. GOLDSMITH, KENT D. RAUSCH, AND HANS H. STEIN

### 10.1 BACKGROUND

As alternatives to petroleum-based energy sources are sought in the US, great attention has been given to renewable fuel sources from agriculturally produced biofeedstocks, for example, miscanthus, switchgrass, sugar cane, rapidly growing tree species, and corn. Renewable fuel sources not only reduce US dependence on foreign sources for energy, but support environmental stewardship through reduction of greenhouse gas production and promote rural development.

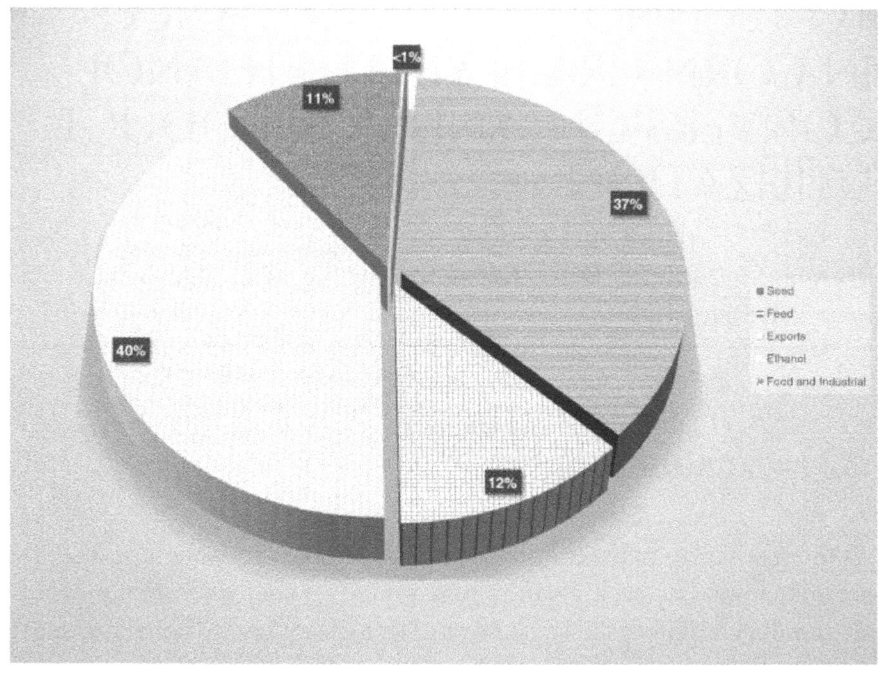


**FIGURE 1:** Disposition (%) among major uses of No. 2 yellow corn harvested in the US in 2011 [1].

The system for producing and processing of corn grain (that is, US No. 2 yellow corn) is well established in the US and corn grain has been used as a source of biofuel for decades. According to the US Department of Agriculture (USDA), of the 12.360 billion bushels of corn grain harvested in 2011, more than 40% (5.007 billion bushels) was processed to produce ethanol while 37% went to livestock feed, 11% to food and industrial uses, and 12% was exported (Figure 1) [1]. The use of corn grain among other biofeedstocks has to be balanced with its longtime predominant purpose for food and feed, and other issues such as economic impacts affecting global food prices, land use, and environmental effects. The Energy Independence and Security Act adopted in 2007 established the Renewable Fuel Standard, recognizing the role of corn grain for ethanol production along with biofeedstocks for cellulosic fermentation [2]. The Renewable Fuel Standard established a limit of 15 billion gallons for the use of corn grain for ethanol. The corn ethanol system produces significant quantities of co-products, including distillers dried grains with solubles (DDGS), corn gluten feed (CGF), and corn gluten meal (CGM). These co-products substitute for corn grain and soybean meal in livestock feed, mitigating to some extent the trade-off between fuel and feed with corn grain channeled to ethanol production.

We examined three key 'supply' variables to quantify the long-term effects and interactions involving corn grain yield, ethanol processing, and livestock feeding to illustrate how these factors affect land area attributed to corn ethanol production. In particular, we assessed the impact of technology as it relates to changes in corn yields due to genetic and agronomic advancements, in ethanol processing due to more efficient fermentation resulting in greater ethanol output, and in changes in livestock feeding practices for beef cattle, dairy cattle, pigs, and poultry. To accomplish this, we first created a model of the US corn ethanol system featuring inputs and outputs involving production of corn grain, processing co-products, livestock feeding, and oil for biofuels.

Secondly, we developed and explored seven scenarios representing various levels of efficiency due to anticipated technological changes in corn production, ethanol processing, and livestock feeding. The seven scenarios feature: 1) corn yield estimates based on historical performance as well as publically shared information about industrial seed corn product pipelines and future product expectations which feature biotechnological advancements i.e. corn yield technology, 2) corn yield estimates (as in Scenario 1) minus 10% to represent a lower-bound estimate of yield with technological advancements, 3) corn yield estimates (as in Scenario 1) plus 10% to represent a upper-bound estimate of yield with technological advancements, and 4) corn yield estimates based on USDA projections which reflect mainly conventional plant breeding practices and take little account of biotechnological advancements; 5) ethanol processing advancements that feature conversion of all starch in the grain and 6) ethanol processing advancements that feature complete fiber conversion; and 7) an average livestock feeding profile associated with a 65%: 35% corn grain to soybean meal substitution ratio representing a shift toward more use of DDGS for dairy cattle, pigs, and poultry.

Thirdly, we considered three time horizons at which to estimate land use attributable to corn ethanol: 2011 (current); the time period at which the 15 billion gallon cap for corn ethanol is achieved; and 2026 (15 years out). With substituting DDGS in livestock diets, we computed land area attributable to corn ethanol accounting for the replacement of soybean oil that would otherwise be derived from DDGS-replaced soybeans, because typically that soybean oil would be directed to biofuel. We also computed land area estimates without accounting for soybean oil replacement.

Thus, we considered effects of three technology factors on the supply of corn and feed co-products within the US corn ethanol system. The analysis employed a 'micro' approach in that the research directly involved discipline specialists and, as a result, integrated discipline-specific insights on the behavior of the production, processing, and feeding components of the corn-ethanol system and the rapid rate of scientific advancement. Furthermore, although other studies have considered the effects of these supply variables individually to some extent (for example, see [3-9]), our view is from a technological perspective and one that considers the interaction among these factors within a particular geography. With this approach, we aim to provide a basis to aid in calibrating parameters for land use models and inform stakeholders of the importance of technological change in biofuel life-cycle analysis.

We limited ethanol demand in the model to a maximum of 15 billion gallons, which corresponds to the maximum allowed for first generation biofuels under the 2007 Renewable Fuels Standard. The capping of demand, while reasonable over the near and medium term, also allows the analysis to solely focus on technological advances corresponding to production, processing, and livestock feeding. Capping demand allows us to limit our scenario analysis to seven but does not allow for analysis of demand or trade scenarios. Moreover, we do not address the implications of corn stover as a biofeedstock as this depends on cellulosic fermentation technology, which is presently in its infancy [10].

## 10.2 RESULTS AND DISCUSSION

Scenarios 1 through 4 explore the impact of increasing corn grain yields over the 15-year period through 2026, with the rate of yield increase influenced by the implementation and farmer adoption of yield technologies. Together, these four scenarios represent the range of growth in corn grain production over the next 15 years. Replacing corn and soybean meal in livestock diets with DDGS, CGF, and CGM has the effect of reducing the land area attributed to corn ethanol production; thus, the land area attributed to corn ethanol production is less than the acreage associated with production of the 40.5% of the corn grain directed to ethanol processing (Table 1). In 2011, not accounting for soybean oil replacement, land area for corn ethanol was 13.9 million acres, 17% of the total 83.98 million total corn acres, compared to 40.5% of all corn grain directed to ethanol processing. Because 2011 yield estimates were based on actual figures, not forecasts, the land area attributed to corn ethanol was the same across all four yield scenarios.

Accounting for soybean oil replacement, land area attributed to corn ethanol production in 2011 was 20.9 million acres, 25% of the total 83.98 million corn acres, instead of 40.5% of all corn grain directed to ethanol processing (Table 1). However, the replacement of soybean oil through canola production contributes positive feedback[a] to the system, meaning that more land area is attributed to corn ethanol than when not accounting for oil replacement.

**TABLE 1:** Estimated land area attributed to corn ethanol production expressed as acreage and percent of US land dedicated to corn grain production for each of seven scenarios, without and with oil replacement, for three time horizons

| | 1 | 2 | 3 | 4 | 5 | 6 | 7 |
|---|---|---|---|---|---|---|---|
| | Corn grain production | | | | Ethanol processing | | Livestock feeding |
| | Yield technology | | | | | | |
| Time horizon | Medium | Low | High | Minimal | Full starch | Complete fiber | 65%:35% substitution ratio |
| Acres without oil replacement (millions) | | | | | | | |
| 2011 | 13.9 | 13.9 | 13.9 | 13.9 | 15.0 | 17.6 | 11.9 |
| Ethanol ceiling | 11.0 (2013) | 13.0 (2015) | 8.8 (2013) | 11.6 (2013) | 11.8 (2013) | 14.0 (2013) | 8.8 (2013) |
| 2026 | 3.6 | 5.3 | 2.0 | 8.2 | 4.4 | 7.0 | 1.2 |
| Percent of US corn land without oil replacement | | | | | | | |
| 2011 | 17% | 17% | 17% | 17% | 18% | 21% | 14% |
| Ethanol ceiling | 13% (2013) | 16% (2015) | 10% (2013) | 14% (2013) | 14% (2013) | 17% (2013) | 10% (2013) |
| 2026 | 4% | 6% | 2 % | 10% | 5% | 8% | 1% |
| Acres with oil replacement (millions) | | | | | | | |
| 2011 | 20.9 | 20.9 | 20.9 | 20.9 | 21.5 | 22.4 | 20.5 |
| Ethanol ceiling | 18.4 (2013) | 20.4 (2015) | 16.2 (2013) | 19.0 (2013) | 18.6 (2013) | 18.8 (2013) | 17.9 (2013) |
| 2026 | 11.0 | 12.9 | 9.4 | 15.6 | 11.3 | 11.8 | 10.3 |
| Percent of US corn land with oil replacement | | | | | | | |
| 2011 | 25% | 25% | 25% | 25% | 26% | 27% | 24% |
| Ethanol ceiling | 22% (2013) | 24 % (2015) | 19 % (2013) | 23% (2013) | 22% (2013) | 22% (2013) | 21% (2013) |
| 2026 | 13% | 15% | 11% | 19% | 13% | 14% | 12% |

As corn yields increase over time, greater quantities of co-products for livestock feeding are produced in the US corn ethanol system, offsetting more land area attributed to ethanol production. In Scenario 1 (medium level of yield technology), land area attributed to corn ethanol production falls to 3.6 million acres by 2026, which is only 4% of all US land area devoted to corn production without oil replacement (Table 1).

However, soybean oil replacement partially offsets the effects higher corn yields have on land usage. Although oil skimming occurring at 50% of the dry grind plants at the rate of 0.24 pounds per bushel of processed corn negatively feeds back into the system to mitigate some of the oil replacement by canola, the benefit is marginal because the volumes skimmed are relatively low. Land usage attributed to corn ethanol production in 2026 (Scenario 1) is more than three times greater when accounting for soybean oil replacement (4% versus 13%). Still, the terminal result dramatically differs from the 40.5% of the grain directed to corn ethanol.

Yield growth assumptions are critical to land usage estimates. There is a difference of 8 percentage points or more than 6 million acres when comparing no/minimal yield technology (Scenario 4) with a high level of yield technology (Scenario 3) in 2026 (with or without oil replacement). Scenario 2 is the only scenario in which the 15 billion gallon ethanol ceiling is reached in 2015; for all other yield scenarios, the ceiling is reached in 2013. Meeting this ceiling occurs when national average corn yields rise to levels in the range of 155.0 to 157.1 bushels per acre (bu/A).

The array of corn yield forecasts provided by the four levels of yield technology implementation and adoption (Tables 2 and 3) is not inconsistent with other predictions of US corn yield averages, especially with technological improvements. Scenario 1 culminates in a 244.3 bu/A average whereas Scenarios 2 and 4, which are similar in depicting a negligible impact of yield technology, average 206.0 bu/A in 2026. A similar range (205 to 242 bu/A) was predicted for the US by 2030 by Miranowski et al., taking into account corn grain yield performance and forecasts by state [11]. Downing et al. predicted that corn yield growth will result in a considerably greater supply of corn grain, with a strong likelihood of a doubling in annual increase due to technology [12]. Even across the entire range of yield estimates provided by the four yield technology scenarios, corn grain yield is shown to be a key factor influencing land usage attributed to corn ethanol production.

Scenarios 5 and 6 reveal the impact of increasing efficacy in ethanol processing on land usage for corn ethanol production. Two effects stand out: the direct effect is that improved ethanol processing increases the amount of ethanol per bushel of grain, which in turn reduces the amount of land needed to meet the 15 billion gallon cap. But improving ethanol output introduces a significant positive feedback force that ceteris paribus raises land usage. Small increases in ethanol output dramatically reduce the volume of co-products available for livestock feeding.

With Scenario 5 involving improved starch conversion efficiency, the ethanol yield increases from 2.759 to 2.829 gallons per bushel of corn. Correspondingly, the quantity of DDGS produced as a co-product in ethanol processing decreases from 17.44 to 16.38 pounds per bushel of processed corn (Tables 4 and 5). Scenario 5 presents an interesting trade-off in keeping with findings by Mueller and Kwik [13]: a net 2.5% increase in ethanol yield per bushel of corn processed for a net 6.1% decrease in the quantity of DDGS produced per bushel. The decrease in feed co-products results in less corn and soybean being substituted, which in turn raises land area attributed to corn ethanol production. In 2011, without oil replacement, land usage is 15.0 million acres, or 18% of the total 83.98 million acres used for US corn grain production, as compared to 17% with Scenario 1 involving baseline ethanol processing efficiencies (Table 1). Including oil replacement raises land area attributed to corn ethanol with full starch conversion to 21.5 million acres, 26% of the US corn grain-producing land. Therefore, increasing ethanol output per bushel of corn by extracting more from the starch component of the grain increases land area attributed to corn ethanol in Scenario 5 compared to Scenario 1, which is a surprising result.

This effect is even more exaggerated in Scenario 6, which features fermentation technology to convert C5 and C6 sugar to ethanol, converting not only residual starch but pericarp and endosperm fiber fractions as well. With complete fiber (and starch) conversion, the ethanol yield per bushel of processed corn increases from 2.759 to 3.078 gallons, compared with baseline ethanol processing efficiencies. Correspondingly, the quantity of DDGS produced as co-product decreases from 17.44 to 12.67 pounds per bushel of processed corn (Table 5). Similar to Scenario 5, Scenario 6 presents an interesting trade-off: an overall 11.6% increase in ethanol yield per bushel of corn for a 27.4% decrease in the quantity of DDGS produced per bushel of corn. The decrease in feed co-products results in less corn and soybean meal being substituted, which in turn raises the land area attributed to corn ethanol. In general, land area for corn ethanol is higher when the system extracts more ethanol per bushel of corn.

**TABLE 2:** Scenario 1 corn yield forecast (bu/A) by year through 2026 and contributing technology factors with associated step changes

| Year of maximum adoption for specified technology factor[a] | Year | Conventional breeding | Advanced breeding technology[b] | Sub-total | Biotechnology traits[c] | Sub-total | Agronomic improvements[d] | Total corn yield |
|---|---|---|---|---|---|---|---|---|
| | 2011[e] | 147.2 | 0 | 147.2 | 0 | 147.2 | 0 | 147.2 |
| | 2012[e] | 123.4 | 0 | 123.4 | 0 | 123.4 | 0 | 123.4 |
| | 2013 | 169.9 | 0.25 | 170.2 | 0 | 170.2 | 0 | 170.2 |
| | 2014 | 171.7 | 0.50 | 172.2 | 0 | 172.2 | 0 | 172.2 |
| 1 | 2015 | 173.5 | 1.00 | 174.5 | 0 | 174.5 | 0 | 174.5 |
| 3 | 2016 | 175.3 | 2.00 | 177.3 | 0 | 177.3 | 3 | 180.3 |
| | 2017 | 177.1 | 3.00 | 180.1 | 0 | 180.1 | 3 | 183.1 |
| 2 | 2018 | 178.9 | 4.00 | 182.9 | 10 | 192.9 | 3 | 195.9 |
| | 2019 | 180.7 | 5.00 | 185.7 | 10 | 195.7 | 3 | 198.7 |
| 3 | 2020 | 182.5 | 6.00 | 188.5 | 10 | 198.5 | 6 | 204.5 |
| | 2021 | 184.3 | 7.00 | 191.3 | 10 | 201.3 | 6 | 207.3 |
| 2 | 2022 | 186.1 | 8.00 | 194.1 | 20 | 214.1 | 6 | 220.1 |
| | 2023 | 187.9 | 9.00 | 196.9 | 20 | 216.9 | 6 | 222.9 |
| 3 | 2024 | 189.7 | 10.00 | 199.7 | 20 | 219.7 | 9 | 228.7 |
| 2 | 2025 | 191.5 | 11.00 | 202.5 | 30 | 232.5 | 9 | 241.5 |
| | 2026 | 193.3 | 12.00 | 205.3 | 30 | 235.3 | 9 | 244.3 |

*[a]Note that multiple waves are anticipated for some technology factors. [b]Advanced breeding technology comprises genomic-based approaches to crop improvement including DNA sequencing, molecular markers, and doubled haploidy. This class of technologies does not include genetic engineering. [c]Biotechnology traits comprises value-added characteristics manifested through genetic modification. [d]Agronomic improvements comprise cultural production practices that relate to the way the corn crop is managed. [e]Actual (not forecasted) US average yields provided for this year.*

**TABLE 3:** Corn yield forecasts (bu/A) by year from 2013 through 2026 for Scenario 1 (medium yield technology), Scenario 2 (low yield technology), Scenario 3 (high yield technology), and Scenario 4 (no/minimal technology change factors)

| Year | Scenario 1 | Scenario 2 | Scenario 3 | Scenario 4 |
|---|---|---|---|---|
| 2011[a] | 147.2 | 147.2 | 147.2 | 147.2 |
| 2012[a] | 123.4 | 123.4 | 123.4 | 123.4 |
| 2013 | 170.2 | 153.1 | 187.2 | 166.0 |
| 2014 | 172.2 | 155.0 | 189.4 | 168.0 |
| 2015 | 174.5 | 157.1 | 192.0 | 170.0 |
| 2016 | 180.3 | 162.3 | 198.3 | 172.0 |
| 2017 | 183.1 | 164.8 | 201.4 | 174.0 |
| 2018 | 195.9 | 176.3 | 215.5 | 176.0 |
| 2019 | 198.7 | 178.8 | 218.6 | 178.0 |
| 2020 | 204.5 | 184.1 | 225.0 | 180.0 |
| 2021 | 207.3 | 186.6 | 228.0 | 182.0 |
| 2022 | 220.1 | 198.1 | 242.1 | 184.0 |
| 2023 | 222.9 | 200.6 | 245.2 | 186.0 |
| 2024 | 228.7 | 205.8 | 251.6 | 188.0 |
| 2025 | 241.5 | 217.4 | 265.7 | 190.0 |
| 2026 | 244.3 | 219.9 | 268.7 | 192.0 |

*[a]All scenarios incorporate actual average yields for 2011 and 2012.*

With both full starch and complete fiber conversion, there is no advantage in terms of reduced land usage attributed to corn ethanol production, even after 15 years. In 2026, 13% (Scenario 5) and 14% (Scenario 6) of US corn-producing land is attributable to corn ethanol, compared to 13% (Scenario 1) without enhanced technological efficiency in ethanol processing to output more ethanol from every bushel of corn processed. The reverse side is that the land usage attributed to corn ethanol is negatively impacted only slightly from as much as an 11.6% increase in ethanol output as corn yields increase to 2026 levels.

Scenario 7 demonstrates the impact of a shift in feeding value for DDGS, CGF, and CGM, with substitution for corn falling from 71% to 65% in livestock diets and substitution for soybean meal increasing from

29% to 35%. This shift reflects a change in the allocation of the feed co-products across livestock types. Specifically, non-ruminants and dairy cattle consume higher proportions of ethanol co-products, whereas beef cattle consume less (Table 6). The shift reflects a tension between offsetting corn grain or soybean meal consumption when feeding ethanol co-products to livestock. Feeding more DDGS to monogastric animals not only has the benefit of replacing more high cost protein from soybean meal with a lower cost alternative, but land usage attributed to corn ethanol production is dramatically reduced as well. This effect is most pronounced with 2026 estimates: only 1.2 million of the total US corn acreage (1%) is attributed to corn ethanol production without oil replacement (Table 1). However, with the shift to 35% replacement of soybean meal in livestock diets, land area attributed to corn ethanol rises to 10.3 million acres (12%) due to higher oil replacement demands.

**TABLE 4:** Composition of feed co-products[a] from ethanol processing scenarios, assuming 86% dry grind with 50% skimming oil and 14% wet milling

| | Distillers dried grains with solubles | | Corn gluten feed | | Corn gluten meal | |
|---|---|---|---|---|---|---|
| | Protein (%) | Fat (%) | Protein (%) | Fat (%) | Protein (%) | Fat (%) |
| Baseline[b] | 27.35 | 9.67 | 17.39 | 4.21 | 58.25 | 4.74 |
| Individual processes: | | | | | | |
| Conventional dry grind, no oil skimming | 27.30 | 10.43 | n/a | n/a | n/a | n/a |
| Conventional dry grind, with oil skimming | 27.40 | 8.90 | n/a | n/a | n/a | n/a |
| Wet milling | n/a | n/a | 17.39 | 4.21 | 58.25 | 4.74 |
| Scenario 5: Full starch conversion | 29.10 | 10.28 | 17.39 | 4.21 | 58.25 | 4.74 |
| Scenario 6: Complete fiber conversion plus full starch conversion | 28.58 | 10.10 | 17.39 | 4.21 | 58.25 | 4.74 |

*[a]Expressed on a commercial or 'as is' basis. Dry matter contents of 89.31%, 87.13%, and 90.04% for DDGS, CGF, and CGM, respectively [40]. [b]Weighted industry averages.*

**TABLE 5:** Ethanol and co-product outputs associated with ethanol processing scenarios, assuming 86% dry grind with 50% skimming oil and 14% wet milling

| | Ethanol yield, L/t (gal/bu) | DDGS yield, kg/t (lb/bu)[a] | CGF yield, kg/t (lb/bu)[a] | CGM yield, kg/t (lb/bu)[a] | Dry grind oil recovery, kg/t (lb/bu) |
|---|---|---|---|---|---|
| Baseline[b] | 410.3 (2.759) | 310.1 (17.44) | 229.3 (12.88) | 49.77 (2.80) | 2.145 (0.12) |
| Individual processes: | | | | | |
| Conventional dry grind, no oil skimming | 414.1 (2.785) | 312.2 (17.56) | n/a | n/a | 0 |
| Conventional dry grind, with oil skimming | 414.1 (2.785) | 308.0 (17.32) | n/a | n/a | 4.288 (0.24) |
| Wet milling | 386.6 (2.600) | n/a | 229.3 (12.88) | 49.77 (2.80) | n/a |
| Scenario 5: Full starch conversion | 420.7 (2.829)# | 291.2 (16.38) | 229.3 (12.88) | 49.77 (2.80) | 2.145 (0.12) |
| Scenario 6: Complete fiber conversion plus full starch conversion | 457.6 (3.078)# | 226.3 (12.67) | 229.3 (12.88) | 49.77 (2.80) | 2.145 (0.12) |

[a]*Commercial or 'as is' basis. Dry matter contents of 89.31%, 87.13%, and 90.04% for DDGS, CGF and CGM, respectively [40].* [b]*Weighted industry averages. DDGS, distillers dried grains with solubles; CGF, corn gluten feed; CGM, corn gluten meal.*

**TABLE 6:** Percentage of feed usage domestically of distillers dried grains with solubles produced from corn grain directed to dry grind processing (4.306 billion bushels) by livestock type with 71%:29% (baseline) and 65%:35% (Scenario 7) corn-to-soybean substitution ratios

| | Percentage of total distillers dried grains with solubles usage (%) | |
|---|---|---|
| Livestock type | 71%:29% substitution ratio | 65%:35% substitution ratio |
| Beef cattle | 50.4 | 30.0 |
| Dairy cattle | 33.5 | 47.3 |
| Pigs | 9.1 | 12.8 |
| Poultry | 7.0 | 9.9 |

Interestingly, feeding a higher percentage of co-products to monogastrics and dairy cattle provides a negative feedback force that reduces the land usage attributed to corn ethanol production. Land usage falls in the terminal period by 0.7 million acres or one percentage point when comparing Scenario 7 with Scenario 1, assuming oil replacement. This occurs because offsetting a low yield crop like soybean through co-product feeding reduces land usage more than a high yield crop like corn. However, this difference would diminish with increases in soybean yield (which were fixed in this analysis), perhaps making corn and soybeans more comparable as land use alternatives in the corn ethanol system.

## 10.3 CONCLUSIONS

Corn grain yield has a profound impact on estimates of land area attributable to corn ethanol production. In 2011, 25% of the acreage used for US production of corn grain was devoted to ethanol fuel production based on the historic corn yield for the year and accounting for replacement of soybean oil with the reduced demand for soybean production. Assuming reasonable increases in corn grain yield with anticipated new yield technologies coming into play in the next 15 years, this percentage could be reduced by nearly half to 13%. Even assuming the most conservative estimate of corn yield growth, that is, Scenario 4, land area attributed to ethanol production drops to 19%. The high rate of technological change in corn production combined with the strong linkage between yield and land use requires biofuel life-cycle analysts to include insightful estimates of yield as well as yield dynamics within models.

Co-product utilization is a powerful force reducing the land usage attributable to corn ethanol in the US corn ethanol system. Thus, the system complementarity between fuel production and livestock nutrition improves because ethanol co-products provide a negative feedback substituting for corn and soybean meal, which in turn reduces land demand for corn and soybean production. System complementarity may be an important element for biofuel life-cycle analysts as they think about the full impacts of the systems under study.

Substitution of soybean meal through co-product feeding removes oil from the market. Accounting for that oil through replacement significantly increases estimates of land area estimates attributed to corn ethanol production. This positive feedback force is especially acute because oil production (soybean or canola) is relatively land intensive compared to starch production (corn). This effect could be ameliorated by substituting an oil crop with a high land use efficiency. We use canola in our model, one of the highest oil-yielding crops on a per acre basis. But perhaps on the horizon, through plant breeding, more superior canola cultivars or other crop alternatives might emerge.

By contrast, improved efficiency in ethanol processing through anticipated fermentation technologies that increase ethanol output has little impact on land area attributed to corn ethanol over time. This is due to the trade-off between volumes of ethanol and co-product outputs, where small improvements in ethanol-processing efficiency significantly reduce amounts of co-product. Thus the need for less corn, and corn land, through advanced processing technologies is offset by reduced negative feedback because there is less co-product to substitute for corn and soybean meal in livestock feeding. The ethanol/DDGS trade-off in dry grind processing may warrant greater scrutiny by ethanol processors from an economic standpoint to maximize returns.

The anticipated change in the overall corn and soybean meal substitution ratio in DDGS livestock feeding only slightly decreases the amount of land area attributed to corn ethanol production. A sizeable (11%) difference in land area with oil replacement versus without oil replacement signifies the importance and prominence of oilseeds to US agriculture.

This analysis demonstrates clearly that, while 40.5% of harvested corn grain in the US goes to ethanol production, the percentage of total corn acreage attributed to corn ethanol production is much less. The estimate of land area attributed to corn ethanol production continues to fall as corn yields increase over time and the 15 billion gallon cap on corn ethanol is reached. As greater output is achieved in corn grain production, the land base required to produce a given volume of grain is reduced. This could translate to either less acreage to produce 15 billion gallons of corn ethanol or an enlarged market opportunity from the same allocation of land. Further research would be helpful to better specify both the demand side impacts of advancing technology, as well as the feedback processes affecting technology research and development investments.

The results challenge other findings about land use and indirect effects on land use change associated with corn ethanol at current and future production levels (for example, see [14-20]) and may be useful in establishing parameters for land use in models that consider a broader view of the biofuel arena. Furthermore, the results cited herein support inferences of others about US agricultural land productivity (for example, [11,12,21]) and the capacity to meet increasing demand in the future. Finally, this analysis also highlights the importance of accounting for technological change to better understand how a particular biofeedstock impacts various aspects of the whole biofuel picture. Technological change challenges discipline specialists when they attempt to analyze an explicitly multidisciplinary system involving numerous technologies outside of their domain expertise. We would argue conceptually, and from our experience coming together on this project, that multidisciplinary teams are essential when researchers choose to explore the macro biofuels system.

## 10.4 METHODS

Technological changes anticipated in each of the three disciplinary areas are described in detail below and related to the seven defined scenarios. In addition, details of the model of the US corn ethanol system created to simulate effects on land usage attributed to corn ethanol production are outlined. Inputs and outputs of this model are varied depending on the scenario.

### *10.4.1 MODEL*

The systems dynamic package, STELLA [22], was used to model current practices and the effects of technological changes related to corn yield, ethanol processing, and livestock feeding on US land usage attributed to corn ethanol production. Our approach was not to mimic real life because the agricultural system in which we are working is extremely complex. Instead, focusing on the three factors of interest, we evaluated critical drivers identified through dialogue with scholars and industry to develop a model that is simultaneously manageable and useful for analysis.

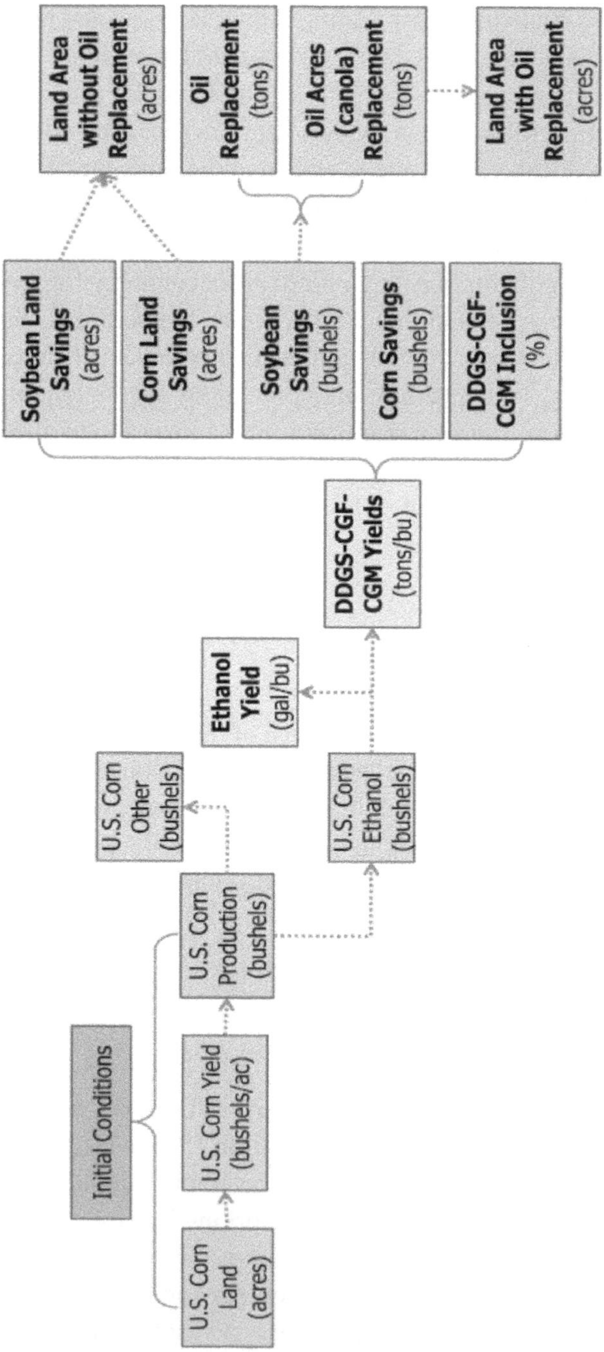

**LFIGURE 2:** Overview of the model to simulate land area attributed to corn ethanol production.

Our simple model (Figure 2) is centered on the US agricultural land base, the proportion of the US corn crop processed for ethanol, soybean and canola crops, livestock population and feeding practices, a 15 billion gallon corn ethanol industry, and a time frame from 2011 to 2026. Estimates of land usage attributed to corn ethanol production are a function of time (future prediction) as well as the variables defined by the scenario. Variables in the model include average US yields of harvested corn grain; ethanol production volumes; availability of corn ethanol processing co-products including DDGS, CGF, and CGM; and inclusion rates of corn grain, soybean meal, and corn ethanol co-products in diets fed to poultry, pigs, beef cattle, and dairy cattle. The model is designed to track both positive and negative feedback[a] processes, for example, the availability of skimmed oil from dry grind processing and replacement of the soybean oil from acres of soybean for meal that are displaced by DDGS feeding. Thus, the model is sensitive to variable inputs as well as internal system feedbacks given the parameters and assumptions specified.

The model produces estimates of land usage attributed to corn ethanol under two conditions: without accounting for replacement of soybean oil from the displaced acres of soybean for meal; and accounting for replacement of soybean oil. With the latter, canola is given as the example of an oil crop replacement for soybean, although others or a mixture of oil crops could be envisioned. Trending on the side of conservatism and with the intent to highlight the three factors under study, yields of soybean and canola crops were held at 2011 levels in the model. Also, the model does not reflect the extra canola meal feed produced through canola replacement of soybean acreage. Fixed in the model are US average soybean yields of 41.9 bu/A (2011 level) [23]; an average 78% yield of soybean meal from soybeans processed [24]; and US canola grain yields at 1,713 pounds per acre (2010/2011 level) [25], resulting in a canola oil yield of 753.7 pounds per acre given an average of 44% oil composition.

The dynamic model computes estimates of land use attributed to corn ethanol in 2011 and 2012 according to actual corn grain acreage and yield figures. The model estimates land use attributed to corn ethanol over the next 15 years (to 2026) under various assumptions related to the above variables and their interactions as defined by the specific scenario. In addition, the model identifies the year in which the 15 billion gallon cap is reached. The model commences with the corn production acreage in the US as a fixed land base according to 2011 levels (83.98 million acres), with harvested grain flowing to either processing for ethanol or other uses (for example, exports, feed). Ethanol production is capped over the 15-year period at no more than 40.5% of the US corn crop or 15 billion gallons of ethanol (or 14.71 billion gallons before denaturation at an assumed 2.2% denaturation rate). This prohibits a greater portion of the total corn grain harvested in any given year that is directed to corn ethanol exceeding either 2011 levels or US policy standards. Holding these factors at a steady state facilitates focus on the factors under investigation.

Over the 15-year window, the seven scenarios build on a baseline involving estimates of US corn grain yields forecasted as a result of a moderate effect of technological influence along with 2011 current practices in ethanol processing and livestock feeding. Only the specified effect (for corn grain yield, ethanol processing, or livestock feeding) is changed for each scenario.

Corn yield forecasts are based on historical trends and in-depth industry analysis and input. Ethanol processing forecasts are based on outputs reported by Mueller [26] and other industry data [27]. Livestock feeding forecasts were linearly extrapolated from the historical trend. Details of the baseline and specified scenarios for corn production, ethanol processing, and livestock feeding are provided in the following sub-sections.

### *10.4.2 CROP YIELD FORECASTS*

Technologies in the area of corn production focus on increases in grain produced per unit of land through both genetic and agronomic improvements. Genetic improvements can be partitioned to identify increases due to quantitative incremental gains through breeding, increased efficiency realized in the seed product pipeline (that is, speed to market) due to new technological innovations, and biotechnology traits (traits manifest through genetic modification) [28]. Agronomic improvements focus on ways to maximally leverage the crop genetic potential through cultural production practices, including spatial arrangement of plants in the field, pre-emptive measures for plant health, soil management and fertility, and seed treatments. In addition, global positioning systems technologies enable increasingly precise application of added nutrients and variable spacing of seed to maximize response to the micro-environment defined by soil type, topography, and other factors.

To derive Scenario 1 yield estimates (Table 2), actual corn yields for 2011 and 2012 were used [1]. Corn grain yields forecast for 2013 through to 2026 in keeping with historical trends, anticipated commercial launch dates, and farmer adoption of technologies, with the following assumptions:

- Based on trends observed from 1930 (development of hybrid corn and adoption replacing open-pollinated varieties) through 2012 [1], an average gain of 1.8 bu/A has been realized per year. This gain coincides with realized gains per year of 1.81 bu/A based on USDA average yields during the single cross hybrid era from 1960 to 1995, before genetically modified traits were commercialized in corn [29]. Thus, an average yearly gain of 1.8 bu/A is assumed to carry forward through 2026 (Table 2) from its historic trend line.
- Advanced breeding technology facilitating genomic-based approaches to choosing parents and identifying superior progeny in breeding populations as well as means to accelerate the breeding process to accumulate genetic gains more rapidly (for example, doubled haploid technology and associated breeding strategies) are expected to contribute additional gains of 1 bu/A per year. Based on widespread adoption of advanced technologies in the early to mid-2000s by the larger seed companies, this step change is likely to be fully realized beginning in 2015, with the launch of corn hybrids developed with these innovations (Table 2). A phase-in period is assumed, with the portion of new hybrid offerings developed using advanced technologies estimated at 25% in 2013 and 50% in 2014.
- Three releases of biotechnology traits are anticipated by 2026, each involving combinations (that is, trait stacks) of novel or improved biotechnology traits, with each combination delivering an estimated yield increase of 10 bu/A (Table 2). The leading biotechnology trait provider in corn, Monsanto Company, together with trait discovery partner, BASF, anticipates that by 2020 new corn hybrids will include >10 biotechnology traits [30] and as many as 20 by 2030 [31]. Monsanto refers to 'yield and stress packages', with enhancements to first-wave trait releases to follow in subsequent trait packages. The package of biotechnology traits would produce a step change in yield by preserving genetic potential through pest and stress tolerances or resistance, enhancing genetic potential, or improving efficiency in the plant utilization of essential requirements such as water and nitrogen. Biotechnology traits in phase III and phase IV stages of development by 2012 were considered to be 2 to 5 years away from market launch [32], for example, drought tolerance I, high yield corn, and CRW (corn rootworm) III featur-

ing RNAi mode of action [33,34]. Biotechnology traits in phase II stage of development by 2012 were considered to be 3 to 7 years away from market launch [32], for example, drought tolerance II, nitrogen use efficiency, and ECB (European corn borer) III [33,34]; this would constitute a second-wave package. The third-wave package of traits is presumed to include next-wave enhancements of the traits in the earlier packages. No phase-in period is accounted for, that is, the effect is not included until the years forecasted for maximal adoption of each release: 2018, 2022, and 2025. Other biotechnology trait providers may also contribute to trait technology to the marketplace; however, product pipeline information from other potential providers was not available publicly for this analysis. Because biotechnology traits are licensed across seed companies in the industry, market penetration does not depend on the market share of any one seed company.

- Agronomic improvements are anticipated in three waves by 2026, each accounting for a 3 bu/A yield increase (Table 2). Seed companies are working with manufacturers of farm machinery to create 'smart systems' to maximize yield through best possible agronomic conditions. For example, Monsanto Company plans the market introduction of IFS (Integrated Farming Systems) I, featuring variable rate planting, as early as 2014 [33,34]. A second wave of agronomic improvements is anticipated, with prescription placement for fertility and water added to IFS II [33]. The development of agricultural biologicals that boost the efficiency of pest controls such as BioDirect™ Technology by Monsanto [33] may factor into a third wave of agronomic improvements. Conservative estimation of yield impact forecasts effect step changes at maximal technology adoption in 2016, 2020, and 2024, with no phase-in period.

Scenario 1 corn yield forecasts serve as a baseline to represent grain input to scenarios that do not consider grain yield (that is, Scenarios 5 to 7). Scenarios 2 and 3 were developed following Lywood et al. [9], based on 10% decrease and 10% increase of Scenario 1 estimates, respectively (Table 3). Note that projections begin in 2013 because actual data are used to represent 2011 and 2012 yields. In contrast to the yield technology scenarios, Scenario 4, which reflects little/no yield gain from technological advancements, was developed using USDA long-term Projection figures that forecast corn yield increases at 2.0 bu/A through 2021 [35]. Extrapolating through 2026, corn yields are predicted to reach a level of 192.0 bu/A with Scenario 4 (Table 3).

Future genetic improvement could be directed to a greater quantity of starch in corn and possibly a higher quality of starch for ethanol production. However, use of such corn grain for improved ethanol yield would require identity preservation of grain destined for ethanol production; such a scenario for corn ethanol raw material supply is unlikely to be widely implemented in the next 15 years [36]. Therefore, this condition was not modeled.

### 10.4.3 FORECASTED CHANGES IN ETHANOL PRODUCTION

There are two methods used to produce ethanol from corn grain: dry grind and wet milling, which account for 86% and 14% of total US production, respectively [26]. Technologies in corn ethanol production focus on increasing efficiencies, leading to greater ethanol output. Advancements, mainly pertaining to dry grind, reflect process modifications that have been developed by researchers but are yet to be adopted at a large scale [37-43]. Enzyme products are being tested and adopted that make more use of the starch in the corn kernel, serving to increase ethanol yields and reduce the residual starch content of the DDGS. Enzymes have been used at a commercial scale that aid liquefaction and saccharification, thus improving the yeast's ability to convert glucose to ethanol [44-47]. Furthermore, experimental enzymes show promise in converting the cellulose and hemicellulose in the kernel to increase ethanol yields as well as reduce these fiber compounds in the DDGS co-product [48-50].

Dry grind and wet milling differ with respect to types and amounts of process outputs. The primary outputs from the dry grind process are ethanol, DDGS, and oil when skimming is practiced (Figures 3 and 4). With wet milling, higher-value co-products result, in the form of CGF, CGM, and corn oil (from the germ) suitable for human use; DDGS is not produced (Figure 5). DDGS, CGF, and CGM are relatively high in protein and thus directly supplant soybean meal, as well as corn grain, in livestock diets. Among dry grind ethanol plants, an estimated 50% practice oil skimming to recover crude oil that is mainly utilized for biodiesel production [26]. In this way, the skimmed oil competes with soy oil as a raw bioenergy feedstock. As soybean meal is supplanted in livestock diets by DDGS from corn ethanol, the soy oil produced along with soybean meal must be accounted for as well. Furthermore, oil skimming has an effect on DDGS composition, lowering the fat (energy) component as well as the protein content. The composition of feed co-products, that is, protein and fat content, depends on the ethanol processing method (Table 4).

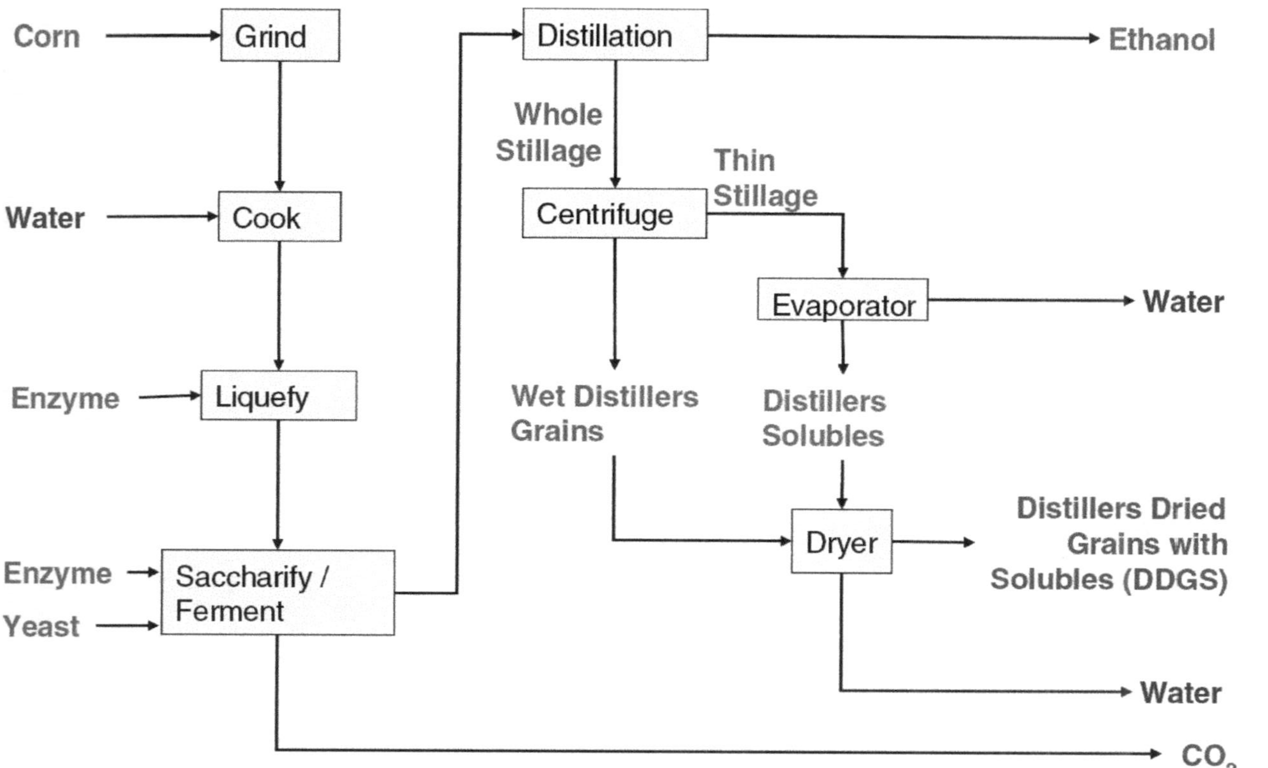


**FIGURE 3:** Conventional dry grind process for production of ethanol and distillers dried grains with solubles.

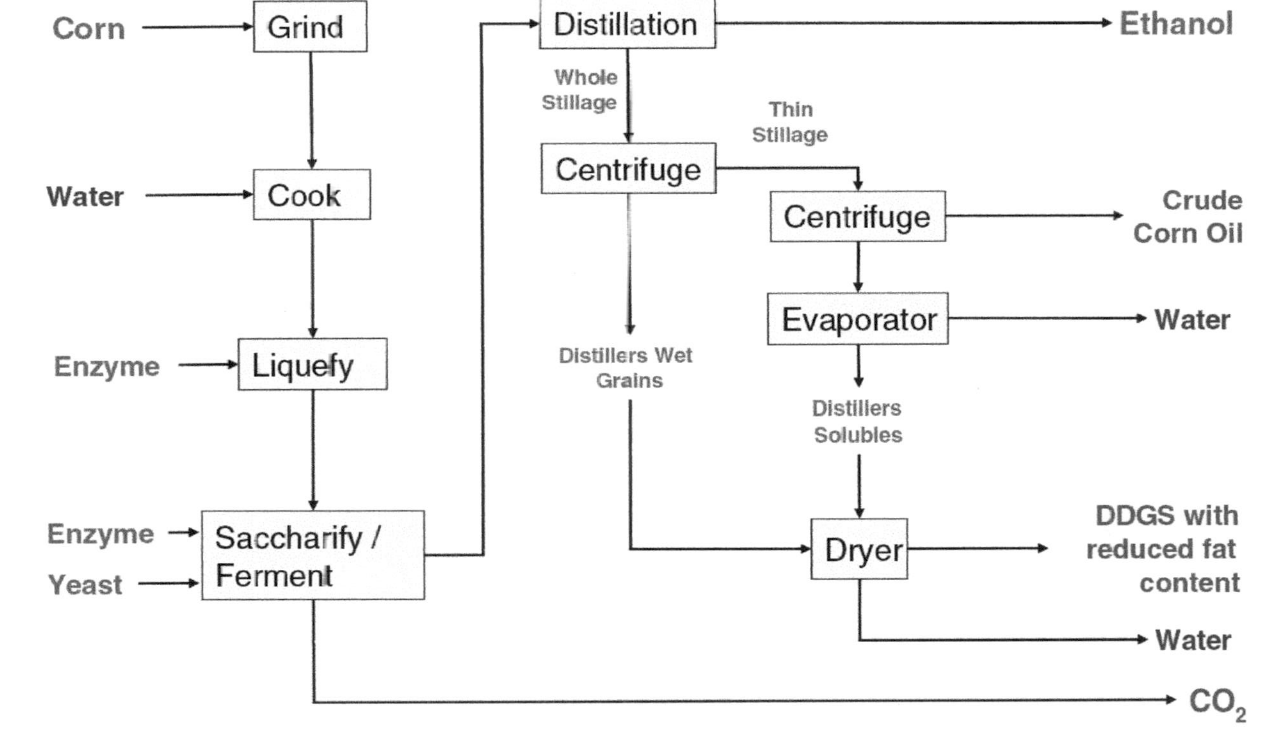


**FIGURE 4:** Dry grind process with oil recovery for production of ethanol, oil, and reduced fat distillers dried grains with solubles.

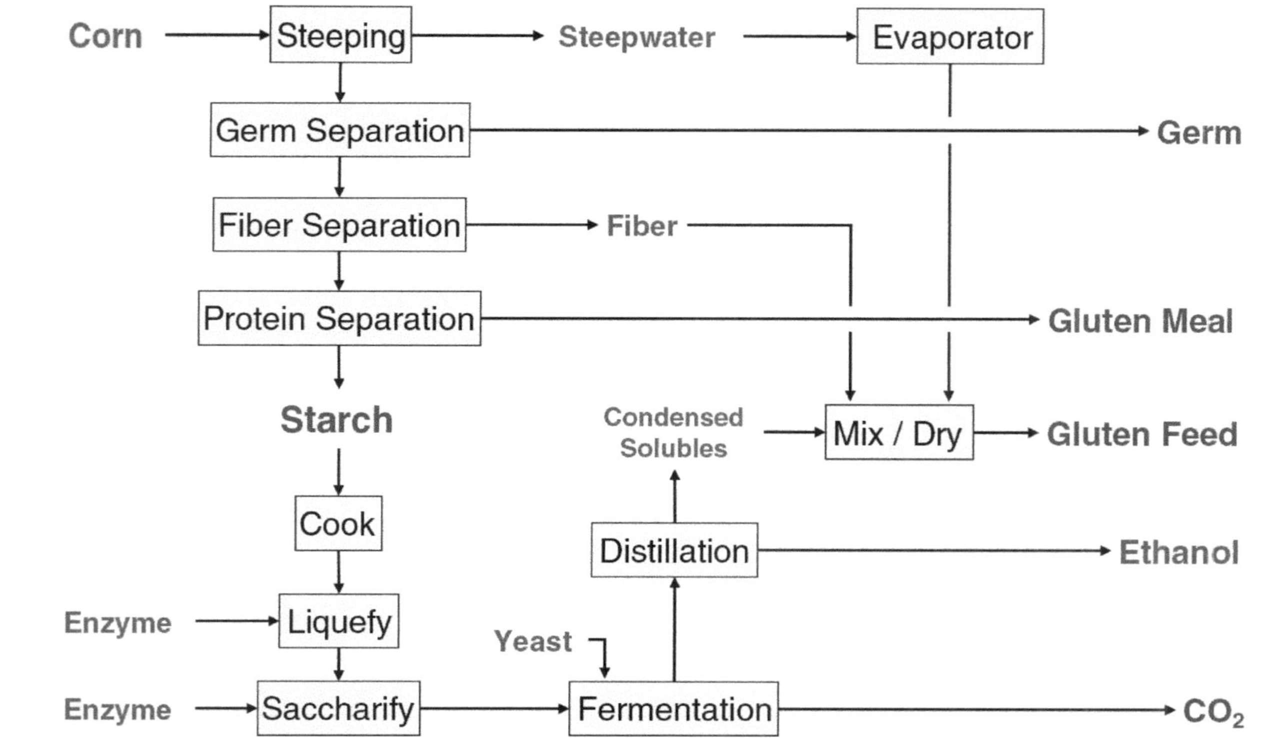


**FIGURE 5:** Wet milling process for production of ethanol, germ from which oil and germ meal are recovered, corn grain feed, and corn grain meal.

A weighted average of ethanol and co-product outputs was computed and used to formulate a baseline to represent ethanol processing in scenarios other than Scenarios 5 and 6 that feature effects of technological changes to the ethanol process (Table 5). Baseline assumptions include:

- A weighted industry yield of 2.759 gallons per bushel reflects dry grind and wet milling ethanol yields of 2.785 and 2.600 gallons per bushel, respectively, with dry grind plants making up 86% of ethanol capacity and wet milling plants 14%.
- Oil skimming, which is done at 50% of dry grind plants, leads to 0.24 pounds of oil per bushel, decreasing DDGS yield to 17.32 pounds per bushel at 89.31% dry matter content (that is, 15.45 pounds dry basis per bushel). Plants that do not skim have DDGS yields of 17.56 pounds (as is) per bushel [26]. Thus, the weighted dry grind average DDGS yield is 17.44 pounds (as is) per bushel.
- Wet milling co-products CGF and CGM represent 23% and 5% output from each bushel of corn processed for ethanol, respectively.
- Based on the current US ethanol production capacity of 14.71 billion undenatured gallons per year [27] and assuming a 50% adoption rate for skimming oil at a rate of 0.24 pounds of oil per bushel, oil production from skimming is estimated at 545.3 million pounds. This can be converted for use as biodiesel at a rate of 9 pounds of biodiesel per 10 pounds of oil, for a total biodiesel volume from skimmed oil of 66.68 million gallons (which is well below the US mandated biodiesel production of 1 billion gallons). It was assumed that biodiesel gallons from skimmed oil would replace gallons produced using soy oil; thus, skimmed oil feeds back into the model to reduce the amount of oil replacement acres by canola. Because oil skimmed from the dry grind process is not economical for use in human food, this prevents the use of some of the soy oil for non-food use.

Scenario 5 anticipates a technological improvement in the dry grind process that allows for fermentation of all starch in the corn grain to ethanol, providing 2.5% more ethanol output per bushel of corn while reducing DDGS production by 6.1%. This scenario is referred to as 'full starch'. At present, a percentage of starch remains unconverted in the DDGS following fermentation, as indicated by average starch content (7.54% starch (dry basis)) of DDGS samples from the National Research Council [51]. If a greater percentage of starch could be converted into ethanol, ethanol yields would increase, DDGS output per bushel of corn would decrease, and DDGS composition would be altered. Dry grind processors continue to use new equipment, enzymes, and process designs in an effort to reduce residual starch content of DDGS. We assume this improved technology would be implemented at all dry grind facilities. Conservatively estimating that 6% residual starch (dry basis) or other convertible sugars in DDGS from dry grind are converted (sources such as the National Research Council report DDGS starch contents of 7.5% to 10.8% (dry basis) [51]), the additional ethanol yield with full starch implementation is estimated at 0.081 gallons per bushel. In Scenario 5, we accounted for changes in yield and composition of DDGS but assumed no impact to CGF or CGM because of the comparatively small fraction of materials in the livestock feeding system and the minimal impact anticipated. Thus, the full starch scenario involves only changes to dry grind co-product outputs; compositions of CGF and CGM were assumed to be unchanged. With more complete starch conversion during dry grind, dry grind ethanol yield increases to 2.866 gallons per bushel, and the improved aggregate industry yield is 2.829 gallons per bushel (Table 5; see Additional file 1 for calculations).

Concomitant with ethanol yield increase, the amount of DDGS produced per bushel decreases due to the complete fermentation of starch. The adjusted rate of DDGS production due to full starch conversion, assuming 50% adoption of oil skimming by the dry grind industry, is 16.38 pounds of DDGS per bushel of corn (Table 5; see Additional file 1 for calculations). At the same time, protein content per bushel of corn increases to 29.10%, although total protein in the DDGS produced per acre of corn remains the same; likewise, fat (oil content) increases per bushel of corn to 10.28% (Table 4).

Scenario 6 anticipates another technological improvement in fermentation in the dry grind process to facilitate conversion of fiber fractions (C5 and C6 sugars) to ethanol, thus leading to 11.6% more ethanol output per bushel of corn while concomitantly reducing DDGS output 27.4%. This scenario, referred to as 'complete fiber', anticipates conversion of fiber portions of the corn grain that are currently unfermented, in addition to conversion of residual starch (as in Scenario 5). Dien et al. document conversion of C5 and C6 sugars to ethanol using conventional yeast and a bacterial strain (*Escherichia coli* FBR5) [48]. More recently, Ha et al. [52] and Bera et al. [53] document the conversion of C5 sugars to ethanol with new fermentation organisms. The fermentation included residual starch along with pericarp and endosperm fiber fractions. Dien et al. reported that use of two types of glucose-consuming organisms increased ethanol yields by 13.3% [48].

In dry grind processing, the 13.3% increase in ethanol yields is accompanied by changes in DDGS yields and composition. Because ethanol is a high-value product, it is assumed that 100% of dry grind producers would adopt this technology (although we do not expect the number and proportion of dry grind facilities to be affected by this technology). The wet milling facilities were not anticipated to adopt this technology because feasible production of CGF relies on a source of fiber that is mixed and dried with process streams such as steepwater and fermentation solids (distillers solubles). In wet milling, fiber is needed as a method to allow removal of steepwater and fermentation solids from the process. Without the fiber stream, drying these solids would not be economical.

The 13.3% increase in ethanol yield with the complete fiber scenario translates to an additional ethanol yield of 0.370 gallons per bushel for dry grind, bringing the aggregate industry yield to 3.078 gallons per bushel (Table 5; see Additional file 1 for calculations). Concomitant with ethanol yield increase, the composition of DDGS output in dry grind processing is altered with complete fiber conversion compared to the baseline. Protein content per bushel of corn increases to 28.58%; likewise, fat (oil content) increases per bushel of corn to 10.10% (Table 4). It could be anticipated that protein and oil contents for Scenario 6 would be much higher than Scenario 5. However, Scenario 5 assumes a 6.1% decrease in the DDGS yield, whereas Scenario 6 is based on actual conversion rates from Dien et al.[48]. With the decreased amount of DDGS produced per bushel, the adjusted rate of DDGS production due to complete fiber conversion, assuming continued 50% adoption of oil skimming by the dry grind industry, is 12.67 pounds of DDGS per bushel of corn (Table 5; see Additional file 1 for calculations).

### *10.4.4 CHANGES IN THE USAGE OF CO-PRODUCTS FOR LIVESTOCK FEED*

Technologies in the area of livestock feeding and nutrition for poultry, pigs, beef cattle, and dairy cattle anticipate increased usage of ethanol co-products, mainly DDGS, by specific groups of livestock. Altered usage of DDGS assumes compliance with dietary nutrition consistent with maintaining high meat and milk quality.

**TABLE 7:** Substitution of corn and soybean meal by distillers dried grains with solubles, corn grain feed, and corn grain meal produced from 5.007 billion bushels (127.440 million metric tons) of corn grain associated with 71%:29% substitution ratio overall

| | | Substitution (%) | | Substitution tonnage (millions) | |
|---|---|---|---|---|---|
| Co-product | Million tons | Corn | Soybean meal | Corn | Soybean meal |
| Export | 9.000 | 51.2 | 48.8 | 4,610 | 4,390 |
| Domestic use | 25.091 | 75.8 | 24.2 | 19.020 | 6.071 |
| Beef usage | 12.646 | 100 | 0 | 12.646 | 0 |
| Dairy usage | 8.405 | 47.0 | 53.0 | 3.951 | 4.455 |
| Pig usage | 2.283 | 60.0 | 40.0 | 1.370 | 0.913 |
| Poultry usage | 1.756 | 60.0 | 40.0 | 1.054 | 0.703 |
| Domestic dairy, pigs, and poultry | 12.445 | 51.2 | 48.8 | 6.374 | 6.071 |
| Distillers dried grain with solubles domestic plus export | 34.091 | 69.3 | 30.7 | 23.630 | 10.461 |
| Corn grain feed | 4.098 | 100 | - | 4.098 | - |
| Corn grain meal | 0.891 | - | 100 | - | 0.891 |
| Total substitution | 39.080 | 71.0 | 29.0 | 27.728 | 11.352 |

In 2011, it is estimated that a total of 127.440 million metric tons (5.007 billion bushels) of corn was directed to ethanol production annually [25], with 86% (109.598 million metric tons; 4.306 billion bushels) being used in dry grind processing and 14% (17.842 million metric tons; 700.917 million bushels) being used in wet milling. Given a weighted average of 310.1 kg of DDGS produced per metric ton of corn grain in the dry grind process (Table 6), a total production of 34.091 million metric tons of DDGS resulted (Table 7). In addition, 4.098 million metric tons of CGF and 0.891 million metric tons of CGM were generated as a result of production of ethanol from corn in the wet milling process, assuming that

23% of the corn grain will end up in CGF and 5% of the grain will end up in CGM. Thus, an estimated total of 39.080 million metric tons of DDGS, CGF, and CGM are produced annually from corn ethanol processing. It is assumed that of the DDGS produced, 9 million metric tons are exported, and the remaining 25.091 million metric tons are used domestically.

Because DDGS, CGF, and CGM supplants different amounts of corn and soybean meal in diets fed to different groups of animals, it is necessary to know the approximate market share for each group of animals for which DDGS is used. There are three recent estimates for the proportion of domestic DDGS fed to beef cattle, dairy, pigs, and poultry [54-56]. The average of these three estimates is 50.4% to beef cattle, 33.5% to dairy cattle, 9.1% to pigs, and 7.0% to poultry (Table 6). Based on these percentages, the total usage of DDGS for each group of animals can be calculated (Table 7).

Inclusion of DDGS, CGF, and CGM in livestock feeding regimes supplants corn or soybean meal in diets for beef cattle, dairy, poultry, and livestock. At present, DDGS replaces approximately 60% corn and 40% soybean meal in the feeding of pigs and poultry [54,57], supporting a 1:1 substitution rate [58,59]. DDGS replaces various amounts of corn and soybean meal in diets for ruminant animals. The rate of substitution per species depends on the requirement for protein and energy. It is assumed that DDGS fed to beef cattle replaces no soybean meal and only corn because beef cattle are usually fed only limited quantities of soybean meal due to their relatively low requirement for protein and because protein equivalents can be obtained less expensively from other ingredients. Thus, it is assumed that DDGS included in diets fed to beef cattle replaces corn at a 1:1 rate, although it is acknowledged that substitution rates of 1.1:1 or 1.2:1 have been proposed [54]. However, currently there are very limited biological data to support substitution rates greater than 1:1 and in the present calculations, the 1:1 rate is used to make sure the DDGS is not overvalued.

It is also assumed that the CGF produced from the wet milling industry is fed exclusively to beef cattle and that it replaces corn on a 1:1 basis and that no soybean meal is replaced by CGF. By contrast, DDGS fed to dairy cows replaces 47% corn and 53% soybean meal and DDGS fed to pigs and poultry replaces 60% corn and 40% soybean meal. As a consequence of these substitution rates, more soybean meal is replaced if DDGS usage is shifted from beef cattle to dairy, pigs, or poultry. It is also assumed that the 9 million metric tons of DDGS that are exported are fed to dairy, pigs, and poultry and that the substitution rates for corn and soybean meal for the exported DDGS are similar to the replacement rates for the domestically used DDGS that is fed to dairy, pigs, and poultry (Table 7). Finally, it is assumed that the CGM that is produced from the wet milling industry is used exclusively in dairy feeding where it replaces soybean meal at a 1:1 rate. Under these assumptions, the total substitution of corn and soybean meal can be calculated.

**TABLE 8:** Substitution of corn and soybean meal by distillers dried grains with solubles, corn grain feed, and corn grain meal produced from 5.007 billion bushels (127.440 million metric tons) of corn grain associated with a 65%:35% substitution ratio overall

| | | Substitution (%) | | Substitution tonnage (M) | |
|---|---|---|---|---|---|
| Co-product | Million tons | Corn | Soybean meal | Corn | Soybean meal |
| Export | 9.0 | 51.2 | 48.8 | 4,610 | 4,390 |
| Domestic use | 25.091 | 75.8 | 24.2 | 19.020 | 6.071 |
| Beef usage | 7.527 | 100 | 0 | 7.527 | 0 |
| Dairy usage | 11.868 | 47.0 | 53.0 | 5.578 | 6.290 |
| Pig usage | 3.212 | 60.0 | 40.0 | 1.927 | 1.285 |
| Poultry usage | 2.484 | 60.0 | 40.0 | 1.490 | 0.994 |
| Domestic dairy, pigs, and poultry | 17.564 | 51.2 | 48.8 | 8.995 | 8.568 |
| Distillers dried grain with solubles domestic plus export | 34.091 | 62.0 | 38.0 | 21.132 | 12.959 |
| Corn grain feed | 4.098 | 100 | - | 4.098 | - |
| Corn grain meal | 0.891 | - | 100 | - | 0.891 |
| Total substitution | 39.080 | 64.6 | 35.4 | 25.231 | 13.850 |

Overall, 27.728 million metric tons of corn and 11.352 million metric tons of soybean meal are replaced across livestock diets, reflecting a substitution ratio of 71% corn to 29% soybean meal in feeding of DDGS, CGM, and CGF (Table 7). This ratio serves as a baseline to represent livestock feeding utilization of corn ethanol co-products in scenarios other than Scenario 7, which reflects an adjusted ratio based on technological changes.

Scenario 7 anticipates a modified overall substitution ratio of corn to soybean meal due to a shift in the utilization of DDGS in feeding diets among livestock types. Because economics favor replacement of soybean meal rather than corn (amount of protein on a per ton basis), it is expected that an increased proportion of DDGS will be consumed by dairy, pigs, and poultry in the future.

Dairy cattle are expected to be a primary target for increased rates of DDGS inclusion in the diet; dairy cattle require a high amount of protein, which is largely provided by soybean meal at present, and can also handle the fiber load. DDGS can be used in diets for dairy cows by at least up to 20% and often up to 30% on a dry matter basis without changing animal performance [60,61]. Furthermore, use of DDGS for dairy favors the lower fat content of DDGS from dry grind ethanol plants that skim oil (Table 4); thus, oil skimming may promote a higher rate of inclusion of DDGS in diets fed to dairy cows [61]. Provided that diets are correctly formulated, there are no indications that milk composition will be changed or that the value components in milk will be reduced [62,63].

With the increased awareness of the benefits of DDGS in diets fed to pigs and poultry and the economic competitiveness of DDGS relative to soybean meal, it is also likely that the penetration of DDGS in the swine and poultry feed markets will increase. This can be easily accomplished without exceeding the maximum recommended rates for inclusion of DDGS in diets fed to pigs or poultry. DDGS can be included in diets for pigs at levels of at least 20% without changing the composition or the nutritional value of the meat that is produced [59,64-67]. Likewise, for poultry, DDGS can be included in the diets by at least 10% to 15% without reducing product quality [68-70]. Because DDGS penetration can be increased greatly without exceeding these thresholds, it is possible to increase DDGS utilization in diets fed to pigs and poultry without negatively impacting pork or poultry meat quality. Furthermore, balancing of DDGS with specific indispensable amino acids will make DDGS more usable in the feeding of pigs and poultry, and use of specific microbial enzymes such as xylanase and phosphatases in livestock diets containing DDGS may help increase the energy and phosphorus value of DDGS. By feeding more DDGS to pigs and poultry, these livestock types will consume an overall greater share of the total DDGS produced.

With greater use of DDGS in feeding of dairy cattle, pigs, and poultry, a decline in the proportional usage by beef cattle is anticipated. In Scenario 7, it is therefore estimated that consumption of DDGS by beef cattle is reduced from the current level of 50.4% to only 30% of total DDGS production, whereas dairy cattle, pigs, and poultry will increase consumption to 47.3%, 12.8%, and 9.9% of the produced DDGS, respectively (Table 6). Reflecting the shift in usage among livestock types, Scenario 7 depicts DDGS substitution of 65% corn and 35% soybean meal across livestock diets (Table 8).

## ENDNOTE

[a]We define feedback drawing on terminology from the field of System Dynamics, http://www.systemdynamics.org/DL-IntroSysDyn/feed.htm. Positive feedback causes systems to grow or expand, and negative feedback causes decline or contraction. For example, in our case, a positive feedback effect results when a perturbation to the system increases land area, such as with higher ethanol yields, whereas a negative feedback effect, such as greater livestock feeding with DDGS, reduces land area.

## REFERENCES

1. United States Department of Agriculture Economic Research Service: Feed Grains Database. http://www.ers.usda.gov/data-products/feed-grains-database/feed-grains-yearbook-tables.aspx#26766
2. US Department of Energy, US Energy Information Administration: Energy Independence and Security Act of 2007: Summary of Provisions. http://www.eia.doe.gov/oiaf/aeo/otheranalysis/aeo_2008analysispapers/eisa.html
3. Wang M, Huo H, Arora S: Methods of dealing with co-products of biofuels in life-cycle analysis and consequent results within the U.S. context. Energy Policy 2011, 39:5726-5736.
4. Arora S, Wu M, Wang M: Update of Distillers Grains Displacement Ratios for Corn Ethanol Life-cycle Analysis. Argonne, IL: Argonne National Laboratory; 2008.
5. Farrell AE, Plevin RJ, Turner BT, Jones AD, O'Hare M, Kammen DM: Ethanol can contribute to energy and environmental goals. Science 2006, 311:506-508.

6. Wang M: Development and Use of GREET 1.6 Fuel-Cycle Model for Transportation Fuels and Vehicle Technologies. Argonne, IL: Tech. Rep. ANL/EDS/TM-163, Argonne National Laboratory; http://www.transportation.anl.gov/pdfs/TA/153.pdf
7. Taheripour F, Hertel TW, Tyner WE: Implications of biofuels mandates for the global livestock industry: a computable general equilibrium analysis. Agr Econ 2010, 42:325-342.
8. Taheripour F, Hertel TW, Tyner WE, Beckman JF, Birur DK: Biofuels and their by-products: Global economic and environmental implications. Biomass Bioenerg 2010, 34:278-289.
9. Lywood W, Pinkney J, Cockerill S: Impact of protein concentrate coproducts on net land requirement for European biofuel production. GCB Bioenerg 2009, 1:346-359.
10. Service RF: Is there a road ahead for cellulosic ethanol? Science 2010, 329:784-785.
11. Miranowski J, Rosburg A, Aukayanagul J: US maize yield growth implications for ethanol and greenhouse gas emissions. Ag Bio Forum 2011, 14:120-132.
12. Downing M, Eaton LM, Graham RL, Langholtz MH, Perlack RD, Turhollow AF Jr, Stokes B, Brandt CC, for US Department of Energy: U.S. Billion-ton Update: Biomass Supply for a Bioenergy and Bioproducts Industry. Oak Ridge TN: Oak Ridge National Laboratory; 2011. http://www1.eere.energy.gov/bioenergy/pdfs/billion_ton_update.pdf
13. Mueller S, Kwik J: Corn Ethanol: Emerging Plant Energy and Environmental Technologies. Chicago, IL: University of Illinois at Chicago; 2013. http://ethanolrfa.org/page/-/PDFs/2012%20Corn%20Ethanol%20FINAL.pdf?nocdn=1
14. Dunn JB, Mueller S, Kwon H, Wang MQ: Land-use change and greenhouse gas emissions from corn and cellulosic ethanol. Biotechnol Biofuels 2013, 6:51.
15. Wallington TJ, Anderson JE, Mueller SA, Kolinski Morris E, Winkler SL, Ginder JM, Nielsen OJ: Corn ethanol production, food exports, and indirect land use change. Environ Sci Tech 2012, 46:6379-6384.
16. Hertel TW, Golub AA, Jones AD, O'Hare M, Plevin RJ, Kammen DM: Effects of US maize ethanol on global land use and greenhouse gas emissions: estimating market-mediated responses. Bioscience 2010, 60:223-231.
17. Hoekman SK: Biofuels in the U.S. – challenges and opportunities. Renew Energy 2009, 34:14-22.
18. Mitchell D: A Note on Rising Food Prices. Washington, DC: Policy Research Working Paper 4682, World Bank Development Prospects Group, World Bank; 2008. https://openknowledge.worldbank.org/handle/10986/6820
19. Havlik P, Schneider UA, Schmid E, Bottcher H, Fritz S, Skalsky R, Aoki K, de Cara S, Kindermann G, Kraxner F, Leduc S, McCallum I, Mosnier A, Sauer T, Obersteiner M: Global land-use implications of first and second generation biofuel targets. Energy Policy 2011, 39:5690-5702.
20. Johansson DJA, Azar C: A scenario based analysis of land competition between food and bioenergy production in the US. Clim Change 2007, 82:267-291.
21. Cai X, Zhang X, Wang D: Land availability for biofuel production. Environ Sci Technol 2011, 45:334-339.
22. Isee Systems: STELLA, v 9.1.4. 2011. http://www.iseesystems.com

23. USDA, National Agricultural Statistics Service: Crop Production: 2012 Summary. http://usda01.library.cornell.edu/usda/nass/CropProdSu//2010s/2013/CropProdSu-01-11-2013.pdf
24. North Carolina Soybean Producers Association, Inc: How soybeans are used. http://www.ncsoy.org/ABOUT-SOYBEANS/Uses-of-Soybeans.aspx
25. USDA, National Agricultural Statistics Service: Crop production and agricultural prices. http://www.nass.usda.gov
26. Mueller S: 2008 national dry mill corn ethanol survey. Biotechnol Lett 2010, 32:1261-1264.
27. Renewable Fuels Association: Ethanol Production Capacity. http://www.ethanolrfa.org/pages/statistics
28. Moose SP, Mumm RH: Molecular plant breeding as the foundation for 21st century crop improvement. Plant Physiol 2008, 147:969-977.
29. Troyer AF: Adaptedness and heterosis in corn and mule hybrids. Crop Sci 2006, 46:528-543.
30. Fraley R: Monsanto Company, Credit Suisse Global AgroChemicals Conference presentation. 2010. http://www.monsanto.com/investors/Pages/presentations.aspx
31. Fraley R, Monsanto Company: A tale of two farms. In Proceedings of the Food & Agricultural Communications; The Next Frontier: 17 February 2012: Champaign, IL. Urbana, IL: University of Illinois Agricultural Communications; 2012.
32. Padgette S, Monsanto Company: Golden Opportunities:Working jointly for higher yields; growth and value in yield and stress. Presentation at Ghent, Belgium on September 16. 2008. http://www.monsanto.com/investors/Pages/presentations.aspx
33. Fraley R, Monsanto Company: Annual R&D pipeline review presentation. 2012. http://www.monsanto.com/investors/Pages/presentations.aspx
34. Fraley R, Monsanto Company: The R&D that drives a yield company: A look at the R&D platforms that define Monsanto. Monmouth: Presented at the Whistle Stop Tour IV; 2012. http://www.monsanto.com/investors/Pages/presentations.aspx
35. USDA, Office of Chief Economist: USDA Agricultural Projections to 2021. http://www.usda.gov/oce/commodity/archive_projections/USDAAgriculturalProjections2021.pdf
36. Goldsmith PD, Bender K: Ten conversations about identity preservation. J Chain Network Sci 2004, 4:111-123.
37. Johnston DB, McAloon AJ, Moreau RA, Hicks KB, Singh V: Composition and economic comparison of germ fractions from modified corn processing technologies. J Am Oil Chem Soc 2005, 82:603-608.
38. Johnston DB, Singh V: Processes for recovery of corn germ and optionally corn coarse fiber (pericarp). US6899910-B2 (Patent); 2005.
39. Moreau RA, Singh V, Eckhoff SR, Powell MJ, Hicks KB, Norton RA: Comparison of yield and composition of oil extracted from corn fiber and corn bran. Cereal Chem 1999, 76:449-451.
40. Singh V, Eckhoff SR: Economics of germ preparation for dry-grind ethanol facilities. Cereal Chem 1997, 74:462-466.
41. Singh V, Eckhoff SR: Effect of soak time, soak temperature, and lactic acid on germ recovery parameters. Cereal Chem 1996, 73:716-720.

42. Singh V, Johnston DB, Naidu K, Rausch KD, Belyea RL, Tumbleson ME: Comparison of modified dry-grind corn processes for fermentation characteristics and DDGS composition. Cereal Chem 2005, 82:187-190.
43. Singh V, Moreau RA, Doner LW, Eckhoff SR, Hicks KB: Recovery of fiber in the corn dry-grind ethanol process: a feedstock for valuable coproducts. Cereal Chem 1999, 76:868-872.
44. Khullar E, Shetty JK, Rausch KD, Tumbleson ME, Singh V: Use of phytases in ethanol production from E-Mill corn processing. Cereal Chem 2011, 88:223-227.
45. Shetty JK, Paulson B, Pepsin M, Chotani G, Dean B, Hruby M: Phytase in fuel ethanol production offers economical and environmental benefits. Int Sugar J 2008, 110:160-167.
46. Vidal BC, Rausch KD, Tumbleson ME, Singh V: Kinetics of granular starch hydrolysis in corn dry-grind process. Starch-Starke 2009, 61:448-456.
47. Wang P, Johnston DB, Rausch KD, Schmidt SJ, Tumbleson ME, Singh V: Effects of protease and urea on a granular starch hydrolyzing process for corn ethanol production. Cereal Chem 2009, 86:319-322.
48. Dien BS, Johnston DB, Hicks KB, Cotta MA, Singh V: Hydrolysis and fermentation of pericarp and endosperm fibers recovered from enzymatic corn dry-grind process. Cereal Chem 2005, 82:616-620.
49. Dien BS, Nagle N, Hicks KB, Singh V, Moreau RA, Tucker MP, Nichols NN, Johnston DB, Cotta MA, Nguyen Q, Bothast RJ: Fermentation of "quick fiber" produced from a modified corn-milling process into ethanol and recovery of corn fiber oil. Appl Biochem Biotechnol 2004, 113:937-949.
50. Dien BS, Hespell RB, Wyckoff HA, Bothas RJ: Fermentation of hexose and pentose sugars using a novel ethanologenic Escherichia coli strain. Enzyme Microb Technol 1998, 23:366-371.
51. National Research Council: Feed ingredient composition. In Nutrient Requirements of Swine. Washington, DC: National Research Council; 2012:265-267.
52. Ha SJ, Kim SR, Choi JH, Park MS, Jin YS: Xylitol does not inhibit xylose fermentation by engineered Saccharomyces cerevisiae expressing xyla as severely as it inhibits xylose isomerase reaction in vitro. Appl Microbiol Biotechnol 2011, 92:77-84.
53. Bera AK, Ho NWY, Khan A, Sedlak M: A genetic overhaul of Saccharomyces cerevisiae 424a(lnh-st) to improve xylose fermentation. J Indust Microbiol Biotechnol 2011, 38:617-626.
54. Hoffman LA, Baker A: Estimating the Substitution of Distillers Grains for Corn and Soybean Meal in the US Feed Complex. USDA ERS; 2011. http://www.ers.usda.gov/publications/fds-feed-outlook/fds11i01.aspx
55. Renewable Fuels Association: Ethanol industry outlook; building bridges to a more sustainable future. 2011. http://www.ethanolrfa.org/page/2011%20ethanol%20industry%20outlook.pdf?nocdn = 1
56. Wisner R: Estimated U.S. dried distillers grains with soluble (DDGS): Production and use. Ames, IA: Agricultural Marketing Resource Center, Iowa State University; 2011. http://www.extension.iastate.edu/agdm/crops/outlook/dgsbalancesheet.pdf
57. Stein HH: Distillers dried grains with solubles (DDGS) in diets fed to swine. In Swine Focus #001. Urbana-Champaign, IL: University of Illinois; 2007.

58. Xu G, Baidoo SK, Johnston LJ, Bibus D, Cannon JE, Shurson GC: Effects of feeding diets containing increasing content of corn distillers dried grains with solubles to grower-finisher pigs on growth performance, carcass composition, and pork fat quality. J Anim Sci 2010, 88:1398-1410.
59. Widmer MR, McGinnis LM, Wulf DM, Stein HH: Effects of feeding distillers dried grains with solubles, high-protein distillers dried grains, and corn germ to growing-finishing pigs on pig performance, carcass quality, and the palatability of pork. J Anim Sci 2008, 86:1819-1831.
60. Schingoethe DJ, Kalscheur KF, Hippen AR, Garcia AD: The use of distillers products in dairy cattle diets. Invited review. J Dairy Sci 2009, 92:5802-5813.
61. Mjourn K, Kalscheur KF, Hippen AR, Schingoethe DJ, Little DE: Lactation performance and amino acid utilization of cows fed increasing amounts of reduced fat dried distillers grains with solubles. J Dairy Sci 2010, 93:288-303.
62. Mpapho GS, Hippen AR, Kalscheur KF, Schingoethe DJ: Lactational performance of dairy cows fed wet corn distillers grain for the entire lactation [abstract]. J Dairy Sci 2006, 90:100.
63. Kalscheur KF: Impact of feeding distillers grains on milk fat, protein, and yield. In Proceedings of the Distillers Grains Technology Council's 9th Annual Symposium: 18-19 May, 2005; Louisville, KY. Ames, IA: DGTC; 2005.
64. Cromwell GL, Azain MJ, Adeola O, Baidoo SK, Carter SD, Crenshaw TD, Kim SW, Mahan DC, Miller PS, Shannon MC: Corn distillers dried grains with solubles in diets for growing-finishing pigs: a cooperative study. J Anim Sci 2011, 89:2801-2811.
65. McDonnell PM, Shea CJ, Callan JJ, O'Doherty JV: The response of growth performance, nitrogen, and phosphorus excretion of growing-finishing pigs to diets containing incremental levels of maize dried distillers grains with solubles. Anim Feed Sci Technol 2011, 169:104-112.
66. Stein HH, Shurson GC: The use and application of distillers dried grains with solubles (DDGS) in swine diets. Board Invited Review. J Anim Sci 2009, 87:1292-1303.
67. Whitney MH, Shurson GC, Johnston LJ, Wulf DM, Shanks BC: Growth performance and carcass characteristics of grower-finisher pigs fed high-quality corn distillers dried grain with solubles originating from a modern Midwestern ethanol plant. J Anim Sci 2006, 84:3356-3363.
68. Shim MY, Pesti GM, Bakalli RI, Tillman PB, Payne RL: Evaluation of corn distillers dried grains with solubles as an alternative ingredient for broilers. Poultry Sci 2011, 90:369-376.
69. Lumpkins BS, Batal AB, Dale NM: Evaluation of distillers dried grains with solubles as a feed ingredient for broilers. Poultry Sci 2004, 83:1891-1896.
70. Roberson KD: Evaluation of distillers dried grains with solubles as a feed ingredient for broilers. Poultry Sci 2003, 83:1891-1896.

# CHAPTER 11

# LAND-USE CHANGE AND GREENHOUSE GAS EMISSIONS FROM CORN AND CELLULOSIC ETHANOL

JENNIFER B. DUNN, STEFFEN MUELLER, HO-YOUNG KWON, AND MICHAEL Q. WANG

## 11.1 BACKGROUND

Biofuels are often considered to be among the technologies that can reduce the greenhouse gas (GHG) impacts of the transportation sector. Yet the changes in land use that could accompany the production of biofuel feedstocks and the subsequent environmental impacts, including GHG emissions, are a potential disadvantage of biofuels. Land-use change (LUC) occurs when land is converted to biofuel feedstock production from other uses or states, including non-feedstock agricultural lands, forests, and grasslands. This type of LUC is sometimes called direct LUC. The resulting change in crop production levels (e.g., an increase in corn production may cause a decrease in soybean production) and exports may shift land uses domestically and abroad through economic linkages. This latter type of LUC is called indirect LUC and can be estimated through the use of economic models.

A change in land use causes a change in carbon stocks aboveground and belowground. As a result, a given LUC scenario may emit or sequester carbon. When an LUC scenario results in a net release of carbon to the atmosphere, it is debated if biofuels result in GHG reductions at all [1,2]. Of particular concern is the conversion of forests [3,4], an inherently carbon-rich land cover that in some cases may be a carbon sink. Their conversion to biofuel feedstock production land could incur a significant carbon penalty [5].

The estimation of LUC and the resulting GHG emissions is accomplished through the marriage of LUC data with aboveground carbon and soil organic carbon (SOC) data for each of the land types affected. The amounts and types of land converted as a result of increased biofuel production can be estimated with an agricultural-economic model, for example, a computable general equilibrium (CGE) model; several recent reports [6,7] provide an overview of CGE models and their application to estimating LUC associated with biofuel production. It is also necessary to know the aboveground and belowground carbon content of the land in its original state and in its future state as feedstock production land. Aboveground carbon content information is provided by databases that are often built on satellite data [8], while SOC content can be modelled with tools such as CENTURY [9].

LUC GHG emissions from biofuel production are typically placed in the context of a biofuel life cycle analysis (LCA), which estimates the GHG emissions of a biofuel on a farm-to-wheels basis [10]. Regulatory bodies, including the U.S. Environmental Protection Agency (EPA), the California Air Resources Board (CARB), and the European Union [11-13], use LCA to evaluate the GHG impacts of biofuels.

When LUC GHG emissions are examined in the context of a biofuel's life cycle, they can be substantive. For example, EPA estimated that LUC GHG emissions were 38% of total life cycle GHG emissions for corn ethanol produced in a natural gas-powered dry mill with dry distillers grains solubles (DGS) as a co-product [11]. LUC GHG emissions are also highly uncertain [14] due to large uncertainties in CGE modelling, aboveground carbon data, and SOC content data [15].

As one of the most prevalent biofuels, corn ethanol has been the subject of most biofuel LUC research [14,16]. Few studies have considered LUC GHG emissions from cellulosic ethanol production. Hill et al. [17] estimated domestic LUC GHG emissions for the production of 3.8 billion litres of ethanol based on conversion of land formerly in the Conservation Reserve Program (CRP) to production of corn, corn stover, switchgrass, prairie grass, and miscanthus. The resulting LUC GHG emissions for corn were between 27 and 35 g $CO_2$e/MJ. These emissions were 0.5 and 0.2 g $CO_2$e/MJ for switchgrass and miscanthus, respectively. Corn stover was assumed to have no LUC GHG emissions associated with its production. Scown et al. [18] considered a number of domestic U.S. scenarios for the production of 39.7 billion liters/year of ethanol from miscanthus, allowing only cropland or CRP lands to be converted to miscanthus production. These authors modelled productivity of miscanthus with Miscanmod at the county level. A model proposed by Matthews and Grogan [19] was used to estimate the SOC content of converted land. SOC changes were aggregated to the county level from a 90-meter resolution. In their calculation of LUC GHG emissions, Scown et al. [18] did not consider the impact of land management history on SOC content. Their study concluded that on net 3.4 to 16 g $CO_2$e/MJ would be sequestered as a result of SOC changes. Separately, Davis et al. [20] considered the conversion of 30% of domestic (U.S.) land currently in corn production to miscanthus or switchgrass (fertilized or unfertilized) production. They used DAYCENT to simulate regional miscanthus and switchgrass cultivation in the central U.S. and identified lower GHG fluxes from cultivation when either crop was grown in place of corn. The reductions after 10 years (1.9% for switchgrass with fertilization and 19% for miscanthus) came from both reduction in fertilizer-derived $N_2O$ emissions and increased carbon sequestration. Similarly, Qin et al. [21] showed that SOC content increases by 50 and 80% when land is converted from corn cultivation to switchgrass and miscanthus, respectively. EPA has estimated LUC GHG emissions for cellulosic ethanol derived from corn stover (−10 g $CO_2$e/MJ) and switchgrass (12 g $CO_2$e/MJ) [11]. CARB has examined forest residue and farmed trees as feedstocks for cellulosic ethanol [22,23]. The agency developed preliminary LUC GHG estimates for the latter feedstock, which is not examined in our current study.

The above literature summary highlights two limitations of previous studies of LUC GHG emissions associated with cellulosic ethanol production. First, application of worldwide CGE modelling to LUC GHG calculations for cellulosic ethanol has been limited to EPA and CARB analyses for switchgrass and corn stover. Second, SOC emission factors have either been developed for very specific lands (e.g., CRP or agricultural lands) or at the national or regional scale for other land types, as in the CARB and EPA analyses. In our study, we sought to address these two limitations of the current literature.

First, we used worldwide LUC results for four biofuel production scenarios (Table 1) as modelled with Purdue University's Global Trade Analysis Project (GTAP) CGE model [24]. The modelling considered domestic U.S. production of ethanol from four feedstocks: corn, corn ethanol, switchgrass, and miscanthus. Second, we applied finer-level SOC emission factors (EF) than have been used in previous analyses for all land categories, including forests. We developed a modelling framework to estimate these EFs at the state-level by utilizing remote sensing data, national statistics databases, and a surrogate model for CENTURY's soil organic C dynamics submodel (SCSOC) [25]. Details of the development of these EFs, which account for both aboveground and belowground carbon content changes, are provided in the Methods section and in a separate publication [26] as is the handling of international carbon EFs [27]. The LUC and carbon EF data were compiled in Argonne National Laboratory's Carbon Calculator for Land Use Change from Biofuels Production (CCLUB) model to enable calculation of LUC GHG emissions [28]. CCLUB is a module of Argonne National Laboratory's Greenhouse Gases, Regulated Emissions, and Energy use in Transportation (GREET™) model which was used to analyse LUC GHG emissions in the context of overall bioethanol life-cycle GHG emissions. GREET covers bioethanol production pathways extensively and is used by Argonne and other researchers to examine GHG emissions from transportation fuels and vehicle technologies [28].

In this paper, we estimate LUC GHG emissions associated with ethanol produced from four feedstocks (corn, corn stover, switchgrass, miscanthus). A sensitivity analysis is conducted to investigate the influence of key carbon content modelling assumptions on results. Addressing CGE model assumptions and their impact on LUC GHG emissions is outside the scope of this paper.

**TABLE 1:** GTAP modelling scenarios[24]

| Scenario | Scenario description | Increase in Ethanol (BL) |
|---|---|---|
| 1 | An increase in corn ethanol production from its 2004 level of 13 billion litres (BL) to 57 BL | 45 |
| 2 | An increase of ethanol from corn stover by 35 BL, in addition to 57 BL corn ethanol | 35 |
| 3 | An increase of ethanol from miscanthus by 27 BL, in addition to 57 BL corn ethanol | 27 |
| 4 | An increase of ethanol from switchgrass by 27 BL, in addition to 57 BL corn ethanol | 27 |

## 11.2 RESULTS AND DISCUSSION

In the following subsections, we describe LUC, domestic U.S. aboveground carbon, and domestic U.S. SOC modelling results. Next, we provide a full discussion of LUC GHG emissions results and place them in the context of life-cycle GHG emissions for each biofuel. The discussion is based on an agro-ecological zone (AEZ) level although SOC EFs for domestic U.S. lands were determined at a state level [27]. Figure 1 provides the distribution of AEZs in the United States for reference.

### *11.2.1 LAND-USE CHANGE*

In this paper, we divide LUC into domestic and international LUC for clarity and simplicity because it is not possible to distinguish between direct and indirect LUC in GTAP results, which are calculated at an AEZ level in the United States and a country/regional level abroad. As described above, types and amounts of converted lands were modelled with GTAP using four scenarios (Table 1) designed to follow the arc of Renewable Fuel Standard (RFS2) implementation. First, corn ethanol production expands until the RFS2 limit of 57 billion litres (BL) is met. Subsequently, cellulosic ethanol feedstocks will be produced on lands that corn does not already occupy. Results for each feedstock are presented in Figure 2. We developed and applied a forest proration factor (FPF) to adjust total domestic forest area converted for production of these feedstocks [27]. We took this approach to align forest land areas in the GTAP land database, the National Land Cover Dataset, and the U.S. Forest Service Forest Inventory Data. This step was necessary for consistency in the analysis because we used the latter to develop emission factors for aboveground and belowground carbon in addition to values for foregone sequestration. GTAP contains significantly more forested land than either of the other two data sources. When applying the FPF reduces the amount of forest converted, the difference is made up with land covered by young, thin trees. In Figure 2, this land type is called Young Forest-Shrub (YFS). The forest emissions factor for YFS is based on the relative height of forest stands in each state compared to shrubland. The relative tree heights for each state were derived from Pflugmacher et al. [29] and Buis [30]. When we apply the FPF, between 20 and 22% of converted land shifts from forests to YFS for all feedstocks.

In the case of corn ethanol (Scenario 1 in Table 1), most of the land converted in the U.S. is cropland-pasture along with some domestic forest (Figure 2). Modelling results indicated that AEZ 10 (temperate subhumid) is most affected by expansion of corn agriculture. Of the cellulosic feedstocks, corn stover has the lowest impact on domestic land use. Although this feedstock has the lowest productivity (Table 2), this result is unsurprising because stover is modelled as a "waste" product of corn production (as opposed to a co-product). Stover harvesting may not fundamentally change corn farming and should not result in significant LUC. Additionally, the greater amount of land converted for switchgrass ethanol production as compared to miscanthus ethanol production in the U.S. can be explained by crop yield, which can be nearly two times higher for miscanthus [31,32]. For both switchgrass and miscanthus ethanol, the majority of the land converted is in AEZ 7 (temperate arid) and is cropland-pasture. Nonetheless, the amount of forest converted for switchgrass is striking.

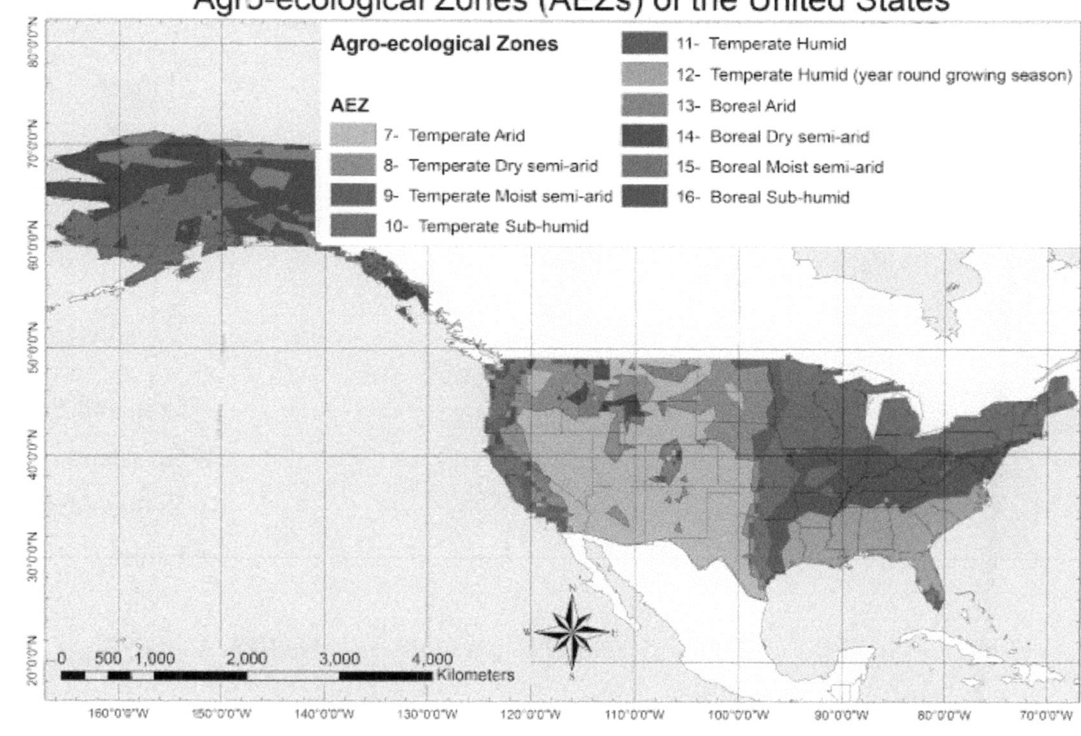


**FIGURE 1:** Distribution of AEZs in the United States.

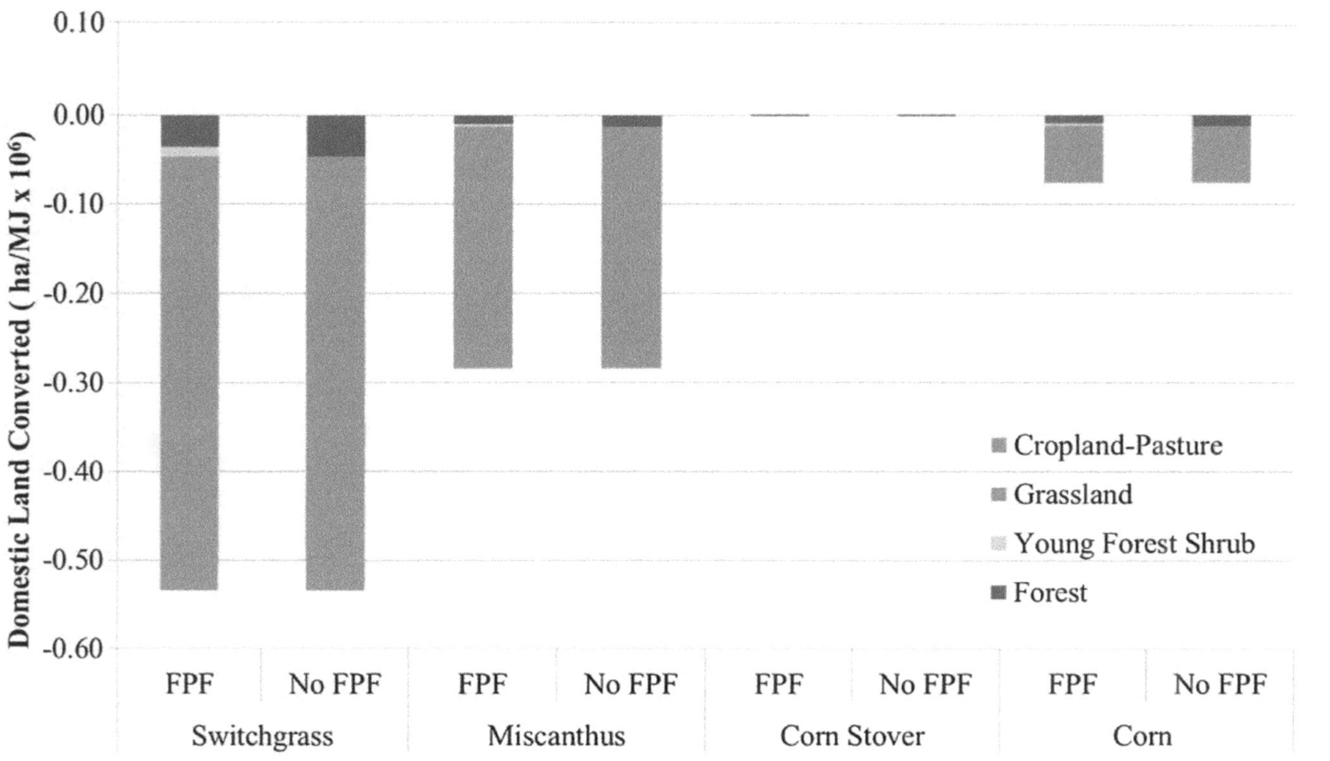


**FIGURE 2:** Domestic LUC for switchgrass, miscanthus, corn stover, and corn ethanol. Legend: Negative values indicate a decrease in land area.

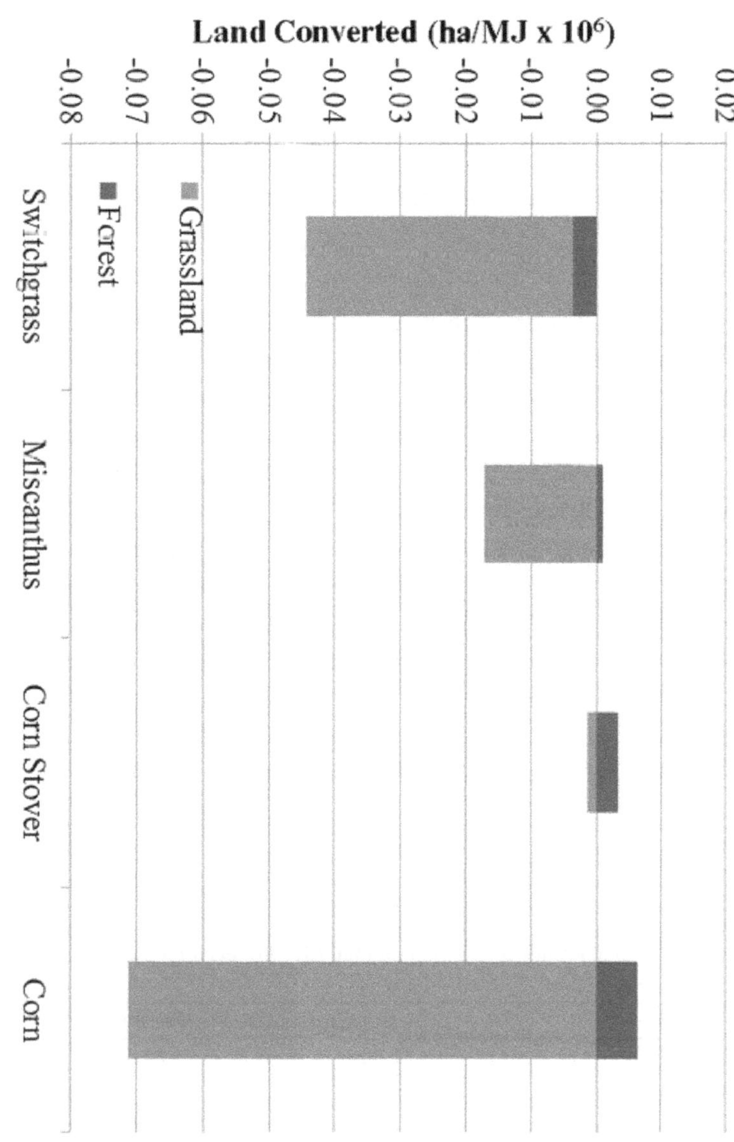


**FIGURE 3:** International LUC for switchgrass, miscanthus, corn stover, and corn ethanol. Legend: Negative and positive values indicate a decrease and increase, respectively, in land area.

**TABLE 2:** Feedstock productivity

| Feedstock | Crop yield (dry metric ton/ha) | Ethanol productivity (L/ha) |
|---|---|---|
| Corn | 7.9[a] | 4,250 L/ha[b] |
| Miscanthus | 24[c] | 6,190 L/ha[d] |
| Switchgrass | 12[c] | 3,200 L/ha[d] |
| Corn stover | 4.1[c] | 1,070 L/ha[d] |

*[a] Yield calculated from [28] and 20% moisture content at harvest. [b] 344 L/dry metric ton [28]. [c] From [24]. [d] Assuming an ethanol yield of 317 L/dry metric ton [24].*

Figure 3 displays international LUC that occurs from production of corn and cellulosic ethanol. Internationally, corn causes more LUC than the other crops because, unlike the cellulosic crops, U.S. corn accounts for a large share of the international corn market and a reduction in U.S. corn exports caused by corn ethanol production increases corn production in other countries. Among cellulosic crops, switchgrass production causes the most land conversion and, as it does domestically, the highest amount of forest conversion. (Note that no FPF was applied for international forest conversions.) Switchgrass production consumes cropland-pasture land in the United States, possibly shifting agricultural production from these lands to other countries. For both corn and corn stover feedstocks, some forest land is recovered internationally, the majority of which is in Russia. Table 3 shows domestic, international, and total LUC for each feedstock.

**TABLE 3:** Total domestic and international LUC for each feedstock (ha/MJ × $10^6$)

| Feedstock | Domestic | International | Total |
|---|---|---|---|
| Switchgrass | −0.54 | −0.04 | −0.58 |
| Miscanthus | −0.29 | −0.02 | −0.30 |
| Corn stover | $5.7 \times 10^{-4}$ | $1.8 \times 10^{-3}$ | $2.4 \times 10^{-3}$ |
| Corn | −0.08 | −0.07 | −0.14 |

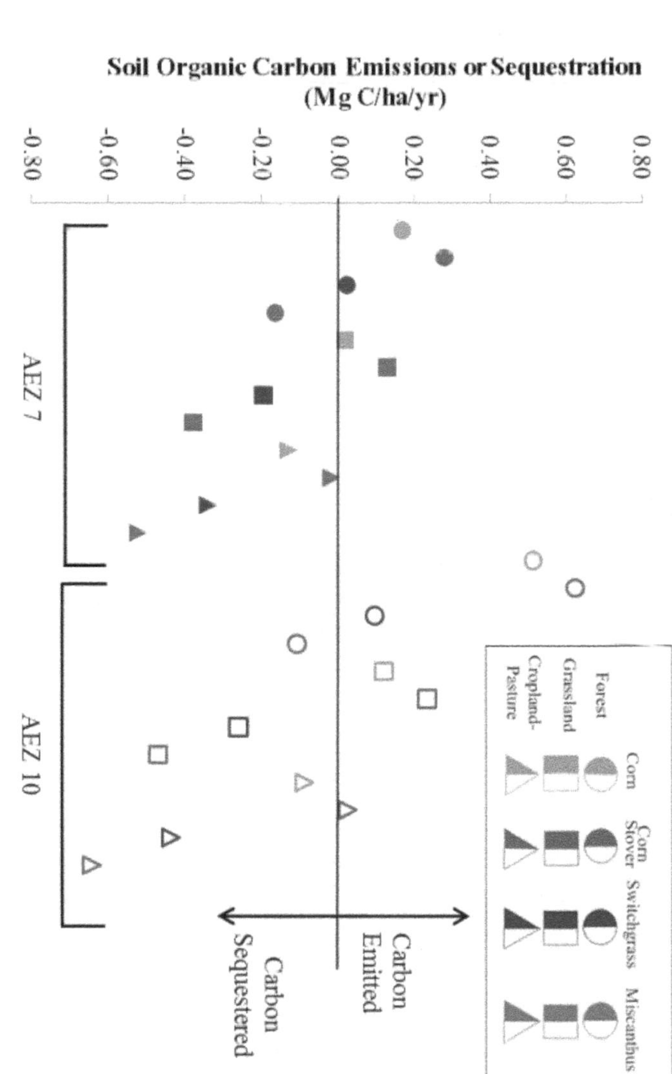


**FIGURE 4:** Soil organic carbon content changes from domestic land-use transitions. Legend: Solid and hollow markers denote transitions in AEZs 7 and 10, respectively. Forest, grassland, and cropland-pasture transitions are denoted by circles, squares, and triangles, respectively. Orange, red, green, and blue markers reflect transitions to corn, corn stover, miscanthus, and switchgrass production, respectively. These results were generated from surrogate CENTURY modelling runs with calibrated soil cultivation effect coefficients, feedstock yields that increase with time, and with erosion effects.

**TABLE 4:** Surrogate CENTURY scenarios in CCLUB

| CCLUB case | Soil cultivation effect coeffecient | | Crop yield | | Erosion | |
|---|---|---|---|---|---|---|
| | CENTURY default | Calibrated | Increase | No increase | Erosion | No erosion |
| sa | X | | | X | | X |
| sb | X | | X | | X | |
| sc | X | | | X | X | |
| sd | | X | X | | X | |
| se | | X | | X | X | |

## 11.2.2 SOIL ORGANIC CARBON EMISSION FACTORS

The development of the SOC EFs used in CCLUB is summarized in the Methods section, a detailed discussion can be found in an earlier publication [26]. Here, we discuss trends in these EFs and the implications for LUC GHG emissions. The variation in SOC EFs with location, a result of soil type and climate differences, is an important feature of this analysis. Although state-level EFs were calculated for each land transition and biofuel scenario, in CCLUB these EFs are rolled up to an AEZ level to match AEZ-level GTAP results. In Figure 4, we present the variation of SOC EFs for two AEZs (7 and 10) that GTAP predicts will experience the largest amount of LUC by feedstock and land conversion type. These results were generated from modelling runs with calibrated surrogate CENTURY soil cultivation effect coefficients, feedstock yields that increase with time, and with erosion effects (surrogate CENTURY case sd in Table 4). (We discuss the influence of surrogate CENTURY modelling choices on LUC GHG emissions in the next section.) Clearly, conversion of forest to produce corn or corn-corn stover results in the greatest amount of carbon emissions. Forest conversion to miscanthus production, however, may not incur a carbon penalty. Carbon sequestration occurs when grassland or cropland-pasture is converted to switchgrass or miscanthus production, which is consistent with other studies [20,21]. The data in Figure 4 consistently show that, of the land use transitions we considered, conversions to miscanthus maximize carbon sequestration. This result is consistent with miscanthus growth generating more aboveground and belowground biomass [26]. The SOC emission factors vary slightly between AEZs 7 and 10 with the exception of forest land converted to corn production. Converting forest to corn or corn stover production in AEZ 10 will produce greater carbon emissions than this transition in AEZ 7.

In estimating GHG emissions from the conversion of forests to biofuel feedstock production lands, we consider two sources of aboveground carbon: carbon contained in aboveground biomass that is cleared and the loss of carbon sequestration that would have occurred if the forest had continued to grow. See Mueller et al. [27] for a full discussion of how these factors were developed. Figure 5 breaks down the total carbon emissions factor applied to converted forest land for each feedstock in AEZs 7 and 10. The largest contributor to these emission factors is aboveground carbon. Both aboveground carbon and carbon sequestered during annual growth are greater in AEZ 10 than in AEZ 7. As expected based on Figure 4, conversion of forest to corn production with stover harvest transitions incur the greatest carbon penalty whereas transition to miscanthus production results in the lowest amount of GHG emissions.

### *11.2.3 LUC GHG EMISSIONS*

CCLUB is populated with carbon EFs generated from surrogate CENTURY modelling under four scenarios outlined in Table 4. The scenarios differ in their treatment of three key parameters: soil erosion, crop yield, and the soil cultivation effect coefficient. The latter was either left at default values or calibrated to real-world data. Additionally, EFs were also produced under different land management practices (conventional till, reduced till, no-till) for corn and corn stover feedstocks. We selected scenario “sd” in Table 4 as the base case for this study. For corn with and without stover harvest, the land management practice of conventional till is the base case setting.

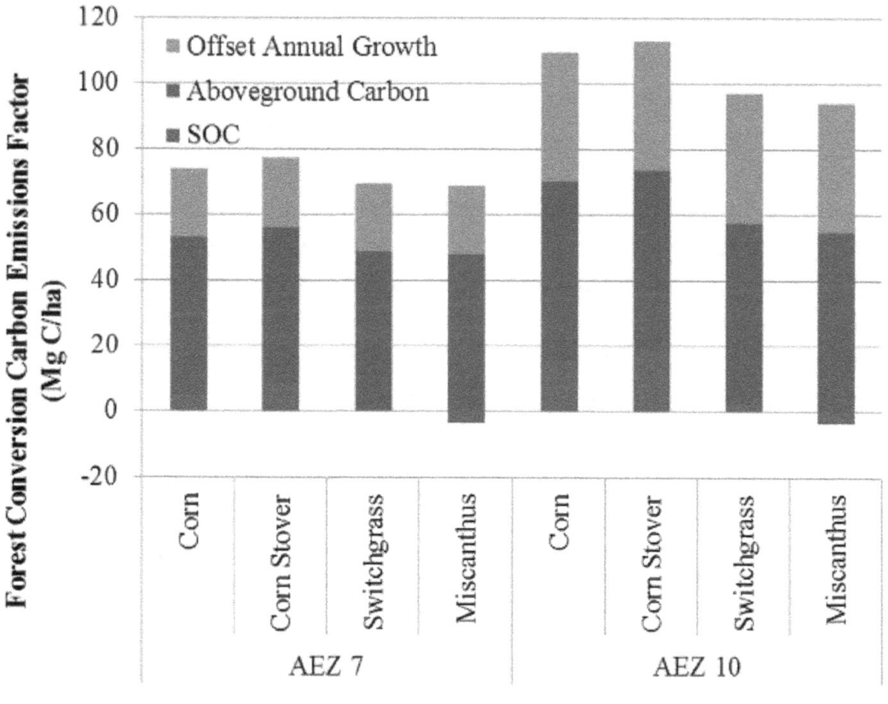


**FIGURE 5:** Forest carbon emission factor for four feedstocks in AEZs 7 and 10. Legend: SOC values were calculated with modelling option "sd" in Table 4.

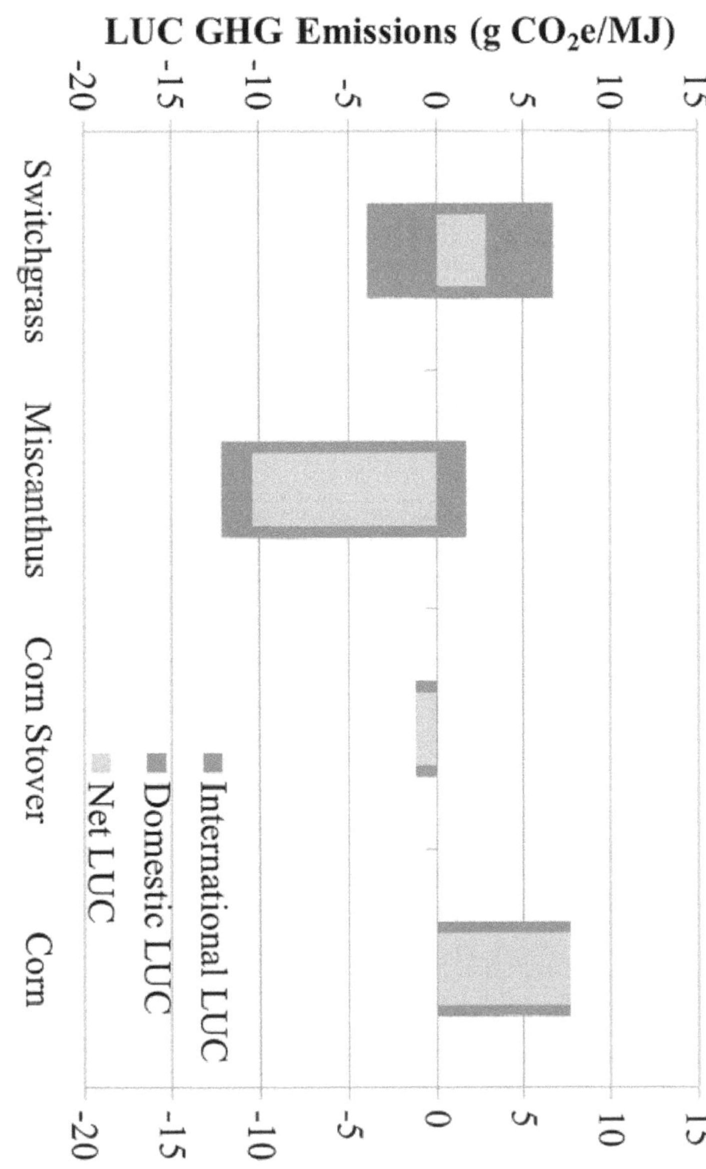


**FIGURE 6:** Base case LUC GHG emissions (g CO2e/MJ) for switchgrass, miscanthus, corn stover, and corn ethanol. Legend: Domestic LUC GHG emissions were calculated with modelling option "sd" in Table4, adopting the FPF, and assuming sequestration of 42% of aboveground live and dead tree carbon in HWP.

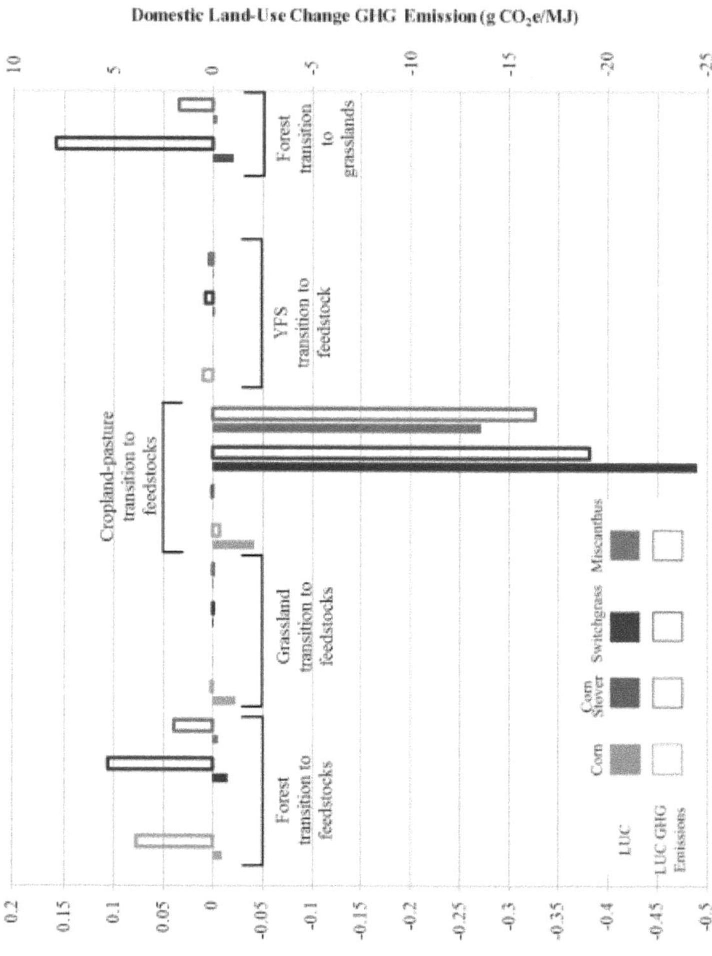


**FIGURE 7:** LUC (ha/MJ × 106) and LUC GHG emissions (g CO2e/MJ) from selected land conversions. Legend: Solid and hollow bars represent LUC amounts and LUC GHG emissions, respectively. Orange, red, blue, and green bars indicate feedstocks of corn, corn stover, switchgrass, and miscanthus, respectively. Results reflect base case modelling conditions.

### 11.2.4 BASE CASE LUC GHG RESULTS

Figure 6 contains the base case LUC GHG emissions results for the four bioethanol production scenarios in Table 1. Figure 7 pairs domestic U.S. LUC for each feedstock with the resulting base case domestic GHG emissions or sequestration. In the U.S., the miscanthus ethanol scenario causes significant SOC increases in the large amount of cropland-pastureland converted for feedstock growth. International LUC GHG emissions associated with this scenario are positive, but minimal. Miscanthus ethanol then exhibits net GHG sequestration from LUC. In the case of switchgrass ethanol, international LUC GHG emissions are significant. As described earlier, switchgrass production converts large areas of domestic cropland-pasture land, triggering conversion of lands abroad, including forests, to agriculture. In the United States, GHG emissions from forest-to-switchgrass conversion cut into gains in soil carbon from conversion of cropland-pasture lands to switchgrass production (Figure 7). The switchgrass ethanol scenario therefore on net emits GHGs as a result of LUC. Less land is converted for corn ethanol production than for switchgrass, yet LUC GHG emissions for corn ethanol exceed those for all cellulosic crops. LUC GHG emissions for corn ethanol are not offset by sequestration elsewhere (Figure 7) because corn reduces or only minimally enhances SOC (Figure 4). The results when corn stover is the ethanol feedstock show a small amount of carbon is sequestered. LUC modelling in this case predicts slight domestic gains in both YFS and forest lands and an increase in international forest lands, which sequester enough carbon to offset the carbon emitted from cropland-pasture conversion. For the most part, however, LUC GHG impacts of corn stover ethanol production can be considered negligible.

### 11.2.5 EFFECT OF KEY SURROGATE CENTURY MODEL PARAMETERS

Next, we consider how three surrogate CENTURY modelling choices affect these base case domestic LUC GHG emission results for corn ethanol (with conventional till) (Figure 8a) and for miscanthus and switchgrass ethanol (Figure 8b). The first modelling choice is whether to use a default or calibrated soil cultivation coefficient. Called clteff, this coefficient represents acceleration in soil carbon decay as a result of cultivation and fertilization under corn-based agriculture. Because it is used to establish the baseline amount of SOC in cropland before switchgrass or miscanthus production begins, it influences results for these feedstocks. Its calibrated value is larger than the default value [25]. Applying the calibrated soil cultivation effect coefficient therefore increases emissions from corn production. On the other hand, emissions decrease slightly from switchgrass and miscanthus production because when more SOC decay occurs prior to establishment of the feedstocks (calibrated clteff), conversion of cropland to produce them yields larger SOC increases. The second modelling choice is whether to assume crop yields are static or increasing. To investigate the influence of assuming crop yields increase, a 1% annual increase in yield for miscanthus and switchgrass was assumed [33]. Corn yield increases were based on historical data [25]. Crop yield increases translate into the production of more belowground carbon, some of which would be incorporated into SOC. Logically, then, assuming crop yields increase with time causes LUC GHG emissions to decline regardless of feedstock. Finally, the impact of soil erosion can be included. Erosion would be expected to decrease SOC, but Figure 8 illustrates that including its impact has a limited effect on domestic LUC GHG emissions.

### *11.2.6 EFFECT OF KEY CCLUB MODEL PARAMETERS*

In addition to containing EFs from surrogate CENTURY modelling under the scenarios in Table 4, CCLUB allows users to explore the effect of two other variables, the fate of carbon in harvested wood products (HWP) (e.g., lumber for buildings) and amount of forested land area in the U.S. (which can be determined with or without the FPF). In the case of HWP, one CCLUB scenario assumes 42% of aboveground live and dead tree carbon is sequestered in HWP [34]. The alternative scenario is that all carbon in these products is emitted. Figures 9a and 9b examine the impact of HWP and FPF for switchgrass and corn ethanol, respectively. We examine switchgrass results because GTAP predicts its production converts the largest amount of forests. In Figure 9a, accounting for sequestration of carbon in HWP reduces LUC GHG emissions by between 3 and 4 g $CO_2$e/MJ when the FPF assumption is held constant. For a given HWP assumption, applying the FPF decreases GHG emissions by between 2 and 3 g $CO_2$e/MJ.

In the case of corn ethanol (Figure 9b), applying the FPF decreases emissions by less than 1 g $CO_2$e/MJ when the type of tillage and the HWP assumption are held constant. Changing the HWP assumption under a constant tillage and FPF scenario decreases emissions by approximately 1 g $CO_2$e/MJ. As expected, for a given HWP and FPF configuration, corn grown under a no-till land management practice emits less carbon because tillage activities do not disturb the soil and release carbon to the atmosphere.

### *11.2.7 BIOFUEL LUC GHG EMISSIONS IN A LIFE-CYCLE CONTEXT*

In Table 5, we provide the range of LUC GHG emissions results that can be obtained by varying the key surrogate CENTURY and CCLUB modelling parameters as described above. We also provide the range of life-cycle GHG emissions assuming the default GREET assumptions for each ethanol pathway [10]. Without the contribution of LUC GHG sequestration, the net life-cycle GHG emissions result for miscanthus ethanol would be positive. Scown et al. [18] reported slightly higher GHG sequestration (between −3 and −16 g $CO_2$e/MJ) from miscanthus production, but limited their study to active cropland or CRP land. LUC GHG emissions could potentially contribute significantly to life-cycle GHG emissions (up to 19 g $CO_2$e/MJ) for switchgrass ethanol. This fuel exhibits the largest sensitivity to changes in modelling parameters in LUC GHG emissions. The area of forest that is predicted to be converted to grow this feedstock makes switchgrass results more sensitive to assumptions about HWP and the FPF than results for the other feedstocks (Figure 9). Corn ethanol LUC results vary considerably, although the base case estimate (7.6 g $CO_2$e/MJ) aligns well with a value in another recent report [35]. At most, LUC GHG emissions contribute 20% of life-cycle GHG emissions for corn ethanol. Regardless of the modelling scenario, corn stover ethanol LUC GHG emissions are essentially negligible.

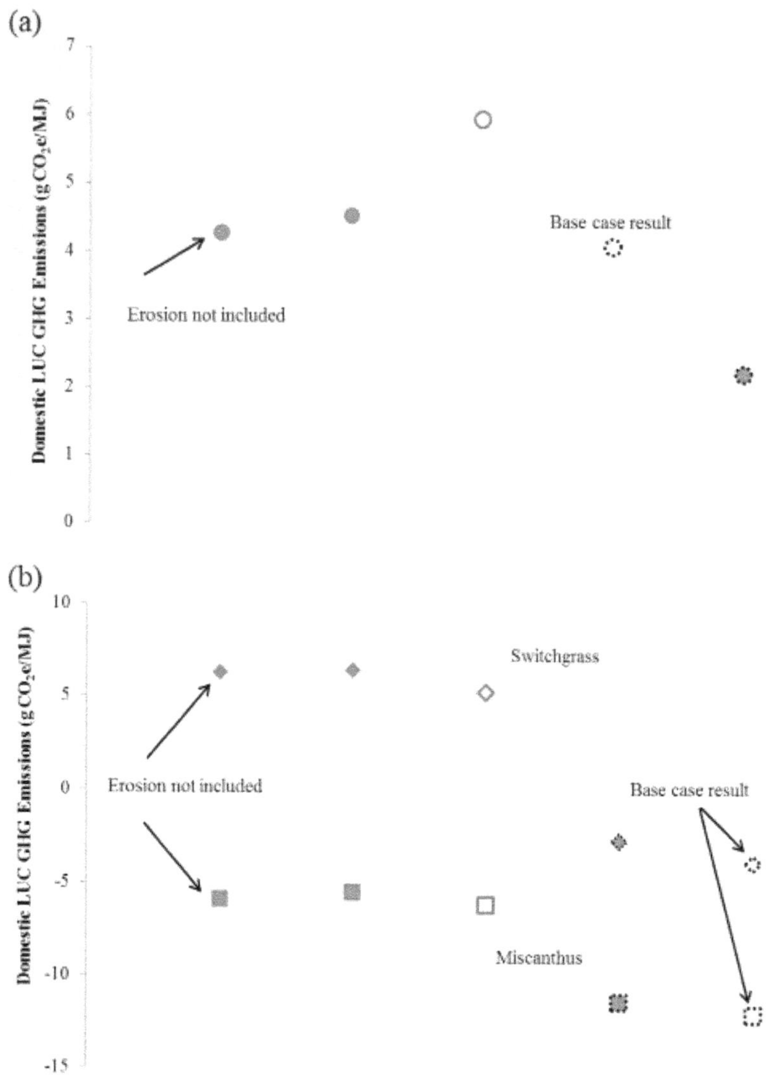


**FIGURE 8:** Surrogate CENTURY parameters' impact on domestic ethanol LUC GHG emissions for (a) conventionally-tilled corn (b) switchgrass and miscanthus. Legend: Solid and hollow shapes indicated surrogate CENTURY modelling with default and calibrated soil cultivation effect coefficients, respectively. Shapes with solid and dashed outlines represent surrogate CENTURY runs with constant and increasing yields, respectively. Diamond markers represent switchgrass results; square markers represent miscanthus results. All results except those indicated include erosion effects. In all cases, HWP sequesters 42% of aboveground live and dead tree carbon and the FPF is applied.

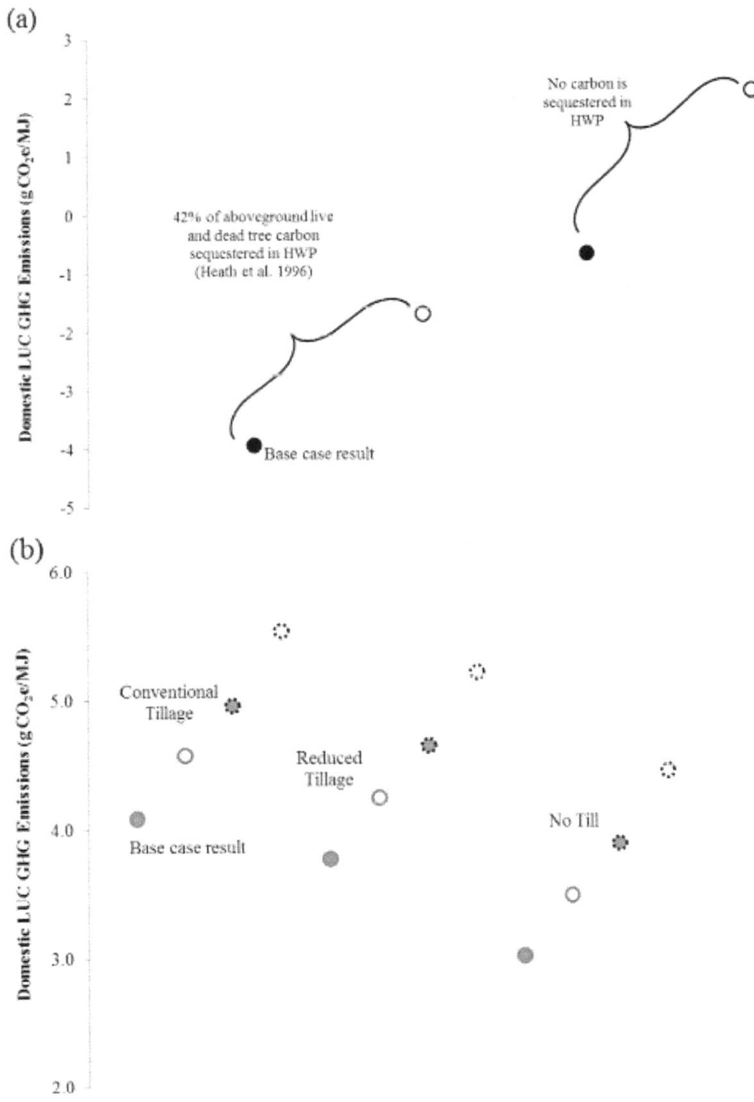


**FIGURE 9:** HWP and FPF impact on domestic ethanol LUC GHG emissions for (a) switchgrass and (b) corn. Legend: In Figures 9a and 9b, solid circles represents results calculated with the FPF applied. Hollow circles represent results using default GTAP results for area of converted forests. In Figure 9b, solid bordered circles represent results that account for some carbon sequestration in HWP. Circles with dashed borders are used for results that assume no carbon is sequestered in HWP.

**TABLE 5:** Range of LUC GHG emissions (g $CO_2e$/MJ)[a]

| | Switchgrass | Miscanthus | Corn stover | Corn |
|---|---|---|---|---|
| Minimum U.S. LUC GHG emissions | −3.9 | −12 | −0.24 | 1.2 |
| Maximum U.S. LUC GHG emissions | 13 | −3.8 | −0.19 | 7.4 |
| International LUC GHG emissions | 6.7 | 1.7 | −0.97 | 3.5 |
| LUC GHG emissions range | 2.7 to 19 | −10 to −2.1 | −1.21 | 4.7 to 11 |
| Lifecycle GHG emissions range[b] | 10 to 26 | −8.5 to −0.20 | 0.97 to 1.0 | 62 to 68 |

[a] *Values presented represent range of results generated at all combinations of surrogate CENTURY (Table 4) and CCLUB modelling parameter settings discussed.* [b] *Using default GREET parameters [10] and varying only LUC GHG emissions.*

## 11.3 CONCLUSIONS AND FUTURE RESEARCH

In this research, we have examined LUC GHG emissions of ethanol from four feedstocks: corn, corn stover, switchgrass, and miscanthus. Of the fuels examined, corn ethanol has the highest LUC GHG emissions. However, the estimate of LUC GHG emissions for this fuel has decreased substantially compared to earlier studies [1,2,11,12,36]. This evolution is due to improvements in CGE modelling such as modifications to the modelling of animal feed, yield responses to price increases, and representation of growth in both supply and demand [24].

Miscanthus ethanol shows the potential to sequester carbon over the course of its life cycle. This result is largely due to its high yield. Scown et al. [18] reached a similar conclusion, although they predict a higher amount of carbon sequestration from miscanthus production-induced LUC. On the other hand, switchgrass exhibits higher emissions than miscanthus because it is produced with a lower yield, necessitating more land, including carbon-rich forests, to be converted for its production. It is important to note that the contrast between switchgrass and miscanthus results is largely due to the difference in their yield. Similar differences may be observed between other high- and low-yield energy crops. LUC GHG emissions associated with corn stover were negligible. As the technology for corn stover's conversion to biofuels and other uses matures, corn stover may evolve into a co-product of corn production rather than a waste product. In that case, future modelling efforts could allocate LUC GHG impacts between the two fuels.

The sensitivity of LUC GHG emissions to key modelling parameters that dictate carbon emissions from converted lands is highlighted from the range of possible results in Table 5, which are affected by belowground and aboveground carbon simulation assumptions and results. As discussed, we did not investigate the influence of key CGE parameters on emissions because we used only one set of GTAP results. The uncertainty associated with these models, including GTAP, is large and difficult to estimate, as Plevin et al. [14] discuss. Improvements to these models, including modelling scenarios in which multiple feedstocks are simultaneously produced, scenarios at higher resolution (state or county-level), and scenarios with dynamic crop yields will shed further light on biofuel-induced LUC and better inform estimates of subsequent GHG emissions.

Improvements to estimates of converted lands' carbon content are also needed. First, SOC content data for soils worldwide is needed, as explained in Smith et al. [8], who provide a vision for developing these data and discuss key sources of uncertainty in their development. Soil organic matter models such as CENTURY would benefit from further calibration of default parameters, including the soil cultivation effect coefficient, with real-world data.

Additionally, it is important to include other factors that accompany LUC beyond soil carbon changes, as we have considered. For example, nitrogen fertilization rates will change, depending on the land use both on the site of feedstock production and at other, indirectly affected agricultural sites, affecting $N_2O$ emissions rates from the soil. The EPA has considered indirect effects like these [11]. Further, Georgescu et al. [37] examine the effects of stored soil water, which can have a regional cooling effect, as impacted by LUC. Additionally, land cover albedo will change with LUC [38]. Because the uncertainty that surrounds biofuel LUC impacts are a key barrier to what otherwise may be a technology that offers environmental and energy security benefits, these impacts certainly merit further study. It is important to realize, however, that the complexity inherent in modelling worldwide phenomena in the future that involve economic, biogeochemical, and biogeophysical effects will likely always lead to large uncertainties and will produce estimates of LUC GHG emissions that vary widely.

Despite the uncertainty and complexity associated with estimating LUC GHG emissions, the continued pursuit of improvement of these estimates will increase understanding of crop management practices that limit GHG emissions from SOC depletion, provide new data for policy formulation that limits LUC impacts through, for example, preventing conversion of carbon-rich lands (forests), and identify crops that minimize LUC GHG emissions when produced on a large scale as biofuel feedstocks.

## 11.4 METHODS

To conduct the modelling for this analysis, we used Argonne National Laboratory's CCLUB and GREET models [28]. The GREET model is developed at Argonne National Laboratory and is widely used to examine GHG emissions of vehicle technologies and transportation fuels on a consistent basis. CCLUB combines land transition data from GTAP modelling [24] with carbon emission factors derived from several sources. Domestic SOC content data were developed with a surrogate model for CENTURY's soil organic carbon submodel (SCSOC) [25,26]. In this modelling, we estimated the forward change in soil C concentration within the 0–30 cm depth and computed the associated EFs for the 2011 to 2040 period for croplands, grasslands or pasture/hay, croplands/conservation reserve, and forests that were suited to produce any of four possible biofuel feedstock systems (corn-corn, corn-corn with stover harvest, switchgrass, and miscanthus). This modelling accounted for prior land-use history in the U.S. dating to 1880. SOC modelling was conducted under a number of parameter settings to examine the effect of soil erosion, crop yield increases, and the calibration of values for a key coefficient that represents the soil cultivation effect. Surrogate CENTURY modelling scenarios are shown in Table 4. Additionally, the effect of three different land management (tillage) scenarios for corn and corn stover production were examined: conventional till, no till, and reduced till. Our modelling of conventional tillage assumes that 95% of surface residues are mixed with soils, whereas no-tillage scenarios assume a converse 5% mixing of surface soils.

International SOC emission factors were adopted from data from the Woods Hole Research Center. The data, available at the biome level, were authored by R. Houghton and provided to CARB and Purdue University to support land-use modelling. Tyner and co-authors [36] reproduced the data set. We incorporated aboveground carbon emissions impacts of forest conversion using data from the U.S. Department of Agriculture's (USDA) Forest Service/National Council for Air and Stream Improvement, Inc. (NCASI) Carbon Online Estimator (COLE) [39]. Technical documentation for CCLUB is available [27]. GREET parameters for feedstock production and growth are provided in several reports [31,32,40]. Other bioethanol life cycle parameters are provided in Wang et al. [10].

## REFERENCES

1. Searchinger T, Heimlich R, Houghton RA, Dong F, Elobeid A, Fabiosa J, Tokgoz S, Hayes D, Yu T-H: Use of U.S. croplands for biofuels increases greenhouse gases through emissions from land-use change. Science 2008, 319:1238-1240.
2. Hertel TW, Golub AA, Jones AD, O'Hare M, Plevin RJ, Kammen DM: Effects of US maize ethanol on global land use and greenhouse gas emissions: estimating market-mediated responses. BioSci 2010, 60:223-231.
3. Gibbs HK, Ruesch AS, Achard F, Clayton MK, Holmgren P, Ramankutty N, Foley JA: Tropical forests were the primary sources of new agricultural land in the 1980s and 1990s. Proc Natl Acad Sci 2010, 107:16732-16737.
4. Popp A, Dietrich JP, Lotze-Campen H, Klein D, Bauer N, Krause M, Beringer T, Gerten D, Edenhofer O: The economic potential of bioenergy for climate change mitigation with special attention given to implications for the land system. Environ Res Lett 2011, 6:034017.
5. Fargione J, Hill J, Tilman D, Polasky S, Hawthorne P: Land clearing and the biofuel carbon debt. Science 2008, 319:1235-1238.
6. Hertel TW, Rose SK, Tol RSJ: Economic analysis of land use in global climate change policy. New York, NY: Taylor & Francis Group; 2009.
7. Delucchi M: A conceptual framework for estimating the climate impacts of land-use change due to energy crop programs. Biomass Bioenerg 2011, 35:2337-2360.
8. Smith P, Davies CA, Ogle S, Zanchi G, Bellarby J, Bird N, Boddey RM, McNamara NP, Powlson D, Cowie A, Noordwijk M, Davis SC, Richter DDB, Kryzanowski L, Wijk MT, Stuart J, Kirton A, Eggar D, Newton-Cross G, Adhya TK, Braimoh AK: Towards an integrated global framework to assess the impacts of land use and management change on soil carbon: current capability and future vision. Glob Chang Biol 2012, 18:2089-2101.
9. Thomas ARC, Bond AJ, Hiscock KM: A multi-criteria based review of models that predict environmental impacts of land use-change for perennial energy

crops on water, carbon and nitrogen cycling. [http://dx.doi.org/10.1111/j.1757-1707.2012.01198.x] GCB Bioenergy 2012.

10. Wang M, Han J, Dunn JB, Cai H, Elgowainy A: Well-to-wheels energy use and greenhouse gas emissions of ethanol from corn, sugarcane and cellulosic biomass for US use. Environ Res Lett 2012, 7:045905.
11. Regulation of Fuel and Fuel Additives: Changes to Renewable Fuel Standard Program. Fed Regist 2010, 75(58):14669-15320.
12. CARB. Sacramento, CA: ; 2009. [Proposed regulation to implement the Low carbon fuel standard, Vol. I, staff report: initial statement of reasons]
13. European Commission: Directive 2009/28/EC of the European Parliament and of the Council of 23 April 2009 on the Promotion of the Use of Energy from Renewable Sources and Amending and Subsequently Repealing Directives 2001/77/EC and 2003/ 30/EC. : ; 2009.
14. Plevin RJ, Jones AD, Torn MS, Gibbs HK: Greenhouse gas emissions from biofuels' indirect land use change are uncertain but may be much greater than previously estimated. Env Sci Technol 2010, 44:8015-8021.
15. Davis SC, House JI, Diaz-Chavez RA, Molnar A, Valin H, DeLucia EH: How can land-use modelling tools inform bioenergy policies? Interface Focus 2011, 1:212-223.
16. Yang Y, Bae J, Kim J, Suh S: Replacing gasoline with corn ethanol results in significant environmental problem-shifting. Env Sci Technol 2012, 46:3671-3678.
17. Hill J, Polasky S, Nelson E, Tilman D, Huo H, Ludwig L, Neumann J, Zheng H, Bonta D: Climate change and health costs of air emissions from biofuels and gasoline. Proc Natl Acad Sci 2009, 106:2077-2082.
18. Scown CD, Nazaroff WW, Mishra U, Strogen B, Lobscheid AB, Masanet E, Santero NJ, Horvath A, McKone TE: Lifecycle greenhouse gas implications of US national scenarios for cellulosic ethanol production. Environ Res Lett 2012, 7:014011.
19. Matthews RB, Grogan P: Potential C sequestration rates under short-rotation coppiced willow and miscanthus biomass crops: a modeling study. Asp Appl Biol 2001, 65:303-312.
20. Davis SC, Parton WJ, Grosso SJD, Keough C, Marx E, Adler PR, DeLucia EH: Impact of second-generation biofuel agriculture on greenhouse-gas emissions in the corn-growing regions of the US. Front Ecol Environ 2012, 10:69-74.
21. Qin Z, Zhuang Q, Chen M: Impacts of land use change due to biofuel crops on carbon balance, bioenergy production, and agricultural yield in the conterminous United States. GCB Bioenergy 2012, 4:277-288.
22. CARB: Detailed California-modified GREET pathway for cellulosic ethanol from farmed trees by fermentation. Sacramento, CA: ; 2009.
23. CARB: Detailed California-modified GREET pathway for cellulosic ethanol from forest waste. Sacramento, CA: ; 2009.
24. Taheripour F, Tyner WE, Wang MQ: Global land use changes due to the U.S. cellulosic biofuel program simulated with the GTAP model. 2011. http://greet.es.anl.gov/publication-luc_ethanol
25. Kwon H-Y, Hudson RJM: Quantifying management-driven changes in organic matter turnover in an agricultural soil: An inverse modeling approach using historical

data and a surrogate CENTURY-type model. Soil Biol Biochem 2010, 42:2241-2253.

26. Kwon H-Y, Wander MM, Mueller S, Dunn JB: Modeling state-level soil carbon emissions factors under various scenarios for direct land use change associated with United States biofuel feedstock production. Biomass Bioenerg 2013.
27. Mueller S, Dunn JB, Wang MQ: Carbon Calculator for Land Use Change from Biofuels Production (CCLUB) users' manual and technical documentation. Argonne National Laboratory; 2012. ANL/ESD/12-5 http://greet.es.anl.gov/publication-cclub-manual
28. Argonne National LaboratoryGREET1_2012. http://greet.es.anl.gov/main
29. Pflugmacher D, Cohen W, Kennedy R, Lefsky M: Regional applicability of forest height and aboveground biomass models for the geoscience laser altimeter system. For Sci 2008, 54:647-657.
30. Buis A: Global map of forest height produced from NASA's ICESAT/GLAS, MODIS and TRMM sensors. 2012. http://www.nasa.gov/topics/earth/features/forest20120217.html
31. Wang Z, Dunn JB, Wang MQ: GREET model miscanthus parameter development. Argonne National Laboratory. 2012. http://greet.es.anl.gov/publication-micanthus-params
32. Dunn JB, Eason J, Wang MQ: Updated sugarcane and switchgrass parameters in the GREET model. Argonne Mational Laboratory; 2011. http://greet.es.anl.gov/publication-hjk5cxlv
33. U. S. Department of Energy: U.S. billion-ton update: biomass supply for a bioenergy and bioproducts Industry. Oak Ridge National Laboratory; 2011.
34. Heath LS, Birdsey RA, Row C, Plantinga AJ: Carbon pools and flux in U.S. forest products. Berlin: Springer; 1996:271-278. [NATO ASI Series I: Global Environmental Change]
35. ATLASS Consortium: Assessing the land use change consequences of European biofuel policies. 2011. http://trade.ec.europa.eu/doclib/html/148289.htm
36. Tyner WE, Taheripour F, Zhuang Q, Birur D, Baldos U: Land use changes and consequent CO2 emissions due to US corn ethanol production. A comprehensive analysis. Purdue University: Department of Agricultural Economics; 2010.
37. Georgescu M, Lobell DB, Field CB: Direct climate effects of perennial bioenergy crops in the United States. Proc Natnl Acad Sci 2011, 108:4307-4312.
38. Pielke RA, Pitman A, Niyogi D, Mahmood R, McAlpine C, Hossain F, Goldewijk KK, Nair U, Betts R, Fall S, Reichstein M, Kabat P, de Noblet N: Land use/land cover changes and climate: modeling analysis and observational evidence. WIRE: Clim Chang 2011, 2:828-850.
39. Van Deusen PC, Heath LS: Weighted analysis methods for mapped plot forest inventory data: tables, regressions, maps and graphs. For Ecol Manag 2010, 260:1607-1612.
40. Han J, Elgowainy E, Palou-Rivera I, Dunn JB, Wang MQ: Well-to-wheels analysis of fast pyrolysis pathways with GREET. Argonne National Laboratory; 2011. http://greet.es.anl.gov/publication-wtw_fast_pyrolysis

# AUTHOR NOTES

## CHAPTER 1

### Acknowledgment

This material is based upon work supported by the U.S. Department of Energy under Award Number: DE-FC26-08NT01923.

### Conflict of Interest

The authors declare no conflict of interest.

## CHAPTER 2

### Funding

The National Key Basic Research Development Program (2011CB707401) (http://program.most.gov.cn/), the National Natural Science Foundation of China (No. 31100440, 21276143, 21376141) (http://www.nsfc.gov.cn/Portal0/default152.htm), and the State Key Laboratory of Microbial Technology (M 201306). The funders had no role in study design, data collection and analysis, decision to publish, or preparation of the manuscript.

### Competing Interests

The authors have declared that no competing interests exist.

### Author Contributions

Conceived and designed the experiments: JL JZ. Performed the experiments: XZL JL. Analyzed the data: XZL JL JZ YBQ. Contributed reagents/materials/analysis tools: XZL JL. Wrote the paper: XZL JL JZ.

## CHAPTER 3

### Competing Interests

The authors declare that they have no competing interests.

### Author Contributions

YQH and JB designed the experiment and drafted the manuscript. JZ designed the equipment. YQH and JZ carried out the experiment. LPZ carried out the CFD calculation. JB conceived of the study. All authors read and approved the final manuscript.

### Acknowledgments

This research was supported by the National Basic Research Program of China (2011CB707406/2013CB733902), the National High-Tech Program of China (2012AA022301), the Natural Science Foundation of China (21306048), the Fundamental Research Funds for the Central Universities of China (WF0913005/1114054/1214025), the Shanghai Leading Academic Discipline Project (B505), and the State Key Laboratory of Motor Vehicle Biofuel Technology (2013012)

## CHAPTER 4

### Competing Interests

The authors declare that they have no competing interests.

### Author Contributions

JL and XZL carried out the experiments, data analyses, and drafted the manuscript. RFY revised the manuscript. JZ designed the work, and participated in manuscript writing and data analysis. YBQ reviewed the paper. All authors read and approved the final manuscript.

### Acknowledgments

This study was financially supported by the National Key Basic Research Development Program (2011CB707401) and the National Natural Science Foundation of China (31100440, 21276143, 21376141) and the State Key Laboratory of Microbial Technology (M 2013–06).

## CHAPTER 5

### Competing Interests

The authors declare that they have no competing interests.

### Author Contributions

All authors contributed jointly to all aspects of the work reported in the manuscript. AP carried out much of the laboratory work, contributed to planning, interpretation of results and drafting of the paper. JH contributed to the purified enzymes used in the study. VA contributed to the planning, interpretation and drafting. JS contributed to the planning, interpretation and writing of the manuscript. All authors read and approved the final manuscript.

### Acknowledgment

The authors would like to gratefully acknowledge Dr. Larry Taylor from the National Renewable Energy Laboratory (NREL) for generously providing some of the antibodies used in this study. The authors also thank Genencor-DuPont and Novozymes for providing the enzymes used in this study. The support of Genome BC Canada and Natural Sciences and Engineering Research Council of Canada (NSERC) are gratefully acknowledged.

## CHAPTER 6

### Competing Interests

The authors declare that they have no competing interests.

### Author Contributions

YC participated in the design of experiments, conducted the work presented here, performed the statistical analysis and drafted the manuscript. YZ participated in the experiments design, performed the work, and helped draft the manuscript. MAS and JH conducted the pretreatment and hydrolysis experiments. HX conceived of the study, supervised the work, and assisted in drafting the manuscript. All authors read and approved of the final manuscript.

### Acknowledgment

The authors thank Drs. Kurt Creamer and Don Higgins for helpful comments on earlier drafts of this paper.

## CHAPTER 7

### Competing Interests

SW, AL were employed by SEKAB E-Technology during the time of this work. EA was employed by Taurus Energy AB during the time of this work. LW is employed by Taurus Energy AB and LO does consultancy for Taurus Energy AB.

### Author Contributions

RK, EA, AL, SW, LW, GZ and LO participated in the conception and design of the study. RK, AL and FN performed the experimental work. RK wrote the manuscript. All the authors commented on the manuscript, read and approved the final manuscript.

### Acknowledgments

This work was part of the project 'Industrial verification of pentose fermenting yeasts' financed by Energimyndigheten, SEKAB E-Technology and Taurus Energy AB, Sweden, grant number: 30805–1. We would like to thank members of Industrial Biotechnology group at Chalmers University of Technology for reading and commenting on the manuscript.

## CHAPTER 8

### Competing Interests

The authors declare that CEW is a cofounder of Mascoma Corporation (Lebanon, NH, USA) and former chair of their Scientific Advisory Board. CEW no longer works with Mascoma. In addition, CEW is founding Editor-in-Chief of Biotechnology for Biofuels. The other authors declare that they have no competing interests.

### Author Contributions

XG performed pretreatment of corn stover with dilute acid, all enzymatic hydrolysis experiments and characterization work reported here, prepared the initial manuscript, and undertook final assembly of the revised paper. CEW developed the initial enzymatic hydrolysis and dilute acid pretreatment experimental design, supervised the work, revised the manuscript, figures, tables, and preparation of correspondence, and approved the final manuscript. RK assisted in the design of experiments, reviewed the results, and prepared and reviewed the manuscript. BED initiated the collaboration, arrangements for supply of corn stover and AFEX pretreated solids, and reviewed the manuscript. VB supplied the corn stover, provision of AFEX pretreated materials, and reviewed the manuscript. BAS supplied the IL pretreated materials and reviewed the manuscript. SS supplied the IL pretreated materials and reviewed the manuscript. All authors read and approved the final manuscript.

### Acknowledgments

We gratefully acknowledge support for this research by the Office of Biological and Environmental Research in the Department of Energy (DOE) Office of Science through the BESC at Oak Ridge National Laboratory (contract DE-PS02-06ER64304), GLBRC (grant DE–FC02–07ER64494), and JBEI (grant DE-AC02-05CH11231). We are also grateful to the Center for Environmental Research and Technology of the Bourns College of Engineering (CE-CERT) at UCR for providing key equipment and facilities. We thank Mr. Charles Donald Jr for preparing AFEX pretreated corn stover for this project and Nirmal Uppugundla for having optimized the commercial enzyme formulations for the three different pretreated solids. We also thank Novozymes and DuPont Genencor Science for providing enzymes for this research. The corresponding author is particularly grateful to the Ford Motor Company for funding the Chair in Environmental Engineering at the CE-CERT at UCR that augments support for many projects such as this.

## CHAPTER 9

### Conflict of Interest

The authors declare that there is no conflict of interests regarding the publication of this paper.

## CHAPTER 10

### Competing Interests

The authors declare that they have no competing interests.

### Author Contributions

RM conceived the study. All authors were involved with designing the study. RM, KR, and HS performed the microanalyses. PG created the model and conducted the system analysis. RM and PG drafted the manuscript, with KR and HS contributing key portions. All authors read and approved the final manuscript.

### Acknowledgments

This study was supported in part by funding from the Illinois Corn Growers' Association and Illinois Corn Marketing Board. We wish to acknowledge David Gustafson of Monsanto Company for verifying public accessibility of information on Monsanto's product pipeline. Furthermore, we wish to thank Steffen Mueller and Geoff Cooper for helpful discussions and review of the manuscript. Finally, we are grateful for the helpful comments of three anonymous reviewers, which improved the quality of this publication.

## CHAPTER 11

### Competing Interests

The authors declare that they have no competing interests.

### Author Contributions

JBD conducted the analysis and writing for this paper with substantial collaboration with SM. MQW and HK also authored and reviewed this paper. All authors read and approved the final manuscript.

### Author Information

HK conducted this research while at the University of Illinois at Urbana Champaign. Recently, he has joined the staff at the International Food Policy Research Institute.

**Acknowledgments**

This study was supported by the Biomass Program of the Energy Efficiency and Renewable Energy Office of the U.S. Department of Energy under Contract No. DE-AC02-06CH11357. The authors thank the support and guidance of Zia Haq and Kristen Johnson of the Biomass Program. The authors acknowledge valuable discussions with Wally Tyner of Purdue University, Michelle Wander of the University of Illinois at Urbana-Champaign, and Joshua Elliott of the University of Chicago. The authors are solely responsible for the contents of this paper.

# INDEX

## X



## Y



## Z

Milton Keynes UK
Ingram Content Group UK Ltd.
UKHW022100141024
449569UK00031B/1716

9 781774 63544